S T E V I E

STEVIE

A BIOGRAPHY OF STEVIE SMITH

Jack Barbera
and William McBrien

New York   Oxford
OXFORD UNIVERSITY PRESS
1987

Oxford University Press

Oxford   New York   Toronto
Delhi   Bombay   Calcutta   Madras   Karachi
Petaling Jaya   Singapore   Hong Kong   Tokyo
Nairobi   Dar es Salaam   Cape Town
Melbourne   Auckland

and associated companies in
Beirut   Berlin   Ibadan   Nicosia

Published in 1985 by William Heinemann Ltd.
10 Upper Grosvenor Street, London

Published in 1987 in the United States by
Oxford University Press, Inc.
200 Madison Avenue, New York, New York 10016

Oxford is a registered trademark of Oxford University Press

Library of Congress Cataloging-in-Publication Data

Barbara, Jack.
Stevie : a biography of Stevie Smith.

Bibliography: p.
Includes index.
1. Smith, Stevie, 1902–1971—Biography.
2. Poets, English—20th century—Biography.
I. McBrien, William.   II. Title.
PR6037.M43Z58   1987   828'.91209 [B]   86-23916
ISBN 0-19-520549-9 (alk. paper)

2 4 6 8 10 9 7 5 3 1

Printed in the United States of America
on acid-free paper

Contents

For Jeanette Barbera, Suzanne Burnett,
and in memory of Mabel York McBrien

Note by the executor of Stevie Smith

This is not an authorized biography of Stevie Smith and her executor is in
no way responsible for it beyond giving permission to the authors to quote
her works. As they had spent some years on their research it would have
been churlish to withhold such permission and the fee they are paying so to
do will benefit The Authors Foundation, the Society of Authors, 84 Drayton
Gardens, London SW10 9SB. Income from the Foundation is used to help
writers, especially younger writers, who need support for the writing of
commissioned books. The authorized biography is in preparation.

Note by the authors

There is a lengthy Acknowledgments section at the back of this book, but
we would like to thank Stevie Smith's executor in this prominent place because of the importance and extent of his cooperation. Not only did he con-
sent to be interviewed when we began work on this biography several years
ago, but he also gave us the full authorization necessary to obtain copies of
Stevie Smith's unpublished writings and drawings. In addition, he has granted
his authorization for quotations from Stevie's writings and for the use of
copyright illustrative material. This is not, however, an authorized biography. James MacGibbon has allowed us the freedom to draw our own conclusions, and to express them as we have done. To our minds, these per-
missions are of equal value, and we are deeply grateful to have them.

I

The Gold Medal for Poetry

On 22 November 1969, Stevie Smith wrote a letter to Eric White, the Literature Director of the Arts Council of Great Britain. "Dear Eric," she began, "Just to thank you very much for a heavenly lunch party, it really was awfully nice of you & please forgive the sadly reprehensible way in which I, as your unworthy guest of honour, behaved." The luncheon had been held the previous day to celebrate Queen Elizabeth's personal presentation to Stevie Smith of The Gold Medal for Poetry. At the time she received it, Stevie was experiencing a second wave of popularity. Literary fame had come to her in 1936 with the publication of her first book, *Novel on Yellow Paper*. But although she continued to write and publish, it was not until the 1960s that she again enjoyed renown, this time for reciting her poems before live audiences and on radio and records.

"The thing about Stevie," said one of her friends in retrospect, "is that she was a professional performer. She appeared on stage, and sang and recited, and was prepared to answer questions." Testimony to Stevie's success as a performer is not hard to find. For example, Karl Miller, in a column for the *New Statesman*, wrote about a poetry festival in 1963, citing two poets for praise: "What I especially liked were Robert Lowell's magnificent grainy readings, and . . . the thunderclap of Stevie Smith, in a pale-pink shift." "Her eccentricity may puzzle," he noted, "yet it also seems to draw . . . a wide affection." Miller concluded, "Stevie Smith is an original and moving poet who hasn't had the appreciation she deserves." Appreciation was swelling, however, especially among the young, something which bemused Stevie: "I say to them, 'I can't see what you see in [the

poems], because on the whole they're a bit deathwards in their
wish . . .' and I should have thought that the attitude of the young was
more courageous than mine."

The incongruity of Stevie Smith captivating a young audience is
conveyed by the poet Christopher Logue, who recalls her performance at the Albert Hall in 1966:

> The audience were quite young, not particularly bookish, and I
> doubt if many of them had heard of her; they were a sexy-looking,
> lively crowd, variously and extravagantly dressed—or so it seems
> in my memory. Onto the stand came this gaunt, plain, stick-legged,
> flat-chested . . . unearthly creature, who, commandingly held up
> her hand until silence reigned, and then, opened her mouth and
> sang, yes, sang her poems in a quavering soprano voice, but with
> such confidence, such grace, and such clarity, that she was given an
> ovation. She was a natural Star.

In February of 1969 *The Financial Times* publicly proclaimed Stevie's
stardom in an article about a poetry gala:

> The mis-firing professionalism of the occasion . . . was put to rights
> by the small . . . figure who darted on to the platform to open the
> second half. This was Stevie Smith, and she is a star. Her delivery,
> level and deadpan, perfectly matches the hilarity of her material,
> and leaves the audience to make up its mind just how desolately sad
> her poems are.

Such acclaim had become common, and made the decision to be-
stow the Gold Medal upon Stevie Smith a popular one. After the
official press release in early November of 1969, reporters and photographers journeyed to her house in the London suburb of Palmers
Green, and the papers printed interviews and notices about the
award. Instituted in 1933 by George V, it put Stevie among a
small company which included W. H. Auden, Siegfried Sassoon and
Robert Graves.

"Darling, I've been offered the Gold Medal!" Stevie exclaimed on
the telephone to Eric White:

> —Now, you know all about these things. I've got to go to the
>   Palace?
> —Yes, they keep the medals in the Palace.
> —They will give me refreshment, won't they? Because I've got to
>   set out from Palmers Green, and I shall be very hungry and
>   thirsty by the time I get to the Palace.
> —No Stevie, they won't. You're going to get a medal. You're not
>   going to get a glass of sherry.

—Oh, but darling, how do they expect me to manage such a long time?

—Well, they do. You will have to manage.

—Are there lavatory accommodations?

After further discussion, White rang off and telephoned Cecil Day Lewis. As Poet Laureate, Day Lewis had chaired the small committee which recommended Stevie for the award. "She needs feeding," White told him. And so it happened that a lunch party was planned for the three of them and Norah Smallwood, Stevie's friend and a director of the publishing firm Chatto & Windus. They would meet at a restaurant in Soho right after the investiture.

In the weeks before her meeting with the Queen, Stevie gave interviews to newspaper reporters, the BBC and the Canadian Broadcasting Corporation. "I'm probably a couple of sherries below par most of the time," she told one reporter, and explained how easily the lines of her poems could go awry. A poem about death, for example, which she thought might "cheer up Her Majesty", has as its first line, "My heart goes out to my Creator in love". "If you're not careful," Stevie said, "your heart goes out to your 'creator in law'." She mentioned that an elderly publisher had congratulated her: "*What* a discerning monarch." And she was busy answering each of the friends who wrote. "I only hope I don't stick at the curtsey," she said to one. "I'm sure H.M. wd. much rather pin it on a doggy-dear than me," she told another. And to a third she joked, "I shall now be terrified of putting pen to paper in case the gold medal says it is not quite what We had in mind!"

Needing a hat for her trip to Buckingham Palace, Stevie went to the jumble sale sponsored by St John's Church in Palmers Green. She acquired a second-hand one, according to the then Vicar of St John's, and "everybody thought it was killingly funny that that hat was going to go up to the Queen." A decade later one of Stevie's neighbours was still marvelling: "She'd actually got a hat from a jumble sale and she wore that to see the Queen!" According to two of Stevie's closest friends, however, on the day she went to the Palace she decided against wearing the hat she'd acquired at St John's. But, as is true of much about Stevie's life, the detail has become the stuff of legend. Even the children of Palmers Green sensed something distinctive about Stevie. They may have been struck by her small stature and nervous bird-like movements, and the incongruity of someone her age dressing as a child. While Stevie pottered in her garden, they would come round and peek through the gate, and when she'd go to talk to them they'd run away shouting, "The witch is coming." "She didn't mind this sort of thing," said the neighbour to whom Stevie related such incidents.

Reporters who went to Palmers Green to interview Stevie about the Gold Medal found themselves describing her residence. As one friend recalls, it was a "dumping ground of furniture from a larger house". To step inside was to enter another era. There was an old stone sink in the kitchen which obviously at one time had had only a cold water tap, "the oldest gas stove in the world", and a larder instead of a refrigerator. The woodwork throughout was dark, and bric-a-brac was everywhere. "Miss Smith likes her surroundings," wrote one reporter: "Sepia grandfather in mutton chops and military uniform in his frame on the dresser . . . And shells along the mantel-shelf, a mirror painted with swallows, a view of pebble-dash across the street, rigid decoration in a flowerpot, dried honesty thin and silvery, portraits in oval frames, a gas fire with a low whistle, and an upright piano swamped in books." Her bedroom, however, was austere, clean, and light. Little squares of carpet and straw, free samples, were on the floor, and pictures of a boy in a Victorian sailor suit, an old mill, and a Madonna and Child were on the walls. It was from this house that Stevie set out for the Palace on the morning of 21 November, nervous about being late.

Her appointment was for 12.40 p.m., but she arrived quite early and found the entrance thronged with people who had come to see the Changing of the Guard. "Well, I've just got time to slip round to the gallery and get some picture postcards," she thought. Returning with her small bag crammed with postcards of da Vinci drawings, her hat pulled "over her dead straight hair", she approached one of the police who appeared not to believe that the figure on foot before him had an appointment with the Queen. "I'd better go for another walk," thought Stevie. When it started to rain, however, she pressed her case and gained entry to the Palace. There she was advised how to enter and leave the room in which she would see the Queen, and told not to worry when to come out—Her Majesty would ring a bell under the table. She thoroughly enjoyed the "slightly giggly time" she had "with the lady-in-waiting & a very decorative young man", and when asked to sing one of her poems, she "hissed it at them" under her breath. Finally, a "staggering-looking woman" who was "tremendously made-up" came out from her audience with the Queen, and Stevie had qualms: "I'm not properly dressed for this."

But the Queen was gracious and impressed Stevie with her "enormous aura of efficiency." "She is absolutely charming," Stevie reported, "a totally different sort of mind in the sense that it's not a poetical mind, but I mean on administration and grasp of politics and economics and all that, she's frightfully good, never a foot wrong." Although things were running late, they spent twenty minutes alone together. "I don't know what you'll do with it," said the Queen, as she presented the medal. "I suppose I could have a hook put on it and wear it round my neck," Stevie replied. The event was "rather like meeting the very best sort of headmistress in the very best sort of mood," Stevie later said. At one point Her Majesty remarked, "I hear you live in a house with nine rooms," and Stevie replied, "Yes, some people think it is wrong for one person to live in a house with nine rooms, but I think it's better for one person to be happy in nine rooms than for nine people to be unhappy in one room." Afterwards she wondered if she'd said the right thing, "as Ma'am has quite a lot of rooms too." For the most part, however, the Queen asked questions which kept them on the topic of poetry. "I couldn't help feeling," Stevie said later, "it wasn't *absolutely* her *very* favourite subject." Stevie got nervous and said, "I don't know why, but I seem to have written rather a lot about murder lately." She had recently completed a poem called "Angel Boley", based on an infamous murder case, and as she started to tell the Queen about it the royal smile "got rather fixed".

The adventure of going to the Palace was, of course, the topic of conversation over lunch, and in Stevie it had a witty teller, whose mask of innocence heightened the humour. But first the menu had to be decided. "We all agreed it was an ideal moment for oysters," recalls Eric White. Four dozen were ordered, the party had martinis, and Stevie passed round the medal. Its obverse bears the crowned effigy of Elizabeth II, and its reverse, conceived by the eminent illustrator Edmund Dulac, shows: "Truth emerging from her well and holding in her right hand the divine flame of inspiration". Conversation flowed, but no oysters emerged from the kitchen. "Nothing came and nothing came," according to Norah Smallwood, "and Stevie got more and more edgy. 'First of all,' she said, 'I don't know why you're having it here. It would have been much more convenient at the Ritz.' Absolutely like that and no soft soap at all. And then no food came, no food came . . . and Stevie said, 'Well, perhaps we'd better change the menu and have tea.' She was very naughty." White's inquiries were deflected by the waiters, with whom Stevie became sharp, and the party had another round of drinks. Looking back, White speculated that the chef discovered one or two of the oysters had gone off, and so the lot was thrown out and a runner sent through Soho to find another four dozen. The oysters eventually arrived and were good, but the party had had too much to drink beforehand, and the luncheon was spoiled by bad timing.

Stevie's distress was extreme. She probably had been afraid to eat earlier in the day. An editor friend recalls that several years earlier he had taken her to the Europa Hotel for lunch and "she was like a little girl . . . excited about a treat." Just before the meal she excused herself from the table. "Are you all right?" he inquired upon her very late return. "Yes," she replied, "but whenever anything exciting happens like this, it always upsets my insides." After the Gold Medal lunch party Stevie went for tea to the home of Anna Browne who was, for much of Stevie's adult life, probably her closest friend. She stayed for dinner and then, exhausted, stayed the night. "She was really angry," because of the luncheon's delay, Mrs Browne recalls. "I think that was a physical thing. She was extremely fragile. She just had been too long without anything to eat." The "physical thing", in fact, had been with Stevie since childhood – as we shall see.

2

A House of Female Habitation

From the time he was a schoolboy, Stevie Smith's father wanted to join the Navy. He was a good student, especially accomplished at languages, and school officials agreed he should continue his education at Osborne, the Royal Naval College. But after his two elder brothers drowned, his mother became terrified at the thought of Charles going to sea. So instead of enrolling at Osborne he went into the family coal exporting business and became a forwarding agent. His heart was not in it. When he met Ethel Rahel Spear, her appearance must have captivated him. A photo, inscribed "With love from Ethel, Xmas 1896", shows her to have been a woman of soft and delicate beauty, with a wistful expression and an hourglass figure. Charles Ward Smith was twenty-six and Ethel Spear twenty-two when, on 1 September 1898, they married in Holy Trinity Church, Hull.

Mr Smith's bride was a woman of fragile health. She was only eleven years old when her mother died and, three years after her marriage to Charles Smith, her father died. In one of Stevie's novels we are told of misunderstanding between the characters based on Ethel Smith and her father, as when he hurt her feelings by laughing at her fancies and the paintings she did at school. Whatever the accuracy of this fictional account, she does seem to have cared for him. When her father died Ethel Smith wrote in her diary, "Dear Daddy passed away very peacefully."

Stevie grew up with no memories of her maternal grandparents, but only stories about them. Born a year after Mr Spear's death, she never forgot the accounts she heard about him in his position as an honorary chief engineer to advise the Admiralty, and of the satisfaction he took in his gold-braided uniform. Characteristically, she could be wry about Mr Spear's position, as well as proud of it: she once remarked that their Lordships at the Admiralty never consulted her grandfather, because they thought they could solve problems without the help of civil engineers. About Mrs Spear Stevie probably heard less. She does have Pompey, the protagonist of her autobiographical *Novel on Yellow Paper*, recall the words of her aunt, that if Pompey's grandmother had lived, her parents never would have met. Where then would Pompey have been? she wonders.

But Mrs Spear was not alive and so Stevie's "romantic girl" of a mother married a man whose hair was "in curl"—or so Stevie wrote in a sardonic poem critical of her father. Stevie's sister always had more sympathy for him. Christened "Ethel Mary Frances Smith" she even took to calling herself, in later years, "Molly Ward-Smith", or sometimes just "Ward-Smith". To Molly her father was a "marker", with a cosy, gay charm, and totally irresponsible.

When the Boer War broke out about a year after the Smiths' wedding, the restless Mr Smith yearned to join British troops in South Africa. Only his wife's pregnancy dissuaded him. In *Novel on Yellow Paper* Pompey is more specific: "There was, chaps, a combination against him. That female dragon my [paternal] grandmother was very much against it, she was very much against it indeed, and she prevailed upon my mama to prevail upon my papa. So sure enough he did not go."

Molly was born on 24 January 1901, two days after the death of Queen Victoria. By the following year Ethel Smith was pregnant again, and Stevie was born 20 September 1902. Neither of the babies made a very good first appearance: in fact, Stevie almost died. If her poem "Infant" is autobiographical—in *Novel on Yellow Paper* Pompey says it is—she was two months premature and, at the time of her birth, her father was in Ostend:

> It was a cynical babe
> Lay in its mother's arms
> Born two months too soon
> After many alarms
> Why is its mother sad
> Weeping without a friend
> Where is its father—say?
> He tarries in Ostend.
> It was a cynical babe. Reader before you condemn, pause,
> It was a cynical babe. Not without cause.

Perhaps Charles Smith was in Belgium in connection with his importing business, but Stevie's verb "tarries" suggests his reluctance to return home. When Stevie was baptized three weeks later, the ceremony took place at the Smiths' house, 34 De la Pole Avenue, Hull. She was given the name "Florence Margaret", although the family always called her "Peggy". Almost two decades passed before she acquired the nickname "Stevie". Ethel Smith recorded the baptism in her diary and added, "The doctor had given up all hope but she began to improve this very night and thank God continued to do so."

Charles Smith had yearned for life on the open sea; instead he found himself saddled with increasing responsibilities. His wife had a weak heart, his daughters were sickly, and then his business ran down and went bankrupt from neglect. Stevie comically depicted herself in a poem as being, at this time, a censorious baby:

> I sat upright in my baby carriage
> And wished mama hadn't made such a foolish marriage.
> I tried to hide it, but it showed in my eyes unfortunately
> And a fortnight later papa ran away to sea.

Inspired by the personal mythology Stevie put forth in these lines, and by the "cynical babe" depiction in the poem "Infant", we might imagine that one day Mr Smith picked up his darling Peggy and, instead of a smiling face, met a reproach which was for him the last straw. Uncoloured by conjecture, the fact is that when Stevie was three and Molly five, Charles Smith bolted to the White Star Line and a life at sea. A photo taken about this time shows him in a white cap, uniform and shoes, standing on the planked deck of what appears to be an ocean liner, a cigarette in his left hand. His smooth face, as he stares into the camera, is sad.

With no income from her husband, Ethel Smith could not afford the substantial four-bedroom house which still stands in De la Pole Avenue. She and her sister, Margaret Annie Spear, immortalized by Stevie in later years as "The Lion of Hull", did have some money left to them by their father. They used it to move to London with Molly and Peggy, thinking the girls would have a better chance there to acquire a good education. In later years Stevie made a point of changing her entry in *Who's Who* to note that she left Hull shortly before her fourth birthday. This was her joke at the expense of regionalists. Where people come from doesn't matter, she insisted.

Ethel Smith and her sister had to work things out carefully, and both Molly and Stevie were to look back on their childhood as a time when they were very poor. Stevie came to think of her mother and aunt as "sheltered ladies", and she celebrated their bravery in "A House of Mercy":

It was a house of female habitation,
Two ladies fair inhabited the house,
And they were brave. For although Fear knocked loud
Upon the door, and said he must come in,
They did not let him in.

There were also two feeble babes, two girls,
That Mrs S. had by her husband had,
He soon left them and went away to sea,
Nor sent them money, nor came home again
Except to borrow back
Her Naval Officer's Wife's Allowance from Mrs S.
Who gave it him at once, she thought she should.

There was also the ladies' aunt
And babes' great aunt, a Mrs Martha Hearn Clode,
And she was elderly.
These ladies put their money all together
And so we lived.

I was the younger of the feeble babes
And when I was a child my mother died
And later Great Aunt Martha Hearn Clode died
And later still my sister went away.

Now I am old I tend my mother's sister
The noble aunt who so long tended us,
Faithful and True her name is. Tranquil.
Also Sardonic. And I tend the house.

It is a house of female habitation
A house expecting strength as it is strong
A house of aristocratic mould that looks apart
When tears fall; counts despair
Derisory. Yet it has kept us well. For all its faults,
If they are faults, of sternness and reserve,
It is a Being of warmth I think; at heart
A house of mercy.

The inheritance from Mr Spear paid earnings quarterly but, according to Molly, sometimes the money did not last. An early draft of "A House of Mercy" had lines, later excised, in which Stevie mentions that her mother and aunt, and the great aunt who eventually came to live with them, often lamented that more of their capital would have to be sold, and then they would flinch from doing so. For the time, the inheritance (£8,557 7s.) should have been quite sufficient, but Stevie reflected that the ladies were innocent about money, and it was not invested wisely.

The move to London took place in the autumn of 1906. Stevie's aunt had gone on ahead, in order to find a house, but she still had not succeeded by the time the contents of the house in Hull were packed. Someone was waiting to move in, and the furniture men were ready to drive away, but Mrs Smith could not give them a destination. Finally a telegram arrived from her sister saying she had found a house in Palmers Green which would do, temporarily. The men drove off with the furniture, and Ethel Smith and her daughters boarded the train to London. She and her sister and Stevie were to occupy that house, 1 Avondale Road, for the rest of their lives.

From her first glimpse Stevie thought the house and garden beautiful. Although Palmers Green is only eight miles from central London, in 1906 it was a country village. Stevie remembered it ever after as illumined in autumnal sunshine and scented with the rich smell of acorns, damp mould, and the large michaelmas daisies that grew in the churchyard. The semi-detached villa found by her Aunt Maggie was part of a little terrace of such houses, surrounded by fields and woods. Red-brick with white facing, it had in front square bay windows and a garden plot. On the ground floor were front and back parlours, and a kitchen which opened on to a small rear garden. Upstairs were three bedrooms and a bathroom. Such houses were inexpensive at that time. When Mrs Smith arrived, she went with her daughters and sister to their landlord's shop around the corner to make some arrangements. The landlord put Stevie on a large machine, the kind used to weigh luggage, and remarked that she made a fine package. When Stevie replied in her Yorkshire accent that she came by train and tram, he called her a foreign package. It was a memorable introduction to Palmers Green.

Two images remained for ever vivid in Stevie's mind from that first year in her new neighbourhood. One was the visit to her home of a gentleman who came in a horse-drawn carriage and wore a top hat. Perhaps both carriage and hat had been hired, she joked decades later. The visitor, Dr Woodcock, became Stevie's family physician. Curiously enough, the second image came from a description of childhood in India. Stevie might have heard it from a relation who had lived there, or maybe the description was in print. When she was too young to read for herself she did enjoy having books read aloud: *Alice in Wonderland*, for example, and another favourite, Charles Kingsley's *The Water Babies*. Her exotic Indian memory was of an ape outside a house, happily swinging back and forth in the moonlight. The image must have fascinated her: years later she incorporated it into a novel as a memory from her protagonist's Indian childhood. The image is also celebrated in Stevie's short poem, *"Le Singe Qui Swing"*, which she sang to the tune of "Greensleeves":

> Outside the house
> The swinging ape
> Swung to and fro,
> Swung to and fro,
> And when midnight shone so clear
> He was still swinging there.
>
> Oh ho the swinging ape,
> The happy peaceful animal,
> Oh ho the swinging ape,
> I love to see him gambol.

Stevie's friend, Anna Browne, notes that it was common to have relations or friends who had been in India and who were indulged by their Indian nurses. For Anna, *"Le Singe Qui Swing"* evokes expulsion from paradise, the Eden lost when British children left to begin school in England.

Stevie wrote that childhood seemed like "a golden age",

> a time untouched by war, a dream of innocent quiet happenings, a dream in which people go quietly about their blameless business, bringing their garden marrows to the Harvest Festival, believing in God, believing in peace, believing in Progress (which of course is always progress in the right direction), believing in the catechism.

Although she endured a kind of exile as a child, it did not begin with Stevie's going to school. The Palmers Green High School and Kindergarten opened the year before she and her family arrived in their new neighbourhood. Popularly called "Miss Hum's School", or even "Missum's School", after its first headmistress, Alice Nellie Hum, it was located three short streets from Stevie's home, at the corner of Osborne Road and Green Lanes—streets then lined with elm trees, hawthorn bushes, an occasional country house and thatched cottages. Stevie attended the kindergarten and once described hearing about the Transvaal there, and using matchboxes and buttons to make covered wagons with wheels. She also recalled the first hymn she sang in kindergarten although, she said, she could not have been older than five when she did so:

Great big wonderful beautiful world
With the wonderful waters around you curled
And the wonderful grass upon your breast
World you are wonderfully beautifully dressed.

Stevie was exiled from the wonderful world, or at least from
wonderful Palmers Green and her beloved mother, when she was still
five, and Dr Woodcock diagnosed her weakness and loss of appetite as
symptoms of tubercular peritonitis. Today doctors can cure the
disease in a few weeks but the necessary drugs did not exist then, and
Stevie had to remain in a convalescent home for sick children which
had been founded at Broadstairs, on the coast of Kent, by the inventor
and philanthropist, Sir Alfred Yarrow. She was there, on and off, for
three years. When Dr Woodcock announced his retirement in 1935,
Stevie sent him a warm letter of thanks for all he had done to help her
make the best of a "rotten inside". To his daughters she said many
times in later years that their father had saved her life.

Yarrow endowed his convalescent home "to the utmost limit that
was wise". It was established for "children of educated people who
could not afford the full expense of a stay at the seaside, so necessary to
a child recovering from illness or an operation". At first Stevie seems
to have enjoyed her stay there. Her poem "The Occasional Yarrow" is
probably based on her happy rambles in the area. And although Stevie
later wrote, "five-year-olds do not have friendships at all, other
children are to them rivals and perils, how can they not be?", she
formed at least one friendship at the time, with another patient, Hester
Raven-Hart, a friendship that continued for many years. Stevie once
recalled that she felt privileged and grand to sit with the boys at a table
for those on a special diet. Hers consisted of three pints of milk a day
and raw eggs. She also enjoyed the prestige of being in the ward for
those most ill: children who left after a short time were considered to
count for less.

But finally, life at the convalescent home became distressing. One
result was Stevie's enduring horror of sleeping other than in complete
darkness, which, she told a friend, stemmed from her childhood in
hospital when lights meant Nurse was coming to visit the ward. To
another friend she mentioned a "ghastly" matron. One thinks of a
convalescent home maid in *Novel on Yellow Paper* who frightens
Pompey by her arbitrary affection, which was not at all the deep love
of a mother but only a terrifying pretence. Stevie's stay at the home
was interrupted by holidays spent with her family, and by visits
from her mother: when the holidays and visits ended, however, the
pain of homesickness and anxiety would be felt all the more. Her

doctors decided that the visits made her too unhappy, so they were forbidden.

While in hospital Stevie received occasional laconic postcards from her father. (If he wrote letters, they have not survived.) Usually the cards were a jot longer than the one cited in *Novel on Yellow Paper* —"Off to Valparaiso love Daddy"—but only a jot. It seems Stevie first went to Broadstairs in the spring of 1908: a postcard from her father dated 23 May of that year, sent to Palmers Green, had to be forwarded to Yarrow Convalescent Home. It read: "Darling Peggy, I hope you are still going on all right & gaining weight. I shall be in York tomorrow & hope to find news of you waiting for me there. Love & Kisses from Daddy." Another card, mailed in New York on 17 August 1908, was sent to Stevie in Pakefield, Lowestoft, where she was on holiday with her mother and sister: "Darling Peggy, I hope you are still continuing on the mend & when I do see you I shall find you strong & well & eating a lot. Love from Daddy."

Of the other cards which have been preserved only one, sent to Molly from New York on 22 June 1909, is of interest. It indicates that Mr Smith visited his family on rare occasions, and it reminds us that the health of Mrs Smith was precarious: "My dear Molly, Just a line to say I hope to see you again next time home. I was sorry that Mama had to go away for a bit, & I hope she is back now & better. Love & Kisses, Daddy." According to Molly, her father once returned from a voyage to America with a gift for Stevie which she loved—one of the first Teddy bears. And friends of Stevie recall her saying that her father showed up one day out of the blue with a parrot on his shoulder, a gift for his daughters. But laconic postcards and occasional visits and presents are thin porridge. We can imagine that Stevie's anxieties while she was in the convalescent home were not allayed by her "unrespected" father's voyages on the high seas, or by her beloved mother's bouts of ill health.

Stevie's prolonged and sporadic exile did have a lengthy break in the autumn of 1910, for in December of that year a local newspaper recorded that she received a prize for attendance in the Upper Transition Class at Palmers Green High School. A story in the same issue of the paper described "the splendid system of teaching carried out by Miss Hum's able assistants", mentioning, for example, a model of the Alps made by children in the geography room. This model was probably supervised by a teacher Stevie mentions in her writings. The form-mistress of her Transition Class, Stevie wrote, combined teaching painting and geography, and had a talent for giving children a vivid impression of the jungle. This instructress, who later left the school to study with the famous Egyptologist, Flinders Petrie, in-

*Stevie's drawing of a parrot which she placed by
her poem "Parrot". One of the motifs in Stevie's
second novel is her autobiographical protagonist's
references to her parrot Joey, "that died, that stood
upright upon his tail feathers, to stand upright and
cry aloud: 'What is come upon me?'"*

spired the children to bring cardboard dress-boxes from home and,
with moss and paint, transform them into tropical forest scenes.

That Stevie remembered with such pleasure constructing those
forest scenes is not surprising. Doreen Diamant (*née* Coke-Smythe),
who was one of her childhood playmates and became a lifelong friend,
recalls that Stevie drew all the time as a child. Often the local papers
carried notices of Honours she gained in the Royal Drawing Society's
examinations. Of course later in life when Stevie's volumes of poetry
were published, she included many of her drawings in them. She also
had a sketch-book published. It could be said that the house at 1
Avondale Road was hospitable to the visual arts: when Stevie died
there still hung on its walls sentimental pictures in the spirit of
Landseer done by her mother.

It seems Mrs Smith's artistic interests extended to literature too, as
she won a copy of Kingsley's *Hereward the Wake* "for a short story on
'Strife' at the Eisteddfod held by the Palmers Green Literary Society,
April 4th 1910". (Publisher Arthur Waugh, father of Evelyn and Alec,
was the adjudicator.) According to a newspaper account, Mrs Smith's
story concerned "the life struggle of a rather unfortunate clerk—a
humdrum hero", and it was the best of five submissions. So one can
also see Stevie's literary career as an emulation. It was from her
mother, at Christmas 1909, that she received an expurgated edition of

Grimm's *Fairy Tales*, stories she loved and which later on she read again and again, in the original German.

1910 was a year of other noteworthy events, in addition to Stevie's prize for perfect school attendance and her mother's literary award. Edward VII died, and George V became king. In Palmers Green the Anglican Church of St John the Evangelist celebrated the induction of the Reverend Roland d'Arcy Preston as its new vicar, the second in the five years the church had been established. The first vicar was a great builder (church records show that Stevie's aunt and great-aunt, Miss Spear and Mrs Clode, were contributors to the Building Fund in 1908), and St John's still owed an enormous debt in 1910. One of the Reverend d'Arcy Preston's early accomplishments was to preside over the elimination of the debt, especially that on the parish hall. In her writings Stevie invented the name "Humphrey d'Aurevilly Cole" for the character based on this vicar, and she recalled d'Arcy Preston affectionately as being scholarly, eccentric, absentminded, and truly kind.* He once himself paid for the Italian holiday of an invalid parishioner. One of Stevie's favourite anecdotes concerned his behaviour at a meeting from which the Lion Aunt, who was a member of the church finance committee, came home "in quite a furious mood".

> "He really is a trying individual," [Aunt] said, "he is really very trying. All the time Mr. Harbottle was giving the estimate for the new cassocks and for the heating plant that is to be installed in the new hall he was laughing quietly to himself. The doctor," she said, "who was sitting beside him on the platform, kept nudging him to stop laughing, and the vicar said, 'What's the matter Blane, why do you keep nudging me?' Everybody could hear what was going on, it was most unfortunate. So then the doctor explained that Mr Harbottle was rather hurt that the vicar was laughing, and the vicar said 'Oh I wasn't laughing at you, Harbottle, it was just something that I was thinking about, just something that was passing through my mind.'"

Despite Aunt Maggie's fury on this occasion, she and the rest of Stevie's family were fond of their vicar. He was a High Churchman, favouring the ceremony and ritual in which Stevie delighted as a child:

---

*The name of the French fiction writer and critic, Jules Barbey d'Aurevilly (1808–89), probably inspired Stevie's invention of the name "Humphrey d'Aurevilly Cole". Barbey's writings fascinated Stevie, and his dandyism and Ultramontanism must have seemed to her a funny parallel of d'Arcy Preston's eccentricities and High Church preferences.

I liked to see the great banners go round the church on festive occasions. I especially liked the great banner that had on the back of it: The gift of Anna Maria Livermore. "Anna Maria amen" (I used to chant) "pray for me and all men." Probably my first steps in verse—in heresy too, I daresay.

But not everyone in the congregation appreciated the vicar's enthusiasm for ecclesiastical display. Trouble began about a year after his induction, when he started wearing white linen vestments at morning celebrations on weekdays. "Falling about in the pulpit" was one thing, but white vestments on weekdays seemed papist to some. Stevie recalled that when the vicar tried "to be a little 'higher', a little more Anglo-Catholic, that is, than his congregation would altogether stand for," her family "was behind him to a man, or rather to a lady."

Church activities played a large part in the life of Stevie and her family. In July of 1910 Mrs Smith and Molly assisted at the church bazaar wool stall. Stevie probably participated in the children's play put on as part of the bazaar and directed by a Mrs Cooper: a photo taken about this time and in the possession of Mrs Cooper's daughter, Olive, shows Stevie wearing a robe with a peaked cowl, her face pudgy and heavy-lidded, kneeling before an imaginary fire with similarly costumed young girls. The following July, at the St John's Flower Show, Molly was at the choir's stall and Mrs Smith and Miss Spear ran the fish pond. Miss Spear, as we have noted, was active also at church meetings, and she made kneeling cushions and mended surplices. According to Molly, however, Aunt Maggie did not often attend services and, when she did go, it was to keep an eagle eye on the choir boys to make sure they didn't put their fingers through holes in the surplices. But Stevie went regularly to church, where she especially enjoyed singing. She would decide beforehand which hymn to sing—usually it was "Once in Royal David's City"—and she sang it, "and it did not matter at all that everybody else was singing something different."

Stevie also enjoyed the lantern-slide lectures sponsored by the Palmers Green Literary Society and held in St John's Hall. On one occasion the Society's Vice-President spoke about "Old London"; another time a Dr H. G. Rosedale, MA, discussed the "Zone of the Zambesi"; and once the Society brought Hilaire Belloc to their parish hall to lecture on "The Men of the French Revolution". But of all the lantern-slide talks Stevie attended she described only one, a "fascinating lecture on the Moon" given by an old clergyman: "He had a slight impediment in his speech, but such was his enthusiasm for the moon and so remarkable were his illustrating slides, that he was one of the

Society's most popular figures and was often required to repeat his lecture."

Many groups not connected with St John's were also able to rent its hall, thereby helping to eliminate the debt on the building and, eventually, to raise funds for other worthy causes. Meetings sponsored by church and non-church groups alike contributed to the neighbourhood's intellectual and cultural liveliness. One year, for example, George Bernard Shaw spoke at the parish hall under the auspices of the Palmers Green Fabian Society. And in November of 1910 Miss Douglas Smith, a suffragette, spoke to a large audience on "Votes for Women". Emotions must have run high, for a local paper reported that the audience listened with "intense interest" and that the Reverend d'Arcy Preston said "wild horses would [not] draw from him his opinion on the subject."

Additional contributions to the St John's Hall Fund came from the proceeds of theatrical productions put on there. In 1910 Molly was praised for being "delightfully serious in her mock dignity" as Little Tommy Tucker in the play *Babes in the Wood. Ali Baba*, produced in January of 1911 and also a success, was reported to be something of a first:

> It is now quite the fashion for children's pantomimes to be given in the parish halls of the local churches. A pantomime is supposed to be performed by adults for the entertainment of the young people, but when 'Ali Baba' was given at St. John's . . . we found that all the performers were children and nearly all the lookers-on were adults.

As Captain of the Forty Thieves Molly "delighted in stamping her feet, which were encased in big top-boots", and her dance towards the end was singled out as "one of the hits of the whole performance". Although Stevie's sister and several of her friends took part in *Ali Baba*, she herself does not seem to have done so.

Quite likely Stevie was back at the Yarrow Home. She refers to being at the convalescent home, and to thinking there of suicide, at the age of eight. In *Novel on Yellow Paper* Pompey, too, thinks of suicide when she is eight, and it is ironic that the thought which Stevie and Pompey found so comforting, that one can take one's life, blunted their desire to do so. As Stevie explained years later, if one can leave the world at any time, why especially now? Psychoanalysts charge a pound an hour, Pompey declares, to tell us that our impressions from childhood make a difference. Rates are higher today, but the truth remains. As a child Stevie loved to sing over and over again the refrain

Thoughts of suicide at 8 - paradox

from an Australian ballad, "Out where the dead men lie", and later in life she acknowledged that many of her poems were inspired by the thought of suicide which cheered her when she was eight. It is not difficult to find such verse, from her first book of poetry, in which "Is it Wise?" implies that suicide is sometimes a wise option, to her last poem, "Come, Death (2)", in which she praises death as a god who comes when called. The latter idea she put into her translation of "Dido's Farewell to Aeneas", although she admitted it is not actually in Virgil. The theme of suicide recurs in her poems, and Stevie mentioned it in essays, and spoke about it in interviews and to friends. Part of her quarrel with Christianity was that it absolutely forbids command over death, a power she found "delicious".

Besides providing her with the ironically fruitful idea of suicide, Stevie's childhood separation from her family no doubt played a part in her fascination with legends of stolen children, and stories of others magically whisked away. Such accounts are found in fairy tales, which she loved to read, and they also surface in several of her poems. Once, reviewing an autobiography, she drew the wry moral that to remain an individual one must arrange to be an invalid as a child. An individual she certainly remained, combining a child's freshness of vision with (as many of those who loved her have testified) a child's need for cosseting and attention. The latter trait is not rare in people who have been institutionalized during childhood. A number of Stevie's friends report that, during visits, she would regard their children as rivals for their attention. Yet she also enjoyed the company of her friends' children and, according to Anna Browne, was desperately stricken by accounts of cruelty to children. She admired D. H. Lawrence's "The Rocking-Horse Winner" because it is about the exploitation of a child, and the heroine of her strange "Angel Boley" poem poisons her mother and husband because they are murdering children. It may seem odd to attribute her rivalry with children and her compassion for them to the same source, yet it is no more odd than that both should have existed in the same person.

Stevie could be proud, too, with the pride which Pompey says is paramount in the rich and spacious thought of suicide. The concept of death which consoled Pompey when she was eight she understood, in her child's way, as a fall from a cliff or out of a window. One thinks of a quotation Stevie loved, Seneca's words to an unhappy slave: "Dost thou see the precipice?" Seneca meant, she would explain, that the slave could always jump off it. True, Stevie's lofty pride would often alternate with tears, but then paradox was the fabric of her character.
The thread of pride can be seen in her poem "A House of Mercy", in

which she praises, among other things, the "aristocratic mould that looks apart / When tears fall."★

Stevie's sporadic periods of exile at the Yarrow Home seem to have ended by the summer of 1911. In June of that year she took an examination given by the London Institute for the Advancement of Plain Needlework, passing first class in the Preliminary Division. Her sister gained a more impressive award that June: although the Palmers Green High School for Girls, as it is now called, no longer stands at the same address, it still has on display a plaque naming Molly as the scholarship girl for 1911, a distinction which provided her free tuition (£10 4s.) for the 1911–1912 school year. Before the autumn session, however, the Smiths were to spend two months of an especially hot summer at Saltfleet in Lincolnshire. They were accustomed to summer holidays: Pakefield in 1908; a week at Clacton in a house miles from the sea, paid for by a grandparent; and other holidays in the years before the Great War, at Heacham, Frinton, Sotwell and Walton-on-the-Naze. Late in life Stevie confessed that she didn't know how they were able to afford those holidays she loved. But her mother always found enough money to take a small house or cottage which had been advertised, and off they went, any doubts about what they would find swallowed up in the sense of adventure.

Stevie recalled the happy summer days of 1911 in her second novel, which describes the first thing she did, or rather the first thing her protagonist, Pompey again, did when she arrived at Saltfleet with her sister. In their joy they ran off from mama and the Lion Aunt, across sand dunes and salt marshes, losing their way on the other side of a broad and dangerous river that flowed between their rented cottage and the sea. Pompey does not remember, but hopes, that her eventual return was accompanied by a proper smack. Stevie also evoked the memory of that Saltfleet holiday in her poem "Archie and Tina". Its speaker recalls the two seaside playmates of the title, and the adventures they all had fishing for crabs, poking anemones, and the like. Where are Archie and Tina now? is the question which opens the poem, and it closes by noting that the *ubi sunt?* is rhetorical, its point being to express joy. Over the poem's course the speaker moves from exclamations of yearning, "oh if only", to what appears to be radiant contentment at the remembrance of past pleasure. The ending of

---

★ The title, "A House of Mercy", is a pun. The poem describes Stevie's *house*, or place of "habitation", as a "Being of warmth" with qualities of "sternness and reserve", in order to celebrate the nobility of her *House*, or family line (traced by its female descendants). Thus Stevie placed a drawing of a house at the poem's beginning, and a drawing of a crowned girl at its end.

"Archie and Tina" is effective not just because the experience it evokes is a happy one (improved by memory, one suspects: never were there so many poppies as then), but also because it is reconciling. Intimations of mortality—animal bones in the sandhills and a dog's earache—are so much a part of the remembered joy as subtly to assist the speaker's shift from yearning for the past to contentment in the present.*

By the autumn of 1911, shortly before her ninth birthday, Stevie was back in attendance at the Palmers Green High School. She acquired there what she subsequently concluded was the best sort of education for a poet. The poems she studied were only the classics of English, "the good poems and the bad-good poems, alike honoured by Time." She was obliged to memorize many of them for homework, and also as a penalty if she broke rules. Far from putting her off, the memorizing put her on. Her English mistress read poems aloud in class, and a phrase would strike Stevie, a narrative grip, an image become vivid. She liked narrative poems, such as "The Rime of the Ancient Mariner", and poems with grand words, such as "Arethusa arose from her couch of snows, in the Acroceraunian mountains" —the very first poem she learned by heart. A tenet of the High School was that good judgment requires a grounding in the classics, and the children were not encouraged to study contemporary poetry. Stevie once wrote that true poets when young are inclined to be derivative, and it is only poems tested by time from which they can safely derive. An imitator of Wordsworth, she argued, will find himself out readily enough, but an imitator of a contemporary poet might not.

During her childhood Stevie did write a poem or two, but she did

---

*Stevie was in her sixties when she wrote "Archie and Tina", and the circumstances of its composition are instructive. A few months before she died, she recalled being thrilled as a child at finding a dog's jawbone, complete with teeth, while digging in the Saltfleet sandhills. It is a fox's jawbone with teeth that the speaker of "Archie and Tina" recalls having found. What precisely Stevie dug up in 1911, of course, hardly matters: it may be, however, that she deliberately substituted "fox" for "dog" in the poem in order to avoid suggesting a meaningless association between the jawbone and the dogs, Bam and Boy. More important, Boy was not a dog of Stevie's childhood. It seems to have been in the spring of 1968, when Stevie was visiting her friends Mungo and Racy Buxton at Cley, Norfolk, that their hunt terrier developed an earache. Stevie had been walking Boy along the beach with the Buxtons' elder daughter, Juliet. The image of his getting an earache in the strong wind, and having to be wrapped in a jersey and taken home, so fitted the mood of "Archie and Tina" —of a cosy world in which calamity is so circumscribed and evokes such care as to be recalled fondly—that Stevie transported Boy to her poem, and the year 1911.

not write poems again until she was in her twenties. One of her essays contains her first, about which as a child she had a high opinion not shared by Aunt:

> Spanky Wanky had a sister
> He said, I'm sure a black man kissed her
> For she's got a spot just here
> Twas a beauty spot my dear
> And it looks most awfully quaint
> Like a blob of jet black paint
> But when he told his sister that
> She threw at him her gorgeous hat
> And with airs that made her swanky
> Said, I hate you Spanky Wanky

This is definitely a child's writing, yet the "gorgeous hat" reminds one of the later Stevie. A letter sent to her in 1920 and signed "F.G." mentions the "Spanky Wanky" poem, and gives Stevie remarkably prescient advice: "I am convinced your forte ought to be quaint, well-made, clear-cut verses. Can you still do the 'Spanky-Wanky' kind? 'She threw at him her gorgeous hat' is a triumph. I believe you could still do that kind of thing a little more advanced, so to speak."

Stevie's years as an invalid, and her early interest in poetry, should not mislead one into imagining that, as a child, she had a quiet, retiring nature. Although she was certainly bookish, she preferred adventure stories—among them works by Rider Haggard, John Buchan, Dumas and Harrison Ainsworth—and she was an adventurer among her friends, too. About her reading of Ainsworth's stories, she once told an anecdote set in a school charabanc going through Windsor Great Park. The driver suddenly turned and said, "This is where Herne the Hunter harried Henry VIII." "I knew what he was talking about," Stevie noted, "because I had read *Windsor Castle*, but I felt that nobody else had or did, because that is not the sort of thing you learn in history books. It was a link between us."

Behind Stevie's house, on the other side of a railway cutting, were privately owned woods where trespassers were forbidden. "This of course made it all the more exciting," she wrote, recalling that having promised her mother she would never cross the railway line, she used to crawl through a large pipe that ran under it, and so play in the forbidden woods. Olive Pain (*née* Cooper) remembers going with Stevie through that drainage pipe, which still runs under the line by Hoppers Road. In the woods Stevie enjoyed "catching tadpoles in the

clever *adventuresome athletes*

energetic

ponds" and outwitting the keeper and his dog, who were there to prevent trespassing. She and her friends used to have picnics by a "witches' pool", and rabbit holes in its sandy banks made a convenient hiding place for their treasure. "I remember," Stevie wrote, "I once left a copy of my Aunt's favourite book—Francis Younghusband's *The Relief of Chitral*—in one of these holes, and never recovered it." Both Olive Pain and Doreen Diamant recall Stevie shooting to the top of a holly tree in those woods on Winchmore Hill. Tied there was a cardboard shoe-box with a slit in it, into which Stevie and her friends would post their "secret letters to each other". To Doreen Diamant, Stevie was a delightful leader in mischief, and the Woodcock sisters recall that she was always saying, "I should have been a boy." On three occasions she sprained her knee playing hockey at school, and once she entered both an eighty-yard race and an egg-and-spoon race at a St John's Flower Show, coming in third each time.

According to Olive Pain, Mrs Smith—whom Olive remembers as being "very pretty, very delicate"—was a bit overwhelmed by Molly and Stevie because they were "so frightfully full of life". Molly once reminisced about a time before the war when her mother and aunt were taken to an exhibition in Earls Court by a suitor from America who wanted to marry Miss Spear. (Aunt never married: Molly recalled her saying, "I can understand liking a man, but not being in love with him.") The children had chicken-pox, but were not really ill, and they stayed at home with their great-aunt, whom they tormented by throwing dolls and clothes out of a bedroom window for her continually to retrieve. Alexandra Palace, a leisure centre not far from Palmers Green, provided a more seemly outlet for the girls' exuberance. (It had a popular skating rink; concerts were held there; and in later years Alexandra Palace became the site of the first public television station in the world.) Stevie used to be taken to the "Ally Pally" where she would "slide down the slopes and go on the roundabouts". Fairs also provided conventional amusement, and Stevie had a great adventure at one when she was about ten:

> The most exciting thing that ever happened to me was when one of those swing things . . . went over the top, and *I did not come off*. How this happened I don't know, and neither does anybody else, but I suppose it was so quick or something, anyway the relief of the fairman who saw it happen was very noticeable indeed. . . . For myself I remember . . . feeling nothing but a slight bewilderment as to where, having inadvertently looped the loop, I now was. The fairman took me behind where his wife gave me a cup of hot tea, in

which, as I distinctly recall "a funny flavour," I fancy there may have been something stronger than tea. This, whatever it was, brought my wandering wits to attention and after gazing at him, the fairman, for what must have seemed to him at least three whole minutes in complete silence (and three minutes is quite a long time for this sort of thing), I brought forth these memorable words: "I ought to have come off."

Stevie showed similar pluck on the occasions when she would climb a wall with a friend, and "crawl along the top of it past other people's gardens right down the whole length of the road". There were perils on this exciting journey, for the girls had to avoid being seen, and they had to dodge overhanging shrubs and trees. Other favourite games required imagination: Beowulf, for example, in which Stevie and her friends would enact, in the forbidden woods and by the lake in Grovelands Park, the story they heard in school of how Beowulf "tore off Grendel's arm in the depths of the lake". Another game they relished was The White Slave Traffic. The Priors were two ladies who lived next door to Stevie all through her childhood. One of these sisters used to warn Stevie: "If a lady comes up to you in a closed cab and leans out of the window and asks you to get in, don't you have anything to do with her," and "If a lady comes up to you and tells you that your dear mama is lying in a faint on the pavement round the corner, don't you believe her, don't have anything to do with her, do not go with her into the cab." "It seemed," Stevie wrote, "that if you yielded to the blandishments of this lady in a cab you would become a White Slave." What that meant Stevie and her friends did not know, but the warnings fired their imaginations and "very far from boggling at such a proposition" they would "play the game with great gusto, trundling [their] dolls downstairs on teatrays—they were the white slaves—and bundling them off in the dolls' pram—the 'cab' of course—to some rather shadowy destination."

School and church theatricals provided opportunities for more organized diversion, in which Stevie followed her elder sister's lead —at least once literally, when she too became Captain of the Thieves in a production of *Ali Baba*. She mentioned to a friend that her first stage appearance had been at the age of nine in *As You Like It*, and this was probably put on at the Palmers Green High School. Doubtless she appeared in other productions there, but no records remain, as they do of plays put on in St John's Hall and at her second school. Many of the plays at St John's were directed by Mrs Zoe Cooper. "My mother was a great person for amateur theatricals," Olive Pain recalls, "and she used to produce a children's play every year, gathering in all the promising people, Flora Robson among them."*

According to Mrs Pain, her mother would start the productions in late autumn:

> She would use the stories found in a publication called *Books for the Bairns*, which provided very good musical scores as well. Mother used to get the Palmers Green Orchestral Society to learn all the parts: we had a full orchestra at St John's Hall every year, and the plays would raise money for various charities. Mother was very friendly with d'Arcy Preston because he used to descend on her and say, "Now Mrs Cooper, what are you going to do this winter?" And she would say, "Well, I don't think I'm going to . . ." And he would say, "Oh, you must!" I remember once Peggy played the Wolf in *Red Riding Hood*—chased me round the kitchen table, terribly good—she terrified me. Molly was mad to go on the stage, and at that time it wasn't easy if you'd got a North Country accent. It wasn't considered the thing at all.

One of Mrs Cooper's more unusual productions was an outdoor pageant of Living Whist she arranged for the St John's Bazaar and Flower Show in June 1913. Mr Cooper spent weeks specially painting sandwich-board playing cards. His daughter Olive, and Stevie, were among the Hearts, while Molly was one of the Clubs. The whole pack marched out and were cut, shuffled, and dealt before the teams began to play.

During Stevie's adolescence she continued to act in plays, both in school and parish productions. These theatricals are of interest, of course, not only because they show her to have been an energetic and clever girl, but also because they foreshadow her later delight in reading and singing her work before live audiences and on the wireless. The theatricals gave her an opportunity to develop a sense of timing, and an ability to project, and they instilled in her at an early age an attitude of professionalism in performance. When in much later years she took to the poetry-reading circuit, the casual and sometimes vulnerable air she so winningly projected concealed her art. Her

---

*Dame Flora Robson, the internationally famous actress, was celebrated in Palmers Green, even at the age of ten or so, for her dramatic talent. That is why Flora Robson, who was not Church of England, was invited by Mrs Cooper to participate in the children's plays put on at St John's. See for example *The Sentinel*, 6 October 1911, p. 2: "little Miss Flora Robson, of local eisteddfod fame, recited and the abandon and dramatic instinct she revealed quite carried away the audience. But she only bowed gravely, each of the three times she reappeared in answer to the clamour for 'more'."

poetry recordings, the BBC scripts she wrote, and reading notes which have been preserved demonstrate, for example, her careful planning of transitions between poems, and her sense of a poetry reading's overall design. She timed her reading of individual poems, making note of exactly how many seconds it took to read each, so she could quickly tailor her performance to any length.

Like all artists, Stevie was captivated to the end of her days by the magic imagination engenders. Frightened by her father's unreliability, her mother's fragile health, and her own illness and exile to hospital, she could in these children's plays safely confront disturbing life circumstances as if they threatened others and not herself. No doubt the role of the crippled boy which she acted in a play significantly called *The Wishing Well* had autobiographical resonances for Stevie. She was praised for being "pathetic" in the part.

Looking back over Stevie's early years, one sees of course the terrors but also those "innocent quiet happenings" of her Edwardian childhood which she described lovingly and often in her writings. "She always envied children," recalls a friend. "They had in her view the ideal life—were looked after and could be themselves. She felt this also about animals who existed for her in a prelapsarian state." And then in a particularly vivid way came the loss of that innocent world for Stevie. *The Wishing Well* was staged at St John's church in July of 1914. Less than three weeks later, when she was eleven years old, the world of her childhood "split with war".

3

Times of Unquietness and Bewilderment

The most immediate effect of the war on Stevie and her family was the cancellation of their usual summer holiday. Four months after war began Mr Smith entered the Royal Naval Reserve as a Temporary Acting Paymaster and, about a year later, he was made an Acting Interpreter (French). Much of his time during the war was spent in the North Sea Patrol. As a naval officer's wife Ethel Smith received an allowance. Periodically Charles Smith would come home to collect it and, as Stevie wrote in "A House of Mercy", her mother "gave it him at once, she thought she should." Although Ethel Smith did not welcome her husband's visits, Aunt had a soft spot for him. He was a man for drinks and women, could "charm the birds off the trees", and "couldn't pass a pub". Molly loved him but Stevie disliked his visits. She thought her mother "immensely loyal" for keeping up appearances and never saying a word against Mr Smith. When he would visit home on leave, Stevie and Molly had to come down from bed and talk to him. She could stand it, Stevie used to tell herself, as it was only for a day and her poor father was spending so much time in the snow and ice. Once Charles Smith was lucky enough to go on a southern patrol and Ethel dutifully sent his "whites" after him. Upon their arrival, however, he had to return to the Baltic.

Mrs Smith's health was in decline during these years, and she had to go about in a Bath chair. "When I saw the suffering of my much loved ma," Pompey says, "I raged against necessity, I raged against my absconding and very absent pa, I raged and fumed and spat." Whatever the situation in the Smith household, Ethel and Stevie were not indifferent to Charles Smith's fate. In the spring of 1916, when Stevie was thirteen and travelling on holiday with relations from Hull, she received a letter from her mother which allayed fears: "Daddy is in Glasgow, so it is some relief to know that he is safe & well. I should have been wondering now one hears of the fight off Lowestoft yesterday, and sometimes it is so difficult to get news."

Enemy action during the Great War also endangered those not in uniform. There were nineteen Zeppelin raids on eastern England and London in 1915, and forty-one such raids in 1916. But the population seemed to take these raids in its stride: after the worst, on 13 October 1915, a Palmers Green newspaper carried the advertisement, "Are your Nerves Upset by the Zeppelins? Then take Dr. Burton's Bontonique No. 2." And when, almost a year later, a Zeppelin was brought down in Cuffley, residents of Palmers Green and Wood Green sallied forth, by tram and motorcar, on cycle and on foot: "Everybody seemed souvenir mad", one paper noted. In contrast to this almost festive report, however, was the shocking account of charred bodies Dr Woodcock, the first physician to arrive on the scene, gave to his family. While still on holiday with her relations from Hull, Stevie wrote home on a card postmarked 25 April 1916: "I see the Zeps have been over the Eastern counties. Does that mean London?" Her mother replied: "We were roused up by the Zepps on Tuesday night but did not hear them for long, several people saw it in the searchlight & we saw the exploding shells from the guns but that was all. I don't think *much* damage was done but that one never knows." On 1 May Mrs Smith wrote again:

> My dearest Peggy
>
> Just a line to send with your fare. I am sending a pound note as I said, you can bring what change you have back with you. We have had a very severe thunderstorm tonight while we were in the Hall at the dramatic entertainment, there was a terrific crack & several people sprang up almost in a panic but someone called out "It is a thunderstorm" & though several went out everything went on quietly after but the lightening [sic] was awful. D. found out that my chair had been out in all the rain & he brought me home in the car before taking the children home. I did think it was good of him. No more just now. Goodnight dear.
>
> Best love from
Mother

Stevie replied on 3 May, thanking her mother for the fare, noting that it was kind of the doctor to take her mother home "in his motor", and expressing the hope that "Aunty Grandma was not upset by the thunderstorm".

As the war continued London residents faced greater danger. German airplanes bombed the city on 28 November 1916, the first of many such raids. In October of the following year a Palmers Green paper complained that a recent meeting of the Literary Society was a "fizzle", that the tendency of Londoners to use their evenings for amusements and useful work had been "rudely interrupted", and that "poor people no longer take their children into the streets as in Zeppelin days".

War also brought changes in routine at the Palmers Green High School and Kindergarten. The girls embroidered the Belgian, French and Russian flags on needle cases, and each had a blanket to drape over her desk so that she could sit under it and be protected from glass splinters. They practised doing this during air-raid drills. Every morning at eleven St John's bell tolled, and all the children stood and prayed for the soldiers at the front. Flora Robson, who entered the High School at the age of thirteen, used to play the piano at daily assembly when a hymn was sung, after which Miss Hum, standing on a dais, would speak on some proverb or moral saying.

"Miss Hum was a very wise and original lady, and rather brave too," Stevie recalled in later years, "because when the war came (the 1914 war) she said that not absolutely *every* German was 'a fiend incarnate' as was the general opinion in our suburb." (Pompey echoes this sentiment, noting that her headmistress "had the strength of her unpopular convictions, to stand up at that time and declare that war on the long view was not a good thing.") Miss Hum, a Quaker, was much loved by her pupils. The motto she chose for the school was, "By Love serve one another", and love and service were demonstrated, not only by the concern and dedication of her staff, but also by the collections taken at school entertainments for such Dickensian-sounding charities as the Ragged School Union and the Poor Children's Boot Fund. One former student recalls that "a Belgian refugee girl came to the school as the guest of Miss Hum". Although it would be impossible to indicate the precise measure of influence this independent and high-principled headmistress had on Stevie's life, it is impossible, as well, to dismiss it. Stevie's admiration for Miss Hum, however, was not pious, as is evident from Pompey's assessment:

But she was very simple too, with not much of an idea of the humour of incongruity. So that well I remember the immensely humorous effect of one line of one of these hymns that said: "Make one sad gigantic failure for eternity" but said it to so lilting and tripping a tune that even a baby must smile, and smile we did, or rather howl and roar, like the children of the devil that we were, that were too much for our singing mistress and our headmistress, that had somehow got a firm grip on the thought that child is but a synonym for sweetness and light.

There were others, besides Miss Hum, whom Stevie remembered from those war years, and whom she incorporated into her writing —Sydney Basil Sheckell, for instance. When Stevie was about ten, Basil used to take her to High Church services in All Saints, Margaret Street. The character based on him in *Novel on Yellow Paper* is called "William", and Pompey notes that the part of the Church of England to which he belongs is "very full of incense and ritual, very attractive, very warm and alive, very alive-o it is there." Basil was wounded during the war, and in *Novel on Yellow Paper* his fictional counterpart, William, is a patient at "Scapelands".* Pompey recalls that her "mother, and Auntie the lion of course, pranced up, my mother in her bath chair, the lion *poussant*, to see William in his hospital bed." There follows a hilarious description of his "arrangement dementia" and the exasperation he caused his nurse by insisting that "everything must be put, it *must be put* just so". Later, when Pompey was about fourteen or fifteen, William and his soldier friends used to visit her house. One of them, Captain Poltallock, whom Stevie based on Sheckell's friend, Captain Joey Porteous, did Pompey's Latin homework for her. Another of Sydney Basil Sheckell's soldier friends, Tommy Meldrum, sent Stevie a letter which she incorporated into *Novel on Yellow Paper* in its entirety, only changing to "William" Tommy's reference to "Sydney". Tommy's mention of Ronald Knox brings Pompey back to the days of the Great War, when Knox's conversion to Roman Catholicism seemed so exciting and important: "those times with their unquietness and bewilderment were very much alive," she says, "and so now I have finished forever with this memorial."†

Although Pompey finished for ever her memorial, Stevie did not. In 1968 her "long long *long* poem about dear Baz" was published, as its last line states, "for a memorial of the soldier dear to us he was". "A Soldier Dear to Us" is, like "A House of Mercy", a poem of reminiscence in which Stevie admires those she loves for their brav-

---

*Stevie's typescript shows that she had originally written "Grovelands", which is a large house near Avondale Road designed by the architect John Nash. Its grounds became a public park in 1913, and the house itself was a hospital from 1916 to 1976.

†Ronald Knox (1888–1957), Anglican priest and theological author, became a Roman Catholic in 1917 and published *A Spiritual Aeneid* the following year to explain his conversion. He was ordained a Roman Catholic priest in 1919 and continued to publish theological and other works.

ery. Basil and his friends from the trenches "came to our house for release", Stevie says. She recalls Basil as a "sweet-tempered laughing man" of High Church preference, remembering his amusement when the Bishop became angry because of the reserved Sacrament in St Alban's, Holborn. But Stevie notices, as well, what Basil did not talk about:

> Basil never spoke of the trenches, but I
> Saw them always, saw the mud, heard the guns, saw the
>   duckboards,
> Saw the men and the horses slipping in the great mud, saw
> The rain falling and never stop, saw the gaunt
> Trees and the rusty frame
> Of the abandoned gun carriages. Because it was the same
> As the poem 'Childe Roland to the Dark Tower Came'
> I was reading at school.

> Basil and Tommy and Joey Porteous who came to our house
> Were too brave even to ask *themselves* if there was any hope
> So I laughed as they laughed, as they laughed when Basil said:
> What will Ronny do now (it was Ronny Knox) will he pope?

> And later, when he had poped, Tommy gave me his book for a
>   present,
> 'The Spiritual Aeneid' and I read of the great torment
> Ronny had had to decide, Which way, this or that?
> But I thought Basil and Tommy and Joey Porteous were more
>   brave than that.

In *Novel on Yellow Paper* Pompey mentions a bishop who spoke of tithes and first fruits, rather than glory and *esprit de corps*, to troops about to go into action. "That took them out of themselves," she says, "and got them thinking there were other problems just as pressing as theirs." And in "A Soldier Dear to Us" Basil talks about "Ronny Knox", not the trenches. His merry heart taught Stevie a secret: "That one must speak lightly, and use fair names like the ladies / They used to call / The Eumenides." Probably she considered that euphemism for the Furies (Eumenides means "the kindly ones") to be appropriately lighthearted.

In July of 1917 Miss Hum awarded a Palmers Green High School prize for literature to Stevie and, that autumn, shortly before her fifteenth birthday, Stevie went to The North London Collegiate School for Girls in Camden Town. There was a marked difference between the two schools. Stevie's days at the High School were "privileged", she recalled, because there never were more than twelve *a compromise w the Latin teacher*

students to a class. As a superior student she was allowed certain licences, such as sitting on one foot which others were forbidden to do. According to Dame Flora Robson, teachers at the High School were surprised when Stevie did not win a scholarship to the North London Collegiate. Molly did win such a scholarship, and she recalled that Stevie also went there because Mrs Smith thought she should.

Miss Buss, who founded the school in 1850, and her contemporary Miss Beale, who founded Cheltenham Ladies' College, were pioneers in higher education for women, immortalized in a well-known rhyme which Stevie once repeated in a letter: "Miss Buss and Miss Beale / Cupid's arrows n'er feel / They are not like us / Miss Beale and Miss Buss." "The universities and schools of the mid-nineteenth century," Stevie wrote, reviewing a biography of Miss Buss and Miss Beale, "the clergy and the parents, seem really to have felt that it was both wrong and dangerous to give middle-class girls any education at all, except a smattering of ornamental subjects, ill-taught and unmastered." One can get a good idea of the tradition of discipline and homework which Miss Buss established by reading M. V. Hughes' memoir, *A London Girl of the 1880s* (1946).

As a girl at the North London Collegiate, however, Stevie was not as admiring of the school as she later came to be. The stiff curriculum and long journey to Camden Town exhausted her, and as she "would rather have been thought naughty than stupid", she rebelled. Celia, the autobiographical protagonist of Stevie's third novel, *The Holiday*, puts it this way:

> it was certainly at school, in my grand second school after I had left behind my dear kindergarten, that I first learnt to be bored and to be sick with boredom, and to resist both the good with the bad, and to resist and be of low moral tone and non-co-operative, with the 'Could do Better' for ever upon my report and the whole of the school-girl strength going into this business of resistance, this Noli-me-tangere, this Come near and I shoot.

Molly once recalled a scene which exmplifies the *noli me tangere* attitude. Her sister was to be deposed to the B set in Latin, and the mistress was taking her to it when Stevie stopped and said, implacably, "I don't want to be in B set!" The mistress gave in and they reached an agreement whereby Stevie would never be asked to construe.

Stevie won only a single prize at the North London Collegiate (for Scripture) and no doubt regretted the stricter standards of her new school when she was asked not to sing because she "put the other little girls off". It wasn't a suitable school for her, Molly explained, because it was too big and too academic: "either you were clever and industrious or no one was interested in you. Stevie had the knack of not doing any work and then managing to get to the heart of the matter in two ticks." The system established by Miss Buss was not designed for such brilliance. Looking back on her days with Stevie at The North London Collegiate, a friend recalls that they "were both a bit bored" and that Stevie's "stimulating mind & unconventional approach to ideas" were, therefore, a special joy: "I shall always be particularly grateful to her for initiating wanderings round London after school. We looked at Chinese scroll paintings in the British Museum, & Italian primitives with the feeling that we had discovered them for ourselves."

In March of 1918, probably on the occasion of her confirmation, the Reverend d'Arcy Preston presented a copy of *Christian Duty* to her with his "prayers and best wishes". Stevie performed a pleasurable "Christian duty" that spring by acting in a production of *Ali Baba*, sponsored by St John's, which raised funds for several charities including the Society for the Prevention of Cruelty to Children and the Central War Work Party. She must have had a sense of contributing to the war effort in another way, as part of the Palmers Green audience consisted of wounded soldiers. The reviewers praised her performance as Captain of the Thieves, for its "really dashing air", and they considered Flora Robson to be terrifying as the Grand Vizier. A droll cast photo shows Flora Robson sitting in a chair with what appears to be a paper lantern on her head. She wears a painted, curled-up moustache, and slippers with curled-up toes. Stevie stands beside her in decorative jacket and brimless cap. Her hair, cut a few inches above the shoulders, flares over her ears and lies in a fringe across her forehead. She looks jaunty, with chin held high and a forearm resting on Flora's shoulder. Perhaps Stevie's friend, Second Lieutenant Basil Sheckell, was among the soldiers in the audience who saw her in this costume. (In July of 1918 he was a witness to Mrs Smith's will.)

Shortly after the Armistice was signed on 11 November 1918, Stevie began preparing for her matriculation examination which she wanted to take in June—a year early. Meanwhile Molly had become infatuated with the actress Sybil Thorndike, to whom she sent a letter on Christmas Eve. Three days later she went to a performance of *The Merchant of Venice*, noting afterwards in her diary, "Sybil simply tophole", and "I do love her." Molly was given a chance to emulate Sybil Thorndike, as it were, when in March the North London Collegiate selected *Merchant* for the next term's dramatic production. Molly was chosen to play Portia and her sister to play Shylock, a role in which Stevie "revelled". But Shakespeare was to be a diversion for the girls after a period of crisis and sadness, for early in 1919 their mother died.

Mrs Smith had been taken ill on 12 January, and for five hours she gasped for breath. On the 15th the Reverend d'Arcy Preston rushed to the house, and Doctor Woodcock advised wiring Mr Smith, who arrived late the next night, "in a fearful state". But he had to leave the morning of the 18th, and his daughters saw him off at Paddington. During the following days Ethel Smith appeared to rally, but then a second tube of oxygen was required, her feet and nose became swollen and discoloured, and the doctor feared gangrene. On 6 February Charles Smith arrived home just in time to witness his wife's death.

The love and anguish Stevie felt when her mother died were expressed years later through her autobiographical protagonist, Pompey, who recalls hating trams, even, for shaking when her mother rode on them. One day Pompey came home from school and found her mother in the midst of an attack: "She was very white in the face, and her lips were not very grey, and she could not breathe at all, she could not breathe." Finally, in February, "darling mama died":

> What can you do? You can do nothing but be there, and go on being there steadily and without a break until the end. There is nothing but that that you can do. My mother was dying, she had heart disease, she could not breathe, already there were the cylinders of oxygen. There was the nurse and the doctor coming day and night. But if you cannot breathe how can you breathe the oxygen? Even, how can the doctors help you. Or? You must suffer and then you must die. And for a week this last suffering leading to death continued.

Mr Smith was "fearfully cut up" at the funeral. He visited home on several occasions before his demobilization from the Royal Naval Reserve (in the rank of Paymaster Lieutenant) in December 1919. Then he "retired to the country with a male friend who decamped with his Naval Gratuity, leaving him high and dry." A few months later Mr Smith was raising poultry with a new wife, Hylda Lingen, a farmer's daughter fifteen years younger than he. She called him "Tootles", and they lived together happily. Although Molly used to stay with them on visits, Stevie never would. She thought her father a "fearful bore". After his death, however, she concluded that if someone could have been so taken with him as to call him "Tootles", he must have had qualities she did not see.

During the period of her mother's final illness and the months that followed, Molly Smith's emotions were stirred by yearnings for Sybil Thorndike and for Roman Catholicism, but Aunt does not seem to have approved. Molly had written to Sybil Thorndike during her mother's illness and, later, telling of her mother's death. Then she received a "letter from Sybil consoling & saying she will come out with me." A subsequent letter "came from Sybil asking me to go *there* to tea", Molly noted in her diary. "I wrote & accepted, then Auntie said I musn't. Utterly fed. . . . I wanted more than I can say to go to tea with Sybil." In the months that followed Aunt did permit her niece to see plays in which Sybil Thorndike acted, and Molly was even able to talk with her heroine backstage on a few occasions. Her "worship" is understandable in one who had "stage-mania", but probably it is also significant that her yearning for Sybil Thorndike, and her yearning for the "safety" of Roman Catholicism, were both intense during the months following her mother's death.

Stevie shared her sister's "Roman Fever"—as Molly once referred to it. Aunt's anger seems mainly to have been directed at Florrie Hook, a Roman Catholic relation who invited the girls to her church. Molly noted, two weeks before her mother's death, an encounter between the Smiths' neighbour, Miss Prior, and Florrie Hook:

> Florrie Hook came in & Miss Prior & argued about R.C.ism. P & I are going to Candlemas at Florrie's Church. They were funny A.P. and F.H. F.H. cold & dignified trying to be, & A.P. muddling along, quite off the point & madly as usual.

Three weeks after the death of Mrs Smith, Molly wrote again in her diary of Florrie Hook:

> Letter from F. Hook saying she wants to give P & I rosaries. *Auntie saw letter & raved. She is awfully bigotted.* . . . wish I was a Roman Catholic. I feel horribly shaken in my old tenets.

Molly began to doubt whether her Vicar "could lawfully administer the Sacraments". "Rome seems so safe," she wrote, "I'm *sure* we need one head on earth & that the Pope is he." On 4 March she noted: "Auntie Madge still railing against F. Hook & her promise: it is *not* her fault at all. We are only using her as a tool to get literature, etc." But two days later there was a "royal row" at 1 Avondale Road, after which "Auntie Granma laughed". All this did not stop Stevie, the following Sunday, from going to tea and to church with Florrie and getting a rosary from her. "She has one for me", Molly noted in her diary, and in subsequent weeks she too went to church with Florrie Hook, Aunt having relented and given her permission. Several years later Molly converted to Roman Catholicism, but Stevie became more and more sceptical of Christian doctrine itself until, eventually, she considered herself an unbeliever.

Through all the turmoil of the months during and after her mother's death, Stevie had been preparing for her matriculation examinations. In July 1919, "to general and spoken astonishment", she passed. That same month Molly received an English scholarship. Stevie went to Scarborough in August, to stay on holiday with relations. Upon her return to the North London Collegiate, she was free to pursue her favourite subjects. Stevie seems most to have liked opportunities for oral performance. She was "a keen member of the Debating, Dramatic and Classical Societies", and in addition to Shylock in *Merchant*, she played Mr Hardcastle in *She Stoops to Conquer*. Although Stevie "did not learn Greek, she acted in the plays produced in 1919 & 1920 in the original Greek, her powers of learning and retentive memory enabling her to be a member of the chorus in 'Alkestis' and a messenger in 'The Bacchae'." About the latter, Stevie recalled that "when there was some suggestion of doing 'Medea'," her classics mistress said "it needed a married woman of 'at least twenty-five' to do justice to the part. So instead we did the sweet peaceful girlish 'Bacchae' which made us laugh rather, that it was supposed to be, you know, 'more suitable'." And, "at the same time I was having to listen to these silly lectures on how babies are born".

Novel on Yellow Paper contains a grand send-up of the character based on Stevie's sex-education lecturer:

> There was once a woman called Miss Hogmanimy. That was certainly a queer name. That was a name you would certainly want to get married out of. But this woman was very queer and wrought up over babies and the way babies are born, and she gave up her whole life going round giving free lectures on how babies are born. And it certainly was queer how ecstatic she got about this way how babies are born, and always she was giving lectures to young girls of school or school-leaving age. And all the time it was mixed up in a way I don't just remember with not drinking, not drinking alcohol, but just carrying on on ginger beer, kola and popgass. . . .
>
> But being all tied up with love and religious sentiment, it was just impossible for her to get the medical side of the question across; she would draw sections on the blackboard and then stand her stout body in front of it, blushing furiously, it was all so holy, and all so terrible if it wasn't legitimate. . . .
>
> Looking back now I have a soft feeling for Miss Hogmanimy, her heart was in the right place but her wits were fuddled. I came away from those lectures with a profound aversion from the subject and a

vaguely sick feeling when I heard of friends and relations about to produce offspring. I used to pray for them and wash my hands of it.

Stevie (and Pompey) found it ironic that she should be memorizing the messenger's speech from *The Bacchae*, in which wine is praised, at the same time she had to attend lectures warning about the danger alcohol poses to virtue.

Almost twenty years after the publication of *Novel on Yellow Paper* Stevie again returned to *The Bacchae*—the only play of all those in which she acted during her childhood and youth which enters her writings. She composed "O Lord!"—a verse translation of the last lines of her messenger's speech, and one of the few translations she ever wrote:

> This god then, O lord whoever he is do receive into the city
> For not only is he great but also as I have heard
> He gave the pain-killing vine to men.
> Take away drink, where's Love?
> Any pleasure come to that, O Lord there is nothing left.

Stevie's world was changing rapidly. With her mother's death, her sister's departure from home for university, and in 1920 her own departure from the North London Collegiate School, her childhood had drawn to a close. Her neighbourhood also was changing, from a country place into a bustling suburb. Building had been effectively limited by the Taylor family of Grovelands, and other great land-owners in the area. They were not pleased when the Great Northern Railway erected stations at Palmers Green and Winchmore Hill, and they refused to sell their lands for development. For years the stations remained isolated among open fields. But growth and change in Palmers Green followed upon the fast electric trams which were put in along Green Lanes from 1908 to 1909, the grinding of their wheels disturbing the night. And after the Great War motor vehicles came into widespread use, also contributing to development. As Pompey says of her neighbourhood in *Novel on Yellow Paper*: "the Armistice already saw the builders at busy work, and the streets of houses going up, and the paving stones going down."

If coming from Hull mattered little to Stevie's life and writing, growing up in Palmers Green certainly meant a great deal. She wrote that as a child she once recorded her name and address in the back of a book, adding "North London, England, North West Europe, The World". One cannot understand the world very well, she said, without having some understanding of that address read backwards. Stevie liked to think that Palmers Green was named "after the holy-man kind of palmer, who might have had his little cell on the green," but she knew that it was not. However unpromising, her neighbourhood afforded Stevie the stuff of a writer's career. The lake in Grovelands Park, for instance, provided her sharp eye with telling details of actual life (young girls swinging arm in arm on its surrounding path, crying out to the boys, "Okey-doke, phone me"), and, as well, inspired many of her fantasy poems "about lakes and people getting bewitched, enchanted, *ensorcelé*". As one of her friends wrote, "An artist, like a child, is happier with just a few old sticks to fire his imagination."

Of course, Stevie's childhood in Palmers Green had not been without pointed sticks—the frail mother and absent father, straitened finances, tuberculosis and exile—but there was love in her home, and her rearing was respectable. The adjective is one she herself used as a term of approbation when discussing the writer William Plomer. "He is not physically strong," she said, "but he is strong in feeling and reserve." And she added, "He has that admirable thing, which his family had also, in spite of eccentricities—respectability." Stevie could have been discussing herself, and her own family.

A delicate child, Stevie was none the less exuberant—romping on stage, climbing trees, and wanting, when she left school, to be an explorer.* Stevie's exuberance was in her case often a triumph of spirit. Sometimes she triumphed in darker ways, as in the aloofness she displayed when bored at the North London Collegiate. When her childhood exile to hospital became painful, Stevie's pride told her Death was at her command, and when her mother's dying choked Stevie with fury and feelings of impotence then, as one commentator has noted, "only intensification of her inner pride could give some stay." She once expressed her admiration of Blake's lines, "little creature, form'd of joy and mirth; / Go, love without the help of anything on earth," and on another occasion she said, "being alive is like being in enemy territory". Courage and style were required, and these had to be earned.

Her tomboy manner, and the cockney vowels which she attributed to her stay at Broadstairs, made her—even at an early age—eccentric, as did her habit in later years, when visiting, of wiggling and twisting in a chair, curling up her legs, and putting her head on the chair arm. Dr Woodcock's wife, not surprisingly, thought such waif-like behaviour strange, and felt protective towards Stevie. Stevie's family,

---

* An advertisement which she placed in *The Times* about her desire to be an explorer was misunderstood, she joked, and became the one occasion on which she was associated—if only by implication—with the white slave traffic.

too, must have seemed distinctive in Palmers Green, a household of women come down from Yorkshire. But they were respectable, were active in church, and had that warmth and reserve Stevie mentions in "A House of Mercy". It is worth noting that "A House of Mercy" and Stevie's two other "long" poems of reminiscence, "Archie and Tina" and "A Soldier Dear to Us", were all written during her last five years, even though all stem from the experiences of her youth. In some of her earliest poems as well—such as "Infant" and "Papa Love Baby"—and in many of the characters and incidents of her novels and essays, Stevie reflected upon the world of her childhood. Out of a respectable girlhood in a respectable suburb came a woman of unconventional perspective, able to find in commonplace neighbours, surroundings, and events a writer's vein of gold.

being alive is like being in enemy territory

4

A Job, Reading and Writing

Stevie enrolled at Mrs Hoster's Secretarial Training College and travelled for six months to Grosvenor Place in central London where the College was then situated. She had expressed an interest in journalism as a career and the possibility of her attending the London School of Journalism was raised, but there was no money for her to do so. The headmistress at the North London Collegiate considered her to be an unsuitable candidate for university, and Stevie later concurred in this opinion: "I do not think I should have done better to go to the university, I was not ready for it, it would have distracted, not fed me." None the less, several of her friends thought she always regretted her lack of a university education. What money there was went to pay for Molly's expenses at the University of Birmingham which she was attending on a scholarship.

In any case, Stevie was bored by the secretarial course and often absented herself in the afternoon to wander around the Serpentine in Hyde Park and feed the ducks. Perhaps it was on these walks that she looked about, like the speaker of her poem "*Rencontres Funestes*", and wondered:

I fear the ladies and gentlemen under the trees,
Could any of them make an affectionate partner and not tease?—
Oh, the affectionate sensitive mind is not easy to please.

In poetry readings Stevie sometimes prefaced "*Rencontres Funestes*" with a description of its speaker walking in Hyde Park, noticing young couples in love. She observed in a cautionary way that people are often so lonely and troubled they may imagine marriage as a solution.

It was around this time, too, when Stevie was nineteen or twenty, that she acquired her nickname. Years later she retained a distinct memory of the incident. She was riding with a friend called Arnold on one of the London commons and, her hired horse being slow to move, she rose in the saddle and kicked it (gently, she said). Little boys who saw her shouted "Come on Steve!"—referring to Steve Donoghue, a favourite jockey at the time. Arnold, who thought it most improper that Stevie was wearing a shirt, said to her: "You look like a jockey, you ride like a jockey, I shall call you Steve." From that it became "Stevie", which gradually caught on among her friends.

Stevie began at a young age to imagine her ordinary suburb as somewhere wondrous, no doubt encouraged by the fairy tales to which she early became devoted. A friend points this out: "When I think of Stevie's poor little house and poor little road . . . I was glad to escape from all this. But to Stevie it was a lovely Victorian house, a lovely road. The loveliness came more from her. She invented it." "She transmogrified everything," another friend said. One finds evidence of similar inventions in most of Stevie's writings. The shift from "Peggy" to "Stevie" was not initiated by her nor did she immediately embrace it. It is notable that throughout the 1920s, in the books she acquired for instance, she would sometimes inscribe herself as "Florence Margaret" or "Peggy" and sometimes as "Stevie". But the change to "Stevie", which by the mid-1930s she used almost exclusively, suggests that in the manner of most writers she began to lead a double life. "Peggy" she remained to a segment of her old Palmers Green world, but another self called "Stevie" was ascendant.

Mrs Hoster's College was thought to offer the best training of its kind in London and had remarkable success in finding positions for its students. Referring to it later on, Stevie called it "the famous Mrs. Hoster's". So successful was the College in placing its students that after completing the course, Stevie, who friends insist could type only with two fingers, found secretarial work in the office of a consulting engineer. He was, she said, "the soul of probity (not so all those who visited him . . .)". Years later she recalled the words of one of the reprobate visitors: "We must get rid of that cursèd Major Hamlet." She said of the consulting engineer that he was "a kind man, a man of a kind of poetry, his words delighted me: 'There was a perfectly horrid little child in my hotel this weekend Miss Snooks, I should not care tuppence if I never saw the little child again.'" Stevie remained in the employ of the consulting engineer for a year before becoming private secretary to the chairman of a large publishing house, in or about 1923.

This was Sir Neville Pearson, whose father had risen from simple beginnings to head his own publishing firm—C. Arthur Pearson, Ltd—and to establish, as well, St Dunstan's hospital for war-blinded servicemen. Sir Neville's first marriage was disastrous, and soon after Stevie came to work for him he was involved in divorce proceedings. The office to which she trekked daily was in Henrietta Street near Covent Garden. According to a woman who worked there, it was "like something out of Dickens, very old & dusty. It overlooked a hospital." Sir Neville she described as "the son of the founder . . . dark & handsome & I think everyone envied Miss Smith."

Meanwhile, changes were taking place at 1 Avondale Road. The great aunt who lived there, Martha Hearn Clode, died in November of 1924 at the age of eighty-four. By this time Molly, having finished university, was earning an annual salary of £100 teaching at Belstead House in Suffolk. She came home only for holidays, and in fact never permanently returned to live at Avondale Road. So, as Stevie wrote, there was only "this darling Lion of Hull and myself" for the more than forty years that followed. Aunt had a broad North Country accent, and has been described by Stevie's friends as "imposing to look at", "regal", and "a woman of majesty". She was frank and plain spoken, had an aquiline nose, dark frizzy hair, and by one account "looked a bit like Gertrude Stein". Her stout figure (others referred to her as "massive", and "a great mountain of a woman") made her a physical foil to Stevie (whose protagonist in *Novel on Yellow Paper* expresses admiration and wonder when watching Aunt enjoy a typical midnight snack of cold game pie, bread, and beer). A photo of them walking out of doors shows Stevie hatless, sleeveless, without jewellery, in a simple dress just covering her knees; Aunt carries a cane, has on a large brimmed hat, a choker and chain, and wears a full-sleeve patterned dress ending a few inches above her ankles. A greater contrast is hard to imagine. Stevie rejoiced that she did not have a literary aunt. She prized Aunt's good sense and was perhaps thinking of Aunt and herself when she wrote, borrowing from Santayana, that "there are the two English animals, you know; always against the Lion of commonsense there stands the Unicorn of fancy". There can be little doubt that the best English qualities for which the Lion Rampant is a symbol were embodied for Stevie in Aunt. In her third novel, *The Holiday*, Stevie describes the domestic condition at home in these years:

But most women, especially in the lower and lower-middle classes, are conditioned early to having 'father' in the centre of the home-life, with father's chair, and father's dinner, and father's *Times* and father says, so they are not brought up like me to be this wicked selfish creature, to have no boring old father-talk, to have no papa at all that one attends to, to have a darling Aunt to come home to, that one admires, that is strong, happy, simple, shrewd, staunch, loving, upright and bossy, to have a darling sister that is working away from home, and to be for my Aunt, with this sister, the one.

Stevie sometimes went on holiday with her "darling sister". In 1926 she and Molly met in Bath and proceeded to Cornwall where they bicycled a lot. The following year the sisters were joined by their Aunt for another Cornish holiday. But whatever harmony existed among these three was soon to be strained.

In 1927 and again in 1928, Church sponsored revisions of the Prayer Book were approved by the House of Lords but defeated in the Commons. Disgusted by the role government exercised in the regulation of Anglican religious practices, Molly became, in 1928, a convert to Catholicism. This greatly grieved the Lion Aunt and did not please Stevie, who anyway agreed with the vote in the Commons opposing alterations to The Book of Common Prayer. Years later she wrote the poem "Admire Cranmer", celebrating the genius of the Archbishop "That wrote the Prayer Book", and in "Why are the Clergy . . . ?" she makes plain her distaste for revisions:

Why are the clergy of the Church of England
Always altering the words of the prayers in the Prayer Book?
Cranmer's touch was surer than theirs, do they not respect him?
. . .

Molly and Stevie were never openly hostile, and indeed they showed more concern for one another than many sisters do. None the less they could be highly critical, and kept up a kind of skirmishing until well into the 1960s, when Molly's failing health and the death of Aunt brought the sisters closer together. Their disagreements often took the form of Cranmer versus Rome. "Pearl" is the name Stevie gave to the character based on Molly in *The Holiday*, and the novel describes Pearl's conversion, as well as her fretful temperament.

Pearl came to breakfast with us after she had been to mass. She is a Roman Catholic, though brought up as I am in the anglican fold. But it was the prayer book controversy that did it, that sent her off. But she is happy, in the religion at least she is happy, though of crocodile birth with me, she cannot be altogether happy, of course that is not possible. She knocked at the door, and I went, and she began to cry.

Oh, Pearl, what is it, darling? I said, and began to cry too. For everything has now gone wrong for Pearl. The priest did not arrive for one hour later, the cows outside her billet had cried all the early morning, and everything was bitter as salt water, and black. We stood in the hall crying softly. Oh, how different is this love for relations from the love for friends. Oh, how it strains and tears.

But when Pearl becomes miserable she also becomes cross, it is the fighting temperament she has, and everything now becomes her enemy.

Most of Stevie's friends describe Molly's temperament in terms that are similar.

"When I left school, at seventeen," Stevie once wrote, "I went to work in an office and I felt this was what I deserved for not having worked better at school, I felt in disgrace, so at once began to be ambitious for learning. So I read and read, and discovered the beautiful Racine, and others, and made friends and went about."

Throughout the 1920s, although Stevie was busy with her work as Sir Neville's secretary, she found time, as she said, for prodigious reading, coupling her quotidian existence with a spirited mental one. (In *Novel on Yellow Paper* Pompey says she sometimes thinks she has read too much.)

Journals Stevie began in 1919 ("Book Notes—Peggy Smith" she entitled the first of the copy books) continue until 1930 and contain copious jottings that document her reading. Very often she copied into them lengthy passages from books she was devouring, many of them apparently library copies, as she records at the start of these journals "Out" and "Rtd." dates. Stevie obviously valued these journals and drew on them later. (We should be grateful to Mary Shelley, she observed in 1960, thanks to whose notebooks "we know what [Shelley] read".) Some of the titles show her continuing interests: Montague Summers's *The Geography of Witchcraft and Demonology*, for instance. The entries show that Stevie was reading steadily and with close attention, often a book every day or two. This confirms her claim that after leaving the North London Collegiate she "suddenly became a real swot, *wanted it all*". The books she read included novels by Huxley and Lawrence (*Sons and Lovers* was a favourite), and French critics on English novelists. Her comments at age twenty-one are intelligent although at times engagingly ingenuous and fervent: How could she have managed to miss Compton Mackenzie's *Sinister Street*? she asks herself.

Among her choices are books about psychology, theology, history, classics, and travel. In the early diaries she makes virtually no mention of poetry. Her tastes are catholic: she praises the sentimental novelist

on Freud

Fanny Hurst and Lord Chesterfield in the same month. Friends from later years remember that Stevie was never intellectually snobbish and approached without prejudice everything she read, practising what she once preached in a broadcast: "Read what you like, capture what you want to bring home, and never pause to think if it is of today or yesterday . . . or ashamed because it comes from the pen . . . of that benefactress of humanity, Miss Agatha Christie." The diaries also reveal that she was paying considerable attention to what *The Sunday Times* and *The Observer* had to say about recently published books. The reviewer in Stevie seems emergent as well.

"A poet reading books is hungry for food", Stevie later said, and during the 1920s she often copied out plots in detail, and passages of dialogue: three pages, for instance, from a book of short stories by Osbert "Ever So" Sitwell (a favourite expression of his, Stevie noted). One senses here the nascent novelist assembling her models. Her enthusiasms include the satiric Stella Benson (though in a later diary entry she calls her "a talented but malicious little bitch") and the unsophisticated Louis Bromfield. Stevie praises Oscar Wilde's *House of Pomegranates*, especially for a virtue she was later to manifest: avoiding quaintness in a fairy tale. *Jurgen* was one of several novels by James Branch Cabell which she read in these years. Regarding Freud's "On Dreams" she says he could deduce anything from nothing. Stevie read the novelist Paul Morand in French and in English, and compared the translation with the original. In connection with a book by Chesterton, she wonders why these RCs (Belloc, Chesterton, Knox and Baring), writers Stevie admired immensely, are so defensive. About Arthur Waugh's *Recent Roman Converts* she resolved to "get this". A potential convert was near to home in Molly, of course, but Stevie herself seems still to have been considering the issue in 1924 when she noted about Knox's *Sanctions: A Frivolity*: "If I could accept this hypothesis [the need of Divine sanctions for a decent conduct of life] I should at once become a Catholic—if they would have me."

A bit later Stevie filled four pages with excerpts from Virginia Woolf's *The Common Reader* and wrote admiringly about *Mrs Dalloway*. Strindberg's *Plays* she scorns in diction perhaps deliberately overblown: "the leitmotif of this lucubration." She comments on Aunt Leonie in *Swann's Way* (could the name of Proust's character have suggested "Lion Aunt"?) and finds the prose charming but precious. Sometimes Stevie's comments are so extensive and shaped that they virtually amount to finished reviews.

In January 1926 Stevie is still reading strenuously. Friends remember her as a quick reader, one who seized on aspects of a book which she immediately made her own. Among the books from which she

copied out long passages that year are Spinoza's *Ethics* and Knox's
Other Eyes Than Ours. (Stevie repeatedly refers to Knox as "Ronny",
although in *Novel on Yellow Paper* Pompey finds such practice funny
and adds about Knox that he is "lost to us, ahem, in the Italian
mission".) Reading a review of a book called *The First Time in History*,
Stevie deplores the critic's lack of humour. Where there is humour,
she says, there will be no Lenin, no Mussolini, and perhaps Christ will
be crowded out. But, Stevie asks, isn't a sense of humour "worth
more" than these? Writing about a book called *False Dawn*, she praises
the chapters on doubt and defeatism and says she thinks she is
sometimes a defeatist. Stevie tells herself to "see to this".

Occasionally one finds observations and *mots* tucked between
entries in these reading diaries. Whether they are Stevie's or not
isn't always clear. Often she copies limericks too. (Limerick contests
were a feature of her youth: one was held at St John's church in 1910.)
Stevie enjoyed bawdy limericks and stories and exchanged them, for
instance, with John Hayward. She quotes part of one in *Novel on
Yellow Paper* but breaks off before the conclusion. But the part either
she or an editor at Cape suppressed in the printed book appears in the
typescript of the novel. *In toto*, the limerick reads:

> There was a young lady of Louth
> Who returned from a trip to the south,
> Her mother said: Nellie,
> There's more in your belly,
> Than ever went in at your mouth.

Although later on Stevie resented the reviewer in the *Daily Mail*
who said that *Novel on Yellow Paper* "suggests what would have
happened if Miss Gertrude Stein had written 'Gentlemen Prefer
Blondes'," Stevie did in fact read Anita Loos's book in 1926 and
thought it shrewd and amusing. She wrote admiringly of Thornton
Wilder's *The Cabala* and of a study of Henry James, and between these
entries finds room to quote the Reverend d'Arcy Preston from the St
John's parish magazine.

Stevie made lists of James's novels with a note to herself to buy
them. Valéry Larbaud and Colette were also on her list of authors to
read. She expressed her dislike of Compton Mackenzie's *Rogues &
Vagabonds*, mentioning that he is a convert to Roman Catholicism
writing propaganda in a book where the "Prots" are all black and the
RCs all white—and she adds, "it ain't facts". Stevie says further that
one must lose balance before toppling over, and especially before
toppling over to Rome. About *Chanson de Bilitis*, which Stevie
describes as a study of Greece in Sappho's time, she scribbled that she

a poets reading a poet wants to be fed.

had written a poem about this which the "gentle reader, hypocritical reader, undersized reader, shall never read".

Stevie's mind seems to have found nearly every subject fascinating in these years between school and her emergence as a writer. She paraphrases for pages in her journal, for instance, Lord Birkenhead's account of the reform of divorce laws in England—a book which may have helped her to look at the complexities of Sir Neville Pearson's marital litigation with a more informed perspective. And she pasted into the journal a cutting from *The Observer* of a paragraph on "the life of dolls" from Francis Thompson's *Essays*. Perhaps she identified with the dolls whom Thompson describes as prey to the whims of their mistresses, blamed or petted for reasons entirely unknown to them.

From Inge's *Philosophy of Plotinus*, Stevie copied out in 1928 the intellectual virtues Plotinus practised. He is the "P." whose identity she makes a mystery in *Novel on Yellow Paper*:

Well and who is this "P." that goes round making these positive assertions? Well this P. is, well have a guess, have a guess, he was a bit A.D. but not so much. And the man I am quoting is that noble Dr. I. that was writing yesterday, today, and forever in the. Well see here Mr. Wedding-Guest Reader, you got to do some work.

"Candour, moral courage, intellectual honesty, scrupulous accuracy, chivalrous fairness, endless docility to facts, disinterested collaboration, unconquerable hopefulness & perseverance, manly renunciation of popularity & easy honours, love of bracing labour & strengthening solitude" are the virtues she cites and which indeed she admired all her life.

Between the years 1928 and 1930, Stevie read and admired David Cecil's life of Cowper, Kafka's *The Castle* (which she thought brilliant) and Baring's *Last Lectures*—the sort of book she likes, Stevie said, because "himself" is not his only interest. She respectfully disagrees with Belloc in his work, *How the Reformation Happened*, and quotes Llewellyn Powys in *Pathetic Fallacy*: "Only in glimpsing moments can we be persuaded to entertain any objective reflections upon our happiness." In these later notebooks she wrote down a mixed lot of poems, among them "Ozymandias", "We'll Go No More A-Roving", "Mariana", George Herbert's "Love", "The Orphan's Song" by Dobell, several of Thomas Hardy's poems, and Bridges's "Angel Spirits of Sleep". A number of the poems she copied were ballads—a form she later tried herself.

Stevie started writing poems "about 1924", she said, although she remembered having written one or two as a young girl. A number of these finger exercises from the 1920s exist in typescript, and Stevie has dated them approximately. One that she calls "Lines addressed to the Fountain in Fuller's Restaurant Regent Street, West side" is written in iambic pentameter and poetic diction that give almost no hint of the original voice later to emerge:

> Pale marble mortuarial muniment
> Of all the spirits that in bondage come
> To sup on silence and to crunch the bones
> Of solitary sad remembering
> While to the sating of base appetite
> Each body heaped upon a gilded chair
> Makes more material munching and takes egg
> Or meat or fish to its necessity.

"1924–27?" Stevie wrote in the margin. She was obviously practising scales here, but does manage to generate some fury: "His teeth tear sausages, pig unto pig / Rears up and happy is that conjugation." Another poem dated "1924–27?"—although excessively literary in diction ("Mourning her bolted star with raucous voice / Engruffened by catarrhal effluence")—is nearer to Stevie's characteristic matter and manner. One hundred and thirty-one lines long, it tells in mock-epic style the tale of "Morbid Maltravers [who] lived in Jermyn Street". (By the mid-1920s Stevie obviously was informed about the streets of London.) He is looked after by Mrs Job, a lady in reduced circumstances:

> Her happier days had never seen her do
> What she did now for gilded gentlemen.
> In those more sumptuous days her doing was
> A little needlework, a little chat,
> A little caring for an unresponsive cat,
> Then came the crash.

The domestic circumstance is faintly a paradigm of Stevie's family history, or perhaps more nearly of her own situation just then—the woman whose star (read "father") had bolted and who must toil away as secretary to Sir Neville Pearson. His function in the firm, according to Stevie and others, was largely decorative. Maltravers, in the poem,

> Did petit point upon a pillow case.
> O the exciting silks,
> The sublimating interplay of colour,
> The lithe soft thread to steel in wedlock given.

The image is inventive and the rhythms nearer to those of Stevie's mature poems. Perhaps like Sir Neville, who was no doubt often unaware of Stevie, Maltravers "Never saw / Poor dingy Mrs. Job in rusty black." His sweet domesticity is threatened when the purple thread runs out. But, braving the elements, he takes a taxi to Liberty's and secures a new supply:

> Maltravers, home by five o'clock
> Had buttered muffins for tea
> How gentle life, how innocent, how mild
> How like a sleeping babe a careless sparrow
> A stretch of grass, a finished petit point
> Maltravers smiled and smiled and almost wept
> He was so happy now he had been so put to
> Old Mrs. Job in the basement boiling with pent in rage
> Made moan for Crash and swallowed heritage.

Stevie's reading obviously inspired much that is in these poems—in "Huderina Elegans" written also "about 1925–27", for instance. The poem is an apostrophe to

> Elegans philoprogenitive,
> Halving your cells that kith and kin may live
> And in their turn dividing so contrive
> That the great race of Elegans may thrive
> And prostituted to the life force merge
> Into Branch Cabell's fearful demiurge.

Not a whiff of vintage Stevie.

"The Spectre of My Ancient Youth" is a poem whose subject and imagery echo the English Romantics: it was "written about 1925". Although Stevie kept the typescript, she marked it with rather an indefinite "No (?)". These verses from the 1920s indicate that Stevie spent a decade and more finding her own very original voice before appearing as a poet in print.

But Stevie's personal life in these years was not only reading books and writing verses. She was in touch with old friends like Hester Raven-Hart, whom she knew from her childhood days at the Yarrow Convalescent Home, and she made new friends. During the 1927 holiday in Cornwall, for instance, Stevie met Joan Prideaux and discovered that they worked near each other in London. Thereafter, and indeed until the beginning of the Second World War, when Joan Prideaux's work put a stop to social life, the two met once a week for lunch and sometimes went out to dinner in the evening.

Whether alone or with such friends, Stevie liked to visit picture galleries. In February 1927, she attended a show of Flemish paintings at Burlington House and listed for six pages in her diary the pictures "I like and want". (To help her remember Memling's "Martyrdom of St Sebastian" she drew a slight sketch of the painting alongside the title.) At another exhibition Stevie admired Toulouse-Lautrec prints. She lists the address of The Soane Museum, suggesting she dropped in at that gallery located in Lincoln's Inn Fields, quite near to her office. Some time between 1928 and 1930 she attended a show of pictures and noticed an El Greco: her early poem "Spanish School" could have originated from that visit:

> The painters of Spain
> Dipped their brushes in pain
> By grief on a gallipot
> Was Spanish tint begot
>
> Just see how Theotocopoulos [El Greco]
> Throws on his canvas
> Colours of hell
>   . . .

In *Novel on Yellow Paper* Pompey distinguishes between those who know about pictures and books "in the rich full way" and "cultured gentlewomen, like you get so many of in America", who earn points in society by talking about the cultural subject currently in vogue. El Greco is suddenly fashionable, and then quite as suddenly old hat: "Oh *my dear*, haven't you heard, why, she's still talking about El Greco?" Pompey cautioned readers to remain "mum-o" when such smarties were about.

Secretarial work, reading, picture galleries, and friends—but what of romance? Certainly it was in the air at the office. Sir Neville first met the actress Gladys Cooper, ten years his senior and herself divorced, in the mid-twenties according to Sheridan Morley, Dame Gladys's grandson and biographer. Although on the sidelines, Stevie must have observed their relationship from its inception. Morley writes:

> That they were both divorcees, at a time when that was still a faintly risqué thing to be, gave them something in common, as did the fact that G had often done charity work for St. Dunstan's . . . A handsome and titled baronet, bowled over by Gladys's beauty and for a while deeply in love with her, Neville Pearson was one of the few people around London at this time actually in a position to marry Gladys. . . .
> Their courtship was a comparatively rapid affair, and broke into print rather too quickly while Gladys was on tour . . . and Sir Neville on holiday in Kenya. Both rapidly denied reports that they

were to be engaged, possibly because his divorce had only just become finalized. On his return to Britain, Sir Neville issued a splendidly high-handed statement to the press: "There is no truth at all in the report of Miss Gladys Cooper's engagement to me, published in an evening paper. Miss Cooper shares with me the deepest resentment at the impertinence and impropriety of papers issuing completely unauthorized statements of this nature. Speaking as one who is intimately connected with the Press, I am disgusted with this type of journalism which savours more of the cheap and sensational methods employed in another continent [could he have meant America?] than of the courtesy and accuracy which we are accustomed to expect from the Press of this country."

Conceivably Stevie typed Sir Neville's statement.

Sheridan Morley quotes the leader from the *Dorking Echo* of 15 June 1928, "MISS GLADYS COOPER MARRIED: BRIDE AND BRIDEGROOM SCRAMBLE OVER FENCE TO OUTWIT DORKING CROWD." The story features "a big closed car", gaping crowds, and Miss Cooper in a beautiful yellow frock and black and yellow picture hat. Soon after their honeymoon, spent at a villa in the South of France, Lady Pearson was back in London appearing in a play with Hermione Baddeley, Nigel Bruce, and others. Although Stevie's early writings depict her own life at this time in some detail, she maintained a discreet silence about the marriage of Sir Neville (or "Naps" as he was known to his friends), and never described the domestic ups and downs of her employer or the fashionable world in which he moved. "She knew all about that," claims a friend who saw her often in those days, "but you couldn't get it out of her." Nor is there any mention of the marriage among her papers and letters which have been discovered to date. But it must have represented dazzle of an exceptional kind to a "Peggy from Palmers Green" and must have touched Sir Neville's Dickensian office with considerable glamour. These were, of course, the Roaring Twenties, the days of The Bright Young Things, and although Lady Pearson was scarcely a flapper, nor any longer young, she and Sir Neville surely represented an astonishingly new and romantic world to Stevie, who was soon to be elated and agonized over two romances of her own.

Travel, Romance and a Reader's Report

On 13 September 1929, Stevie sent her sister a postcard announcing that she had been in Weimar the day before and that the next day she would go to Berlin. She was in Germany for two weeks that year, and returned in July 1931 for just under two weeks. Not more than a month in all Stevie later said, but she gathered ample literary gain from these journeys, especially for *Novel on Yellow Paper*. The autobiographies of John Lehmann and Stephen Spender and *The Berlin Stories* of Christopher Isherwood, among other writings, record memorably the compelling yet menacing atmosphere of Germany in these years. Briefly, Stevie dipped into this world. She did once send a copy of Isherwood's *Goodbye to Berlin* to friends and in an accompanying note said it depicted just the time she was in Berlin, so that his novel brought all that back to her. She added that she never visited the neighbourhoods in Isherwood but stayed in boring districts such as Wilmersdorf and Charlottenburg. Later she said that she was on the "first last train" to leave the city.

Stevie describes in *Novel on Yellow Paper* and elsewhere the German holiday her heroine Pompey had, and from that account we can infer many of Stevie's experiences there. She did admit to a friend, after all, that "everything in the novel is true". And she must have told the same to a fan, who said in his letter of reply: "I was gratified and relieved to hear [from you] that 'Novel on Yellow Paper' was autobiographical." In the typescript of *Novel on Yellow Paper* Pompey is first called "Miss Smith", then "Miss Smart", and finally Pompey Casmilus. The novel begins with the 1931 journey and then flashes back to 1929. The interior monologue shifts frequently between these two trips to Germany. Aunt sees Pompey off on the boat train to the Hook where she boards another train to Berlin. Throughout the journey Pompey thinks of Aunt at home in "Bottle Green". Pompey's sadness over leaving Aunt is tempered by the pleasure she takes in even the simplest German phrases—*Schokolade* and *Zigaretten und bitte wie lange ist hier Aufenthalt* (Chocolates and cigarettes and how long, please, do we stop here?); or those steady favourites of Stevie, nursery rhymes: *Spieglein, Spieglein an der Wand* (Mirror, mirror on the wall).

The last time she was in Berlin, Pompey recollects, she came from Dessau where she had stayed with friends. Years later Stevie described her first German holiday in a letter. She recalled "staying with a dotty family called Studders in Halle" whom she had picked from a students' exchange bureau, although she was not a student:

I travelled about with Frau S. to Weimar, Leipzig, where I purchased that little paperbacked copy of *Der Lebende Leichnam* by Leo Nikolajewitsch [i.e. Tolstoy's play, *The Living Corpse*] which proved so useful in Novel on Y.P. . . . , Dresden, & so on. I had not learnt German at school but before going to Germany I worked through Hugo and Otto Sauer [i.e. the book *German Conversation-Grammar*] and of course these Studders spoke no English so I had to speak German. I also had some riding lessons at the riding school in Halle which were very funny. A nice riding master but when we were going round the ring he used to shout at me "wieder falsch" [wrong again] because I couldn't ride "flat", as they do. We also used to wander round in the woods near Halle and once we saw a hunting party in pink coats—in August.

Pompey dislikes the "*männlicher Protest*" she encounters in Berlin during her second visit: the Hitler campaign, she tells us, was commencing in 1931 and "Jew friends" she visits in Charlottenburg had had the swastika scrawled on their gateposts. She goes to a museum and quite by chance meets Karl—a young Swiss–German whom Stevie presumably met first in London. (She once told a friend "It was of course my peculiarly mixed feelings for Karl that drove me deutsch-wards.") Together Pompey and Karl toured Weimar and held hands on the bus. A card Stevie sent from Berlin on 12 July 1931 to her sister, then teaching in Ipswich, confirms the autobiographical character of some of these events. Stevie tells Molly that she is remaining in Berlin, a city she loves and in which she feels at home, until "next Wednesday". She has been to tea with friends in Charlottenburg and is enjoying the cafés where she spent several evenings dancing. "*I met Karl* (!) in the National Gallery this morning," she tells Molly, and comments that this is amazing but true. She later, in a letter to a friend, remarked once more on this chance meeting:

> In 1931 I went again to stay with the Studders for another two weeks. They were in Berlin by then—Wilmersdorf—and even more screwy. I went with them to Neuhäuser near Pillau, on the Baltic, but I came back to Berlin pretty soon, stayed with some friends of my family in Charlottenburg and went round a bit with Karl whom I had run into in the Kaiser Friedrich museum. But I think all this is in the book.

By now, Pompey says in *Novel on Yellow Paper*, she and Karl had become "*ganz platonisch*", dancing, swimming, and sight-seeing together. Apparently shopping for Tauchnitz editions was part of Stevie's programme because she owned many of these, inscribed "Berlin 1931". On a postcard sent to 1 Avondale Road from aboard the ship *Hansestadt Danzig*, Stevie reports she has had three lovely days in Berlin with Karl and she plans to return there the following Saturday.

Stevie introduced her childhood friend, Doreen Diamant, to Karl. Doreen had married in 1930 and moved to Mayfair, and Stevie used often to drop in for a visit. The two had seen little of each other after Stevie began school at the North London Collegiate and Doreen went to study the violin in London and on the Continent. Now they resumed their friendship and saw each other frequently. Stevie was, in Doreen's words, "very good company indeed, vital, witty, fun." Doreen's husband liked Stevie very much and they often took her riding in Richmond Park.

Doreen met Karl in London, perhaps on the very outing Stevie has Pompey recall in *Novel on Yellow Paper*. It is the account of a day she and Karl spend walking in Hertfordshire. Mrs Diamant met him also in Germany and says that the description of Karl in the novel is lifelike. Six foot two and with bright blue eyes, Karl was bespectacled and formal in manner. Stevie kept his calling card, "Karl Eckinger, stud. phil.," in her copy of *Faust*. Eckinger studied history at universities in Basel, Munich and Berlin and, while in London, he did research for his dissertation at the Public Record Office and the British Museum. To Pompey, Karl seemed a sweet boy, although his habit of commenting on the superiority of things German irritated her. On their stroll they discovered in the Hertfordshire countryside a vacant house and, as it began to pour with rain, the slim and agile Pompey slipped through a broken window. Karl climbed to the first floor and entered through another window. Inside they found some sacking and lay down together in each other's arms discussing the struggles of Luther and St

Anthony against sexual temptations. He is "darling Karl" and "sweet Karl".

Pompey and Karl went about London, pausing for tea at Gunter's, and talked, it seems, unceasingly. But he always managed to vex her with his benighted views about the non-German world, and sometimes she was reduced to tears by them. Reflecting Stevie's ambivalence about life in central London, Pompey found it preferable to see Karl in the country. She says that London isn't always sympathetic, although she and Karl embrace tenderly at the Duke of York's steps when he tells her he loves her. Their love oscillates between fierce quarrels and laughter and kisses, but as time passes it grows perceptibly darker. Pompey describes their "winter campaign of love and strife", tragically developing against the countryside she loved so especially—Monks Green Woods, the woods of Smith-Bosanquet. The contest becomes for the reader an image of the deepening struggles between the English and Germans. "An icy crackling wrath", she says, "lies in rimy ridges on us both."

Nothing for these lovers to do but part. For all time, however, he is that sweet boy she refers to having known so long ago. Pompey says that she herself would have championed an authentic Germany that cherished all Karl held dear: philosophy, music, landscape. But this dream changed; and how, Pompey wonders, can one explain the nightmare it became by 1936?

Her friend, Sally Chilver, remembers Stevie chatting about "the German chap" and, in the context of a conversation Stevie had with her, about the impossibility of romantic love. Stevie also discussed her later friendship with "Freddy", giving the impression that Karl was far more exciting. It is Sally Chilver's view that the Karl affair foundered finally on politics—"a kind of ethnic intellectual antipathy, just the way he thought, and a certain lack of lightness." Many years later, Stevie agreed with a critic's assessment of "poor Karl" as "pedantic & self-righteous", although she hoped Karl would not see himself so characterized in the Swiss newspaper where this description appeared. Karl's "lack of lightness" is Stevie's complaint in her poem "Dear Karl": she cannot afford to send Karl *Leaves of Grass*, indeed, she cannot afford a copy of her own. So she sends him a sixpenny edition of Whitman and is defensive, for she knows he will be indignant at the imposition of an editor's selection:

And now sending it to you I say:
Fare out, Karl, on an afternoon's excursion, on a sixpenny
    unexplored uncharted road,
Over sixpennyworth of tarmac, blistered by an American sun,
    over irrupted boulders,

And a hundred freakish geology's superimpositions. Fare out on a
    strange road
Between lunchtime and dinner. Bon voyage, Karl, bon voyage.

"Dear Karl" is not one of Stevie's better poems, marred as it often is by
cumbersome diction ("freakish geology's superimpositions"), but it
does convey the rub of two temperaments, and in Stevie's lyrical "Fare
out, Karl" one senses, also, her farewell. She later told a friend that by
1931 "feelings were all nicely worked out . . . and [Karl and I] haven't
written or seen each other since."*

Suzannah Jacobson, a Jewish friend whom Sydney Basil Sheckell
had introduced to Stevie around 1928, and on whom the character
Rosa is based in *Novel on Yellow Paper*, thought of Karl as Stevie's
"Nazi boyfriend" and believes that in those days Stevie herself was
anti-Semitic. It is true that early in *Novel on Yellow Paper* Pompey is
socially snobbish about Jews. The only "goy" at a party, she feels
superior. On her second visit to Germany, however, Pompey stays at
the home of "Jew friends" in Berlin. While there she wonders what
will become of the country once it has driven out the gifted Jews.
Cruelty was always terrifying to Stevie, and finally Pompey's loathing
of the cruelty she observes in Germany forces her to leave. On board
the departing train, Pompey experiences a moment of self-awareness
and repents the wicked way she had looked at Jews in the past. She sees
herself contributing to the evil and cruelty engulfing the world. Then
it is that she wonders how the Nazis would treat "dear Rosa". She
thinks of Rosa and her husband, Herman, and another Jewish friend,
Leonie, with love, and begins to weep for them and for all victims of
cruelty.

Pompey reaches unflattering conclusions about the German people
—that not only the political leaders but the people themselves are
neurotic and dangerously possessed of cravings that eventually can be
curtailed only by the British Admiralty and the War Office. Stevie said
about her second visit to Germany that only when she boarded the
Channel ferry and headed for England did she feel safe. And indeed
later in the 1930s Stevie tried to discourage the playwright Denis
Johnston, with whom she had become friends, from travelling to
Vienna and spending money there. Her friend Helen Fowler remem-

---

*Karl Eckinger became a newspaper editor in Switzerland after the Second World
War, and was active for many years in the Liberal Freisinnig-Demokratischen Partei
(the centrist Swiss political party). The disappearance of the Weimar Republic had
made an important impression on him, and he became an opponent of totalitarianism.
Although Stevie Smith never heard from Karl Eckinger after the war, she did learn he
married a woman named Elfrieda. From that marriage Eckinger had two daughters.

bers that during and after the Second World War, Stevie "absolutely loathed Germany", and once reacted very aggressively when Helen took Stevie to the German food centre in Knightsbridge. "I never want to go to Germany again," Stevie declared in 1953, and she made synonyms of "cruelty" and "German" in her radio play of 1959. Two years later she criticized Dean Inge for his pre-war claim that Jewish influence was impeding an understanding with Germany, and she linked "vanity" and "Germany" in "Have Done, Gudrun", a poem published shortly before her death.

In 1931, probably before her second visit to Germany, Stevie advertised in *The Observer* for a German-language exchange with someone who might want to improve her English by meeting regularly for conversation. Gertrude Wirth, a German whose husband worked for an American firm, responded. They began to meet weekly and in time became friends. As far as Mrs Wirth can remember, Stevie spoke hardly any German at all and they never conversed in that language. Stevie was energetic, though, in criticizing Gertrude Wirth for the Americanisms she had acquired. "You mustn't say this," Stevie would tell her firmly. She did not present herself in 1931 as a writer; and Mrs Wirth recalls that, at one of their meetings, when "I told Margaret that my piece on a journey with the Flying Scotsman had been accepted by the *Frankfurter Algemeine Zeitung*, she was so excited I wondered later whether that had not been the moment when she started writing herself." During the early 1930s, Stevie and Mrs Wirth went together to art exhibitions, and she remembers Stevie speaking passionately about ideas and sometimes "maliciously" about people. Mrs Wirth thought a young Swiss man, who escorted Stevie to a dance they attended together, "pale" (this was probably Karl Eckinger). Stevie she remembers as "spinsterish" despite her youth, a woman whose teeth protruded slightly (a feature others mentioned and early photos reveal), unsophisticated, and lonely.

Stevie's eccentric vicar, the Reverend d'Arcy Preston, may also have been lonely in these years. His sister, who resided with him, died, and some time afterwards he resigned the vicariate of St John's. Several months later, in December of 1931, Palmers Green was startled to learn of his marriage. The local paper reported the event a few days after it took place, playing up the wedding's "secrecy", telling of "exhaustive enquiries", and noting that, "As far as can be ascertained none of the bride or bridegroom's friends were present, and few of them are even aware of the wedding." In her novel, *The Holiday*, Stevie's protagonist speculates about the reason for d'Arcy Preston's long delay before marrying:

This vicar—the Rev. Humphrey d'Aurevilly Cole—had been for many years in love with the organist, a lady of high stomach. There was some trouble between this lady and Humphrey's sister, and the lady said that never would she cross the vicarage doorstep until the sister departed. So in the end the sister died and Humphrey married the organist, and the London papers had a paragraph about it, 'Vicar Marries Organist After Twenty Years' (for twenty years it was).*

While d'Arcy Preston defied convention by the secrecy of his wedding, Stevie defied it another way, for whatever her personal anxieties, she did feel social pressure to marry. "At that period," she said, referring to the early 1930s, "I thought [marrying] was the right thing to do, one ought to—that it was the natural thing to do, hey-ho—but I wasn't very keen on it." Stevie probably realized then what she later told Sally Chilver: that she didn't "quite catch success-fully (to be infected by, I mean) that troubled stirring world of Two's . . . I cannot ever get the *à deux* fix." Nevertheless her protagonist in *The Holiday* appears to express a cherished notion of Stevie when she says "love is everything, it is the only thing, one looks for it. Yes, everybody is hankering after it."

And so, after her romantic friendship with Karl, Stevie at thirty or thereabouts found a new friend, but this time someone much closer to home. It seems reasonable to consider autobiographical much of what Stevie attributes to Pompey as she describes her feelings about Freddy. As David Garnett said about Freddy in his review of *Novel on Yellow Paper*, "Stevie Smith shows that she loved him." According to Anna Browne, Stevie often referred to "Freddy" in later life, and Anna had strongly the impression that he represented the tenderest of Stevie's romantic friendships. She never referred to Karl in Anna's presence as anyone she really cared about, but repeatedly spoke with great fondness of "Freddy".

His name in reality was Eric Armitage, and in the typescript of *Novel on Yellow Paper* Stevie called him "Ricky". "I've been curious as to how Eric and Stevie met", his niece writes. "The great meeting grounds of the suburbs in those days seem to have been the amateur dramatic groups, and the tennis clubs . . . I certainly can't imagine Stevie playing tennis, and even the amateur dramatics seem a little unlikely." Probably they met in the early 1930s at the "Sixteen Plus"

---

* Stevie's friend, Nina Woodcock, who has close knowledge of the circumstances surrounding d'Arcy Preston's marriage, reports that the organist's love for the vicar had for many years been unrequited. She suggests that it was less love on the vicar's part than his need for someone to look after him, and the organist's persistence, that brought about the marriage.

Club socials held at St John's church. It was at one of these socials that Rosemary Cooper, a neighbour and friend of Stevie from 1932, remembers meeting Eric in company with Stevie. At that time she struck Rosemary (later an assistant editor of *Vogue*) as unattractive, a person who went unnoticed by others. Eric she recalls as "weak and unimpressive, the only boyfriend Stevie ever had." Rosemary felt sorry for Stevie, and once with her beau walked Stevie home after a dance because Eric had neglected to do so. Another time Rosemary and her escort disturbed Stevie and Eric when they entered a derelict mansion in the neighbourhood where couples often went "for a flirtation". She thinks Stevie and Eric weren't "doing anything interesting. They were just sort of crawling around."

Doreen Diamant clearly remembers Stevie saying she had become engaged, and in *Novel on Yellow Paper* Stevie bases Pompey's engagement on her own, and reveals her ambivalence about it:

> Now we are engaged to be married. That is a grand expression, we are engaged to be married, I have the *éclat* of the fiancée. But somehow marriage and being engaged to be married does not bite upon my consciousness at all, it is not right, perhaps it is not the right sort of acid, it has no bite. And so when we are engaged I move my mother's engagement ring that I had from her when she died when I was sixteen, this ring I shift to my engagement finger, it is like a game that has no significance but to play we are engaged. But it galls and wounds us this marriage game, and in our hearts we are beginning to think: Never never can we marry.

Both fear of libel and concern for Eric's feelings may have impelled Stevie to delete from *Novel on Yellow Paper* her candid and at times sharp comments about their friendship. In these deleted sections she describes him as sometimes churlish in the courtship, and hurt and angry when Pompey refused to marry him. He then takes to sending letters in which he hurls reproaches and accusations at her and claims he has lost weight and is neglecting his work. Pompey feels remorse for having hurt him so deeply. In one of the deleted passages Ricky refers to himself as her lover, but Pompey tells the reader he never has been her lover though he might have been had he not procrastinated. She says, "He had not the ferociousness and the single-mindedness and the courage to bring together the thing the place and the time." And where Pompey criticizes "Sir People" and "people" in *Novel on Yellow Paper* for being priggish and frightened, one finds "Ricky" in the typescript.

Not deleted from *Novel on Yellow Paper* is Pompey's remark that her sad love for Freddy "maybe never had a meaning, except in that ivory tower, that made such a good poem they all chose it first." This is an allusion to Stevie's brilliant early poem "Freddy", in which she evokes the happiness she was unable to sustain with Eric:

> Nobody knows what I feel about Freddy
> I cannot make anyone understand
> I love him sub specie aeternitatis
> I love him out of hand.
> I don't love him so much in the restaurants that's a fact
> To get him hobnob with my old pub chums needs too much tact
> He don't love them and they don't love him
> In the pub lub lights they say Freddy very dim.
> But get him alone on the open saltings
> Where the sea licks up to the fen
> He is his and my own heart's best
> World without end ahem.
> People who say we ought to get married ought to get smacked:
> Why should we do it when we can't afford it and have ourselves
>   whacked?
> Thank you kind friends and relations thank *you*,
> We do very well as we do.
> Oh what do I care for the pub lub lights
> And the friends I love so well—
> There's more in the way I feel about Freddy
> Than a friend can tell.
> But all the same I don't care much for his meelyoo I mean
> I don't anheimate mich in the ha-ha well-off suburban scene
> Where men are few and hearts go tumptytum
> In the tennis club lub lights poet very dumb.
> But there never was a boy like Freddy
> For a haystack's ivory tower of bliss
> Where speaking sub specie humanitatis
> Freddy and me can kiss.
> Exiled from his meelyoo
> Exiled from mine
> There's all Tom Tiddler's time pocket
> For his love and mine.

"Freddy" is built upon a simple contrast between the social world of friends and relations, of tennis clubs and public houses, and the lovers' private world which is a refuge out of time, a "time pocket".★ Like

---

★Anna Browne writes: "I think [time pocket] must have been formed on the analogy of *air pocket*, a small area of low pressure into which an aeroplane suddenly

Pompey, the speaker surrenders none of her wit and worldliness as she spots the comic amid immortal longings. Her pub chums find Freddy "dim", and in his suburban "meelyoo" she is rendered dumb. She knows that marriage, despite the chorus who recommend it, would be a mistake. The poem is a celebration, however, of the lovers' sweet moments alone. Stevie's use of Latin phrases and echoes of religious refrain ("World without end ahem") deftly emphasizes the timelessness to which both religion and love aspire, although "ahem" alerts us to her simultaneous awareness of the intrusive temporal world. But her closing reference to the game of Tom Tiddler's Ground suggests bliss, for children who trespass on that fairy land gather silver and gold. Scenery and slang of the period balance both this image, and the romantic *tour d'ivoire*-out-of-haystack. In *Novel on Yellow Paper* Pompey too speaks of "a hearth-rug's ivory tower of bliss, a little space, a time-pocket for love and play", to describe her moments alone with Freddy.

In appearance Eric was "tall, dark haired, and a good looking man", writes his niece, with "somewhat aquiline features". She continues:

> He talked very little about [Stevie], perhaps because of his loyalty to his wife, Catherine. He worked in an insurance company, I believe —his bad stammer handicapped him from achieving the success he might have had, I'm sure—his father was an attorney. He married a quiet, very devoted woman who made him a very happy home, and after her death he totally lost interest in living and died about a year later. They had no children. I remember his sending me a cutting about Stevie on her death and saying he didn't care for her poetry.

Essentially, the incompatibility between Stevie and Eric, as gleaned from *Novel on Yellow Paper* at least, is his having been a "foot-on-the-ground" person and she, famously, "foot-off-the-ground": "in his extreme moods of exasperation he has said to me: Bring your ideas down to earth. You want sense knocked into you. Keep your feet on the ground." Anna Browne says that in retrospect Stevie felt her romantic friendship with Eric had been unlikely and she thought of him as part of her youth. "Freddy", as she spoke of him to Mrs

---

drops. It was a term very much used in the days of light aircraft flying at low altitudes. . . . The time pocket is a timelessness within time, just as an air pocket is an (apparent) airlessness within the air. The haystack is an ivory tower, Tom Tiddler's ground fairy ground (with Fairyland's timelessness). Stevie's imagery is both beautiful and exact. She and Freddy are not meeting in his real world or hers, or even on some common (and real) ground outside. They are meeting in little spaces of non place and non time, within the real world but not continuous with it, all the more precious because cut off from reality."

Browne, never understood that Stevie could not have been the conventional wife he wanted. "Freddy's heart", as Pompey says, "is in these little homes."

Mrs Browne's impression is supported quite independently by Eric's niece when she writes about her uncle and Stevie: "I can see why he would have been attracted to her, and also that marriage would have been impossible. Eric needed a stable, conventional marriage, and he found it with Catherine." Eric's niece thinks the film *Stevie* quite authentic except for the portrait of her uncle:

> He wasn't anything like that hearty soul that bounded on saying "Anyone for tennis?" Eric was an enormously intelligent man with a great love of literature and music . . . I can see they would have a lot in common, but it's impossible to imagine Stevie married. . . [Stevie] used to drop in on Sunday mornings before lunch and have a drink, after walking in Grovelands Park, which she loved to do . . . I suppose we would give her a small goal on her Sunday walks, and provide a glass of sherry!

Talking with young friends years later, Stevie said that she had considered marriage at one time, decided no, and never thought about it again. And she told Anna Browne that Eric had not realized she *never* in fact wished to marry.

None of Stevie's friends questions her fondness for men. Helen Fowler recalls that Stevie enjoyed male company and could be quite flirtatious. This is confirmed by the writer Jeni Couzyn, who describes Stevie's having flirted "with my escort [the poet, George MacBeth] with the most alarming success." "I don't feel happy in love", Stevie reflected late in life: "I think I'm much happier in sort of friendships, you know. Having someone you can giggle with and have fun. I mean men. I adore that. As soon as I get on to the other side—it's a bit odd somehow, but I get sort of frightened."

Although she had a lifelong dread of snakes—and one knows what an unsophisticated Freudian would make of that—fear of sex does not seem to describe accurately her fright. Helen Fowler elicited a "probably not" from Molly when she asked her directly if Stevie had ever gone to bed with a man, but that was perhaps the polite sisterly response. Mrs Fowler said that in the 1930s it would not have been extraordinary for a woman in Stevie's set to have done so. Stevie did once say to her that girls in her day declined from fear of pregnancy to do things that their daughters now do with impunity. Stevie's friend, the composer Elisabeth Lutyens, concurred: "When my generation was young, nobody did the last fence, we were too terrified before birth control. And men were too terrified of paternity suits." Miss

Lutyens said she sympathized with any men Stevie might have slept with. She thought Stevie would consider it all "a ridiculous gymnastic act" and with her intelligent black eyes would note their capers unblinkingly. Stevie did tell Liz Lutyens with amusement about a young man with whom she went to bed inquiring of her: "Are you enjoying yourself, dear?" But Miss Lutyens kept to her view that Stevie never "took the last fence".

Stevie told her friend Lady Lawrence that she'd early in life gone to bed with someone as an experiment, and she said to Sally Chilver "she couldn't understand why she'd done some terrible things". It was clear to Sally from the context that she meant sexual things. Stevie also told Sally that one morning after a party she found herself in bed with a man and, unable to remember anything about what had happened, crept quietly out of the bedroom. Sally thinks that Stevie's few sexual experiences convinced her sex was "ridiculous", although she suffered no real guilt or dismay about the experiences.

As her writings reveal, Stevie could be shrewd, and also funny, about sex. The shrewdness flashes in her judgment that D. H. Lawrence was "a bit of a sentimentalist in his idea of the satisfactions of animal life". "I doubt", Stevie adds, "if the tomcat is ever satisfied; in the hands of Nature, sex is a tyrant's weapon." She saw what was behind a spate of novels which she described as examples of female authors "falling in love with themselves when young": we shall have many more such novels, Stevie noted, "until perhaps the ladies are chased out of it by recognising it for what it is—the waif as object of sexual desire." And she could be astute about the limitations of Colette's subject: "It is a matter of painful experience that in no field does the law of diminishing returns more dreadfully operate than in the physical." But she could be humorous as well as perspicacious about sexuality, and human attempts to deal with it:

This Englishwoman
This Englishwoman is so refined
She has no bosom and no behind.

The Rehearsal
I always admire a beautiful woman
And I've brought you some flowers
for your beautiful bosom.

According to Sally Chilver, Stevie was good at discussing the tangles of her friends' love lives. She felt able to criticize so sophisticated a novelist as Sybille Bedford for writing, Stevie said, like a schoolgirl about love. Yet, on another occasion, Stevie admitted she

wasn't much good on the subject of romantic love, and speculated that
something to do with it was missing in her. Love mapped a country
she knew nothing about, Stevie told Helen Fowler, and Mrs Fowler
traces to that Stevie's almost exclusive interest, later on, in writers like
Angela Thirkell, E. F. Benson, and even Agatha Christie who largely
omit any love interest in their novels.

And "*quelque chose de Lesbos*" as Stevie wrote in her second novel,
Over the Frontier? Not a bit of it, according to Elisabeth Lutyens.
Another friend, whom Stevie asked to lunch in Palmers Green quite
often in her later years, recalls Stevie telling him of a day when
Elizabeth Sprigge, the writer, walked with her in Grovelands Park and
the two lay down together. He inferred from the account that
Elizabeth Sprigge made some sexual overture to Stevie, but he did not
believe Stevie responded. Once in a theatre with Helen Fowler, Stevie
noted a woman making a frightful fuss because she was unable to
secure leave from military service at the same time as her girlfriend,
whereas husbands and wives were always allowed to get leave
together. Those in the audience who could hear the woman's conversation were startled. Stevie considered this hilarious, although it is
likely her hilarity had more to do with her strong sense of social
incongruity than with defensiveness about or hostility towards les-
bianism. Helen Fowler wondered at one time if Stevie were lesbian,
but concluded she had no sexual interest in women—a view confirmed by nearly all Stevie's women friends.

Stevie did have good friends who were lesbians, however, and
seems to have been comfortable and affectionate with them. Of
Radclyffe Hall and Lady Troubridge she wrote these affecting lines:

> But there was this love between them, and no tyranny on the one
> side and no servility on the other. Now she must miss her very
> much. But they were both Catholics, so hope to meet again. This
> book [Lady Troubridge's biography of Radclyffe Hall] may seem to
> some a study in self-delusion. It does not seem so to me because out
> of whatever dreams they had came the realities of love and
> friendship. Good luck to them, one thinks. In a desperate world, in
> the fearful business of being a human creature, they made a corner
> for themselves and were happy.

And in a review of Mary Renault's *The Charioteer*, Stevie objected to
"the repulsive current euphemism 'queer'", and quoted from *Mädchen
in Uniform*, "Die Liebe ist tausendförmig", adding, "love of any sort
being held preferable to an absence of love".

In *Novel on Yellow Paper* Pompey says, "how I have enjoyed sex",
and proposes three cheers for Venus, coupling cenobites with dic-

tators as the troublers of the modern world. Stevie seems to have shared these sentiments and at the least was free from a puritan view of sex, but she could hardly be called an adept at it. The fear she felt, when flirting seemed it might turn into something more serious, had probably to do with her wish for a love which would combine romance and security, her doubts about finding such a love outside the *tour d'ivoire*, and her fear of the demands it would make, if found. Hadn't her "romantic" mother married a man who abandoned her with a weak heart and two children? Stevie was fond of saying that "for many, marriage is a chance clutch upon a hen-coop in mid-Atlantic".

"I love life," she said, "I adore it, but only because I keep myself well on the edge. I wouldn't commit myself to anything." When asked how she felt about children, Stevie said she liked them but was always thankful they belonged to someone else. She told a man who wondered if writing a poem isn't like having a child that, unlike a mother, the poet doesn't have a poem on her hands for twenty years. She would have made a bad wife, Stevie insisted, being too selfish and interested more in her own thoughts. She probably wouldn't recognize her husband in the street, she joked. And she could not have endured watching while someone she adored turned to hating her as, with increasing exhaustion, she became less successful as a spouse. One of the stories she wrote, "The Herriots", may be read as an unhappy fantasy of what marriage would have been like for her, and her eye was keen at detecting sadness in the marriages of others. The poem "Goodnight", for example, she based on a vivid incident —exactly what a man said in front of Stevie to his wife:

Miriam and Horlick spend a great deal of time putting off going to
    bed.
This is the thought that came to me in my bedroom where they
    both were, and she said:
Horlick, look at Tuggers, he is getting quite excited in his head.

Tuggers was the dog. And he was getting excited. So.
Miriam had taken her stockings off and you know
Tuggers was getting excited licking her legs, slow, slow.

It's funny Tuggers should be so enthusiastic, said Horlick nastily,
It must be nice to be able to get so excited about nothing really,
Try a little higher up old chap, you're acting puppily.

I yawned. Miriam and Horlick said Goodnight
And went. It was 2 o'clock and Miriam was quite white
With sorrow. Very well then, Goodnight.

"'Goodnight'" shows, Stevie wrote, "how awful marriage can get and yet go on."

Once Stevie told friends how at a "stuffy black-tie dinner", rather to her own surprise, she found herself saying to the company she thought it a good thing she'd never married, and that had she done so her husband would have to have been undersexed. In the view of Anna Browne, Stevie never wanted to marry but instead "to warm her hands at other people's hearths". She and others who knew Stevie are of the opinion that the chronic and profound physical fatigue which afflicted her would have made her ill-fitted for marriage, its physical tasks, and perhaps even for sex.

"Lovable people are often rather indecisive", wrote Stevie, expressing a favourite sentiment in her review of a book whose heroine was unable to choose between two suitors. Stevie's dilemma was different: well into the 1930s she wondered whether or not to marry. Then she decided against it. Although often lonely in later life, she seems not to have regretted this decision. And all the evidence validates her judgment.

One of the strongest impressions her German friend, Gertrude Wirth, had of Stevie was that she seemed "married to her Aunt". Stevie certainly relished her domestic life with Aunt, who cooked for her and dealt with tradespeople. Like Pompey, Stevie would have dismissed the advice of anyone who told her, "You want to get right away from that old-fashioned aunt of yours." She seems never to have shared the view of the girls in Palmers Green, as Pompey reports it in *Novel on Yellow Paper*: "The unmarried girls have an idea, that if only they were married it would be all right, and the married women think, Well now I am married, so it *is* all right." Pompey adds that "sometimes they have to work very hard saying all the time: So now I am married, so now it is all right, so Miss So-and-So is not married, so that is not all right." By marrying, Stevie would have exchanged Aunt's dependable love for what she called "the witch-craft spider-web of love and affection that seems such a fair house of security". Such an exchange must always have seemed foolish to her, although for a time she toyed with making it. "In some ways I'm romantic but my basic root is profoundly sensible—profoundly sensible. About everything." With Aunt, Stevie could be *à deux* in a way that put few constraints on her life and yet was enduring and deeply comforting. "People think because I never married, I know nothing about the emotions", she told a friend some months before her death. "When I am dead you must put them right. I loved my aunt."

The final section of *Novel on Yellow Paper* describes the Freddy Affair and the sad ending it came to. Stevie's second novel, *Over the*

Frontier, is scarcely under way before Pompey, still Stevie's protagonist, confesses her anguish over the break-up. There is briefly a reconciliation, but the tides of farewell are engulfing Pompey, and soon the relationship ends. Stevie describes the pain of parting in *Novel on Yellow Paper*:

> But when it is over, it is over, then it is tearing inside, it is 'tearing in the belly' one would wish oneself dead and unborn. And one does little things and goes to see friends and does one's work and fusses with this and that and feels in one's heart the drift and dribble of penultimate things, and thinks: To-morrow I shall be dead.

Pompey still thinks longingly of "ineffable Freddy", but vows that "no man ever again shall scour the heart of Pompey that is now numb and ripe for death." In *Over the Frontier*, Pompey is able to distinguish between the Freddy-chimera she fell in love with and the genuine Freddy, though when she lies down she still imagines herself in his arms. Later in this novel, when she is in Germany, she is surprised by the ghostly appearance of Freddy in her dreams—Freddy whom she'd left behind in England "with the rights and the wrongs of it as between the two of us unsolved but shelved"—and she affirms her farewell. But underneath the pain and the resolve one senses gaining in Stevie the realization: "Nothing that produces suffering can be nothing."

In 1933 Stevie travelled with her sister to Bandol, Var, in the South of France, where Aunt, who was in Eastbourne on holiday, wrote to them. It is a characteristically loving letter in the course of which Aunt's discreet solicitude for her nieces is revealed when she tells them about a woman who nearly drowned while swimming. We do not have Stevie's reply, but she did write a witty poem about the French spa:

> Bandol (Var)
> In the south of France, my dear,
> Is full of most awfully queer
> Majors of the British Army, retired.
>
> They live in boats tied up to the quay
> (No income tax, no port dues here you see).
>
> And they're always trying to buy or to sell each other things
> And they hope what they lose on the roundabouts they'll make on
>   the swings.

And they say
Quite in the best stage-army traditional way:
'England was quite a good place to live in before the War
Hawkahaw.'
They all seem to have got catarrh.

Bandol (Var)
How picturesque you are.

Stevie returned from her holiday to continue, as she had throughout the stormy years of her affair with Eric, the daily round at Pearson's. In 1933 the Piccadilly underground had been extended to Southgate and into the open land of Cockfosters, affording her an alternate means of travelling to central London. The underground changed Southgate more radically than had anything in its earlier history: only the park lands were saved from the buildings which rapidly rose and crowded the landscape, making the district part of the middle-class commuter belt. Stevie's work at Pearson's was not "excessive", she said, and she provides glimpses of it in *Novel on Yellow Paper*. Often bored by their tasks, Pompey and Sir Phoebus Ullwater, Bt., the character based on Sir Neville, comfort themselves with tea prepared by the office girl.

Pompey has speeches to write, accounts to handle, servants to pay, manuscripts to read, dogs to walk. Once Sir Neville's wife, Gladys Cooper, swept into his office desperate to find a certain breed of dog for a play in which she was appearing, and Stevie went and found one. Dogs seem very much to have occupied Stevie in those days. Sir Neville's daughter believes that the dog "Belvoir" (pronounced Beaver) in Stevie's poems "To the Dog Belvoir" and "Death of the Dog Belvoir" is "the same Beaver who in fact was my nursery dog when I was a baby." It seems that several members of Sir Neville's family were Stevie's concern in these years. Mabel Lady Ullwater, for example, is based on Sir Neville's mother. Pompey describes her visits to the home of Lady Ullwater whose book, *Twenty Thousand Miles Up the Amazon*, she is helping to write. The author gladly allows Pompey to exercise her fantasy in composing the work, and so she conjures up with profit her memory of the cardboard jungle populated with exotic animals she constructed as a child.

By 1934, Sheridan Morley writes, the marriage of Sir Neville and Gladys Cooper had foundered and she was determined to divorce him:

It didn't prove easy: Pearson may not himself have been the soul of fidelity in these past months, but he took the somewhat unusual view that for him to be divorced would do irreparable damage to both his social life as a baronet and his professional life as a magazine publisher. He therefore insisted, against the custom and practice of the day, that if Gladys was to get custody of their daughter (which was he knew the one thing she still needed from him) he would have to divorce her . . . G's other menfolk were . . . appalled. People simply didn't do such things in 1934; it was a gentleman's duty and obligation to let his wife divorce him, whatever the circumstances, and Pearson's decision now to play the game by an altogether new and different set of rules was a source of continuing amazement for the three-year period that the divorce process then demanded.

Stevie loyally omits any of this from her novels, although among the letters she typed for Sir Neville there must have been some that touched at least obliquely on the matter. Of course the danger of libel, to which she was always alert, would have deterred her.

It was also in 1934 that Stevie tried to publish her poetry, probably for the first time. She submitted to the literary agents Curtis Brown Ltd a collection of verses of which a number were later to appear in her first book of poems. "E. B.", whom Curtis Brown in a letter to "Miss F. M. Smith" called "One of our best readers", wrote a three-page assessment of the poems. Curtis Brown, whose office was near Stevie's at 6 Henrietta Street, said that the reader's report was "by no means intended for your eye, and will probably make you furious." None the less he sent it because of some suggestions that the reader made. (Brown discouraged a response by saying that he was "just off to America.")

E.B. remarked with some astonishment on the size and mix of Stevie's submission: "there is more verse here than is put out by most poets in half a life time", she rather wearily said, and described how she "clipped together all the poems which show some imitative semblance of form and . . . put in another clip the majority of the poems which seem entirely incoherent." Among the poems E.B. found imitations of Milton, imitations of hymns, "ecstatic praise of Lawrence and Huxley and several valiant attempts to out-Sitwell the Sitwells"—this last referring to "The Bereaved Swan", a poem Stevie published in her first book of poems but later said she too disliked. The reader thought "Lord Barrenstock", also included in Stevie's first book of poems, illustrative of "a kind of fixation about sexual perversion". But E.B. was "left gasping" by

> Never for ever, for ever never, O
> Say not, aeonial, I must for ever go,
> Sib to eternity, to confraternity
> Of Time's commensurate multiples a foe.

(This appeared as "Intimation of Immortality" in Stevie's first book of poems.) She was "shocked" too, after such poems, to come upon "a long and careful imitation of *Paradise Lost*" (Stevie put a question mark after that mention). Probably this was "Satan Speaks", an accomplished poem Stevie composed "about 1925", as she indicated on the typescript. In it, Satan addresses the blind poet as the one who has most powerfully sung of hell. At age twenty-three or so when she wrote it, and still *pratiquant*, Stevie imagined the arguments Satan would use to persuade Milton to leave heaven and choose an infernal home. Interestingly enough, Satan's manner of argument, playing upon conundrums, became her own years later in poems which reveal her impatience with Christianity.

Other "imitations" Curtis Brown's reader alluded to in her report were one of Lawrence's "Look! We Have Come Through" ("not read", Stevie comments) and "a melancholy little echo" of Wilde's "Ballad of Reading Gaol" ("?" notes Stevie, and adds "not up in the nineties").

E.B. noted what she took to be a "decoration" for verses she admired: "Casual copulation / Cheats the heart and the imagination." But Stevie, who apparently intended even in 1934 to submit drawings along with her poems, jots in the margin that the "decoration in pencil depicting a melancholy gentleman in a double breasted coat standing in a yard full of washing" is meant to illustrate not "Casual copulation" but (as it does now in the *Collected Poems*) "Alfred the Great". The reader not altogether approvingly notes "similar illustrations of queerly shaped infants on other pages"

E.B. is amazed, after "having ploughed solemnly through an immense quantity of vers libres", to find a poem she oddly thought "far and away the best verse in this huge collection of extraordinarily mixed stuff":

> "Give me the cup, Soldier,
> I'll drain the last drop away,
> But a thought runs in my head,
> Soldier, let the cup stay.
> I remember Christ would not taste this wine
> I remember Christ put away this cup
> And I think, What the Saviour put down
> Shall a man take up?
>
> Drop the cup, Soldier,
> Break it if you can,
> For the cup cries like a harlot
> In the ears of man."

When E.B. faults some of the poems for taking "blasphemous digs at the Christ", Stevie notes "not blasphemous"; and Stevie counters the reader's objection to "small but vicious attacks on British Army officers for whom the writer seems to have a superb contempt", by something now indecipherable, although the word "not" is clear to see. "Ignes Fatui"—a self-directed exhortation, no doubt, about the time lost in contemplating religion and philosophy—E.B. understandably cites as an illustration of the "rubbish" Stevie had written:

> . . .
>
> Do not consider Philosophy
> Do not consider Religion
> Do not think about the Thoughts
> *Zeit verschwendet*
>
> Consider rather
> The Milk Can
> How – the Milk Can
> What – the Milk Can
> Why the Milk Can
> Nice-nice – the Milk Can.

(Stevie obviously modified this view when she wrote in 1951, replying to a questioner, that poetry must be rooted in religion and philosophy.)

The reader sums up Stevie's work as "bitty" and "very ultra-1934" and "*so many*", from which we can at least infer how industrious she had been in the previous years. E.B. finds them "the outpourings of a neurotic type of mind" and objects (twelve years after publication of "The Waste Land") to Stevie's having mixed lines written in Greek and Latin with English as "a form of classic snobbism which seems quite deplorable" ("why," notes Stevie, "if it's apt?"). E.B. goes on to say that "Ugliness of everything under the sun is dragged out and commented on—particularly sexual ugliness." Decades later Stevie sided in some ways with this reader when she said she didn't like very well the first poems she published. She found them in retrospect composed of "fear and pain and disgust and dislike and all those negative things." "At the same time," Curtis Brown's reader added, "the writer can write verse on occasion." She ended by saying that "The reader very much doubts the literary quality of most of the poems but feels there may be some power in them which she has failed to find." All in all Stevie may have felt as she finished reading the report rather like the woman she called "The Songster":

Miss Pauncefort sang at the top of her voice
(Sing tirry-lirry down the lane)
And nobody knew what she sang about
(Sing tirry-lirry-lirry all the same).

In the following year, Stevie was to be read more sympathetically
by several editors and publishers and finally to see her poems in print.
Meanwhile she went about her work as Sir Neville's secretary. Like
Pompey in *Novel on Yellow Paper*, Stevie had to finish off letters to her
employer's old school chums, write to boring relatives, write
speeches,

> writing writing writing; seeing government officials "about land
> tenure in the Argentine", about "mining rights in Alaska", about
> shipment of religious statue home to France—a gesture (generous
> too the thing cost him 100 guineas), on Sir Phoebus's part "to make
> his soul"; paying servants, doing accounts, coding and decoding,
> walking dogs, writing charitable publicity for firm's pet charity
> (103 different appeals to the season); reading manuscripts. Signing
> my letters: Private Secretary—on one never to be forgotten and
> glorious occasion receiving a reply from the Minister of Guess
> What's secretary addressed to me: Dear Private Secretary, and
> signed Private Secretary; deep baying to deep, private secretaries on
> the line . . . Signing my letters: Phoebus Ullwater *tout court*,
> without the hedging chivvying invaliant p. p. Phoebus Ullwater, in
> slanting ruffling swashbuckling forgery. Forgery.

Besides all this she answered other letters sent to the agony columns of
the Pearson publications and shopped for her employer.

Stevie often spoke warmly of her job and employer to friends and
told a soldier who wrote to her during the war that it was a "lovely
mixture" which she had described in her books *ad nauseam*. More
evenings than not, she left her office and made her way back to the
snug suburban world of 1 Avondale Road—a woman, in 1934, aged
thirty-two and unmarried; a poet as yet unpublished, but about to
make a move that would fix her permanently in literary history.

6

In the London Literary Swim

Few outside the highest court and government circles knew in 1935 of the liaison between the Prince of Wales and Wallis Warfield Simpson, an American woman already twice married. But by 1936 when George V died and the popular prince succeeded him as Edward VIII, their affair was the subject of international speculation. "A pity you were not saying that the King was *affiancé*", wrote Stevie in *Novel on Yellow Paper*. Residents of Palmers Green no doubt followed the story with fascination and concern, Stevie and Aunt among them. But, of course, none of this would have disturbed their peaceful life in Avondale Road. Stevie loved her home and felt safe there with Aunt, although the dissatisfaction she expresses in an early poem did sometimes possess her:

> Suburbs are not so bad I think
> When their inhabitants can not be seen
> Even Palmers Green.

Suzannah Jacobson remembers that Stevie had, from the time they met in 1928, the wish to transform herself into a "go-ey girl" like Pompey in *Novel on Yellow Paper*. Stevie wrote there that "in this world of catch-as-catch-can we are so often being in the place that is certainly *not* our place at all, and so being unhappy." Throughout the 1920s and early 1930s some part of Stevie must have longed to forsake her genteel suburb, filled with single ladies looking after their aging mothers, and seek out the pleasures of an interesting social life in town. She says in *Novel on Yellow Paper* that it isn't at all chic to live with an aunt in the suburbs, but her longings for another and different world had probably deep down more to do with loneliness. Pompey describes the people of Bottle Green as uncommunicative and with little awareness of themselves. She tells us that her being goes out to precious friends who live in town or in the countryside where she wearies herself by visiting them. Even the trip to town is fatiguing for Pompey, but she prefers tiring travel to a friendless life in Bottle Green. Reading had allowed Stevie some further escape, but finally the passport she acquired was authorship.

In retrospect Stevie acknowledged it was the company more than the fame that she sought in authorship. She describes in *Novel on Yellow Paper* an existence which allows her to link London with Palmers Green, and to move freely in both among her friends. As Pompey says, she wanted Bottle Green and the pub-lub life in London as well. Throughout her life, Stevie continued to impress friends by her capacity to bridge the two: in the 1950s and 1960s, when the art historian Jonathan Mayne knew Stevie, he considered her ability to reside in Palmers Green while moving in London literary society the most compelling thing about her.

Stevie finally broke into print in 1935 when David Garnett, then literary editor of the *New Statesman*, accepted six of her poems, among them "Lament of a Slug-a-bed's Wife", which may have been one of those Curtis Brown's reader thought "blasphemous":

> Get up thou lazy lump thou log get up
> For it is very nearly time to sup
> And did the Saviour die that thou should'st be
> In bed for breakfast, dinner lunch and tea?

"Freddy" was also among the poems accepted by Garnett, as were "The Hound of Ulster" and "The Zoo"—with their echoes of Belloc and Blake. The latter two are cautionary tales addressed to unwary boys who fail to realize about the caged lion in the zoo, for instance, that

> God gave him lovely teeth and claws
> So that he might eat little boys.
>
> So that he might
> In anger slay
> The little lambs
> That skip and play
> Pounce down upon their placid dams
> And make dams flesh to pad his hams.

Some time in 1935, Stevie tried again to have a book of her poems published and she submitted a collection to Chatto and Windus. It was

Ian Parsons, then a young editor and junior partner there, to whom she spoke. She had sent Parsons her manuscript at the suggestion of the novelist Alice Ritchie, his wife's sister. She was "Liz" Ritchie to Stevie who based a character (Topaz) in *Novel on Yellow Paper* on her. Liz Ritchie lent her the copy of Dorothy Parker's stories from which Stevie took many of the Americanisms she used in *Novel on Yellow Paper*, and, more importantly, the rhythm which got into her head. Stevie ultimately regretted that so much from Dorothy Parker found its way into that novel and thought it "pseudo-American" coming from her, and mannered in a way she disliked. But she obviously found Parker's style attractive in these years and adopted it elsewhere, most evidently in her letters.

Ian Parsons told Stevie to "go away and write a novel" before publishing poems. It pleased her later on to describe how in fact she went and did what Parsons suggested: "He also said (to himself, *sotto voce*, ho ho): 'I'm sure they'll never come back.' But I really did go away and write a novel." "Go away" is perhaps misleading: Stevie wrote at home and in her office, typing on the yellow paper used at Pearson's for carbon copies. One night she wrote six thousand words—"in a dream state". *Novel on Yellow Paper* was completed in six weeks, and Chatto turned it down.

The reader at Chatto and Windus had a blind spot about *Tristram Shandy*, according to Ian Parsons, and so was intransigent about any suggestion that they publish a novel in the spirit of Sterne, as *Novel on Yellow Paper* seemed to be. His opinion convinced a majority of the editorial board. In February of 1936, doleful and embarrassed, Ian Parsons wrote by hand and on personal stationery a letter addressed to "Dear Margaret Smith," saying that "the depressing, and to me humiliating fact is that I have failed to persuade my partners that 'Pompey' is the quite remarkably good book which I firmly believe it to be, or to convince them that—good or not—it would find a public big enough to cover its production costs." He continued:

> They insist on seeing the book as a series of isolated incidents, some 'funnier' than others, and not—as I see it—as a continuously developing theme, a kind of spiritual Quixote's Progress with a very definite unity and shape. And in this they are supported by our editorial readers, so that I am outnumbered by four to one . . .
>
> And so now I find myself in the hateful position of having to hand back to you a book which I encouraged you to write, praised you for when you had written, and even insisted on making you attune to my suggestions. What you will think of me after all this I dread to think, though it is quite clear that I am now a first-class bar none

candidate for exposure in your next novel. And it certainly won't be any consolation to you to know that I, at any rate, still think the book a masterpiece in its way. But there it is. I think C. & W. are being incredibly foolish in turning it down, and I've told them so. That's why this letter is written in my own illegible hand and not on office paper. And as I hear a rumour that Mr Cape may do your poems, I can't for a minute doubt that he will also have the good sense to do "Pompey". There will be a morose satisfaction for me when it is hailed as a classic and sells a hundred thousand.

I hardly have the face to add "with best wishes."

Yours dejectedly,

Stevie and Ian Parsons remained friends through the years (although she complained in 1955 that he had by then become unresponsive to her work). From time to time they lunched together, and Stevie would often regale him with imitations of Sir Neville Pearson, whom Parsons heartily disliked. Stevie did love to repeat the flourish with which Sir Neville finished his talks, "As Lady Godiva said, thank heavens I am drawing to my close," and perhaps amused Ian Parsons with this tale.

Cape had seen a poem of Stevie's in the *New Statesman* and had asked if she had a manuscript to show them. She sent *Novel on Yellow Paper* and they took it. This led Stevie to say years later that she was "very grateful to the old N.S. as it was under David Garnett and Raymond Mortimer and have reason to be, as if they hadn't published my poems Cape would never have asked for Novel on Y.P." Rupert Hart-Davis looked after Stevie at Jonathan Cape Ltd, and she valued his advice and that of Hamish Miles, Cape's distinguished reader. Some time in mid-1936 Miles went to Stevie's office in Henrietta Street to work with her on the copy-editing of her first novel. Stevie "amusingly described" the event in her second novel, already well under way, and the description vividly conveys her temperament as she juggles the odd mix of her day-to-day chores in the office. "You must imagine a very *very* hot afternoon", says Pompey, the autobiographical protagonist carried over from *Novel on Yellow Paper*. Pompey has the 'flu and is irritable, but she none the less feels sorry for the man who comes "to help me through this ordeal". While they attempt proof-correcting, messages for Pompey's employer interrupt them:

> But there is the message from his lady aunt to say to remind him to buy some smoked cod's roe. But I have tried, I say, I have rung up Fortnum's and Harrods, and cookie has rung up Selfridge's, it is absolutely not the season for smoked cod's roe, it is but absolutely impossible. For instance it is possible to force asparagus out of

season, but the female cod will not be so constrained. She will not have it. It is impossible.

And there is the message from my uncle, that Lord Snooks wants to know how much bloody well longer he is to hold on to Korpsdorf Prefs. And then I must think very carefully and cleverly, Now what were Korpsdorf Prefs. this morning? And last week? And so there is a possibility that they will shoot higher? Oh yes, I certainly think they will touch two and seven-eighths, so it is perhaps worth holding on, and holding on till the cows come home? Because really Lord Snooks holds a packet and if he starts to unload—no that must be prevented. My Uncle Stockbroker says so.

So I come back to my proofs in rather a scattered frame of mind. And here before me on the page is another problem: Was Semele the aunt or the sister of Pentheus? But how abominable I have been, how frightful, how could I have left this important relationship-question to the chance of my mendacious memory? But this is a point of the utmost urgent importance. But it is already running round in my mind with Lord Snooks and his Korpsdorf Prefs. And it is running round with the offended female cod that must not, must not be thought of to be mishandled in this out of season way. It is an outrage even to think of this quiet female cod—Well now look, she is sitting at rest upon the lovely soft warm green mud of some steep rock pool. And do I not wish that I was so happily situated as she? I do wish this.

Hot and distraught, Pompey yearns for an open window and for a cigarette. For twenty pages of type-slips, misquotations, misspellings in five languages, "and heavily sub-surface libel actions" she remains silent. Going to the window she encounters another "frightfulness":

For there on the iron balcony lies a dead bird. But it is something that is so horrible, so sad. It is a bird that has fallen out of its eggshell, it lies there splosh on the iron ledge, it has absolutely no feathers on at all. But no feathers—that is horrible. For a bird, that is not possible. But is it my fault? Did I ask that the bird should fall out of its eggshell? To lie untimely upon the balcony? Without a feather upon it? An *embryo*?

Oh now do you see how frightful the mechanics of authorship are when one is involved at once in so many disturbances and cannot keep that peace and integrity of the inner life of the soul that is for the whole health and strength of body and mind?

So I got sick of it, sick at heart with a great world pain, and sick of the sight of page proofs and sick of the irritation of a stimulated imagination that must go to having so many thoughts, so many many thoughts.

Stevie inscribed in Miles's copy of *Novel on Yellow Paper* verse in which she thanks Dr Hamish for being so beamish. She says in part, "If it hadn't been for you, believe me / My child would never have seen the light of day. Stevie." Alongside her inscription she drew a child atop a copy of the novel. Looking back at their relationship years after both were dead, Miles's son wrote, "Miss Smith was . . . diseased with diffidence; I believe my father did much to give her confidence."

The success of *Novel on Yellow Paper* allowed Stevie at thirty-four to reach outside Palmers Green and enter literary society in London. When it appeared in September 1936, Raymond Mortimer, successor to David Garnett at the *New Statesman*, wrote to congratulate her:

> I think it is much the best book which has appeared this year. I like the poetry of it as much as I like the humour. And in my cottage this week end I was continually disturbed by peals of laughter from Eddy Sackville-West on one side of me, and Desmond Shawe-Taylor on the other, reading simultaneously two copies. I don't at all agree with Garnett's view that not many people will like it—I think quite low brow people will. And really it is a book which does one good, & makes one feel one has got out of one's bed on the right side.

This early and enthusiastic response to the novel by wits so exquisite and discriminating must have exhilarated Stevie. And it certainly suggests how quickly the publication catapulted her to fame.

In October John Hayward, already distinguished as a critic and man of letters, invited Stevie to lunch with Raymond Mortimer, Eddy Sackville-West, Rosamond Lehmann and Joe Ackerley. This luncheon came about as the result of a letter Hayward wrote to "Dear Miss Smith" to say that he had fallen under the "strange and delicious spell" of *Novel on Yellow Paper*:

> Since most of my time is spent in reading and in writing about other people's books, it is seldom that I have time or inclination to feel really excited by a new novel. Pompey's enchanted me. Please allow an obscure stranger to thank you for it; and to say how much he hopes that it will have a sequel.

Stevie replied, telling Hayward that she had already written half of a new novel about Pompey but stopped, thinking that readers would not really care for Miss Casmilus. His letter, she said, encouraged her to continue. In the "London Letter" column Hayward wrote for *The*

New York Sun, he had praised *Novel on Yellow Paper* and referred to it as "the talk of the town". "Reminiscent of *Tristram Shandy*, Virginia Woolf, Clarence Day, Anita Loos and the Joyce of the last chapter of 'Ulysses'", the book none the less showed, said Hayward, "striking originality". Stevie was grateful and mentioned that Pompey had already travelled to America and returned with a nice contract from New York.

The contract was issued by William Morrow & Co. At their invitation, Noel Coward sent the publishers an account of his re-actions to the novel. An editor at Morrow told Stevie that Coward "rarely endorses a book", but his response to *Novel on Yellow Paper* was thoughtful and largely laudatory:

> I think it is a brilliant and original book, although at moments a trifle self-conscious. As an exposé of Stevie Smith's mind it is extraordinary and she has a wise, witty and entrancing mind. The self-consciousness I spoke of appears every now and then when she too often reiterates her effects. In the strange formula she has evolved this is permissible but I think unnecessary as it is liable to be a trifle irritating.

"It held me fast from beginning to end", Coward added, and Morrow quoted his words in a *New Yorker* advertisement for the novel. Coward's praise, along with that of Raymond Mortimer and John Hayward, testifies to the mandarin approval with which Stevie was launched.

The day before the book appeared, Stevie dined with Suzannah and Maurice Jacobson at their home in St John's Wood and, according to Mrs Jacobson, "got a bit tipsy". Perhaps this was owing to what she called "publication nerves" in replying to a fan letter the playwright Denis Johnston sent. In this first of her letters to him, written early in September 1936, she added that she was "very sad-dog and dippy at the thought of the number of people . . . who will now know a great deal more about [me] than they did . . . I admit I have a slightly let's-get-out-of-here feeling." The Jacobsons were shocked, Suzannah said, to read a few days later what they considered to be Stevie's anti-Semitic account of them as Rosa and Herman in the novel. Mrs Jacobson resented as well the reference to her husband's "fat little behind on the music stool". "Stevie", she says, "had a warped way of looking at things." And she insists, "you can't get drunk on other people's sherry and then say such things about them." "Pagan" was a term the Jacobsons applied to Stevie: by it they meant that "she didn't think about other people's feelings."

Mrs Jacobson recalls that Stevie was "very young and unknown in

those days" before the publication of *Novel on Yellow Paper*. The
Jacobsons took Stevie to parties and to the theatre; or she would come
to them and stay the night, frequently leaving something behind
which had to be sent on next day to Palmers Green. Occasionally
Stevie would ask Suzannah to lunch in Palmers Green where she met
Aunt and marvelled at a house which looked exactly as it must have
done when Aunt moved in. Suzannah introduced Stevie to many of
her friends, among them the "Larry-party-crowd" in the novel. In
this most autobiographical book, Stevie bothered to change the names
of her friends and acquaintances, but little else in depicting them.
"Larry", though, was a woman. Stevie refers to Larry as a "ladylike
boy" in her novel, and in the typescript Stevie called her "Hesione"
before she changed the name and sex. Suzannah Jacobson writes that
"Larry" gave "fabulous parties where one met everyone—Radclyffe
Hall, Galsworthy, [Paul] Robeson . . . an older generation than
Stevie." Some time before the War, Stevie went to a party the
Jacobsons gave at which the pianist Louis Kentner consented to play.
Stevie talked all through his performance. It may have been Suzan-
nah's displeasure over this that led Stevie to say when she brought
Sally Chilver to a musical party there later on: "We've got to mind our
P's and Q's." Mrs Chilver remembers that Stevie bubbled over with
ideas at parties and elsewhere when one saw her, and attributes this to
her having lived in rather a solitary way. "Sorry I've been talking so
much", she once heard Stevie say, "but my vents have been blocked."

Beginning *Novel on Yellow Paper* with a farewell to all her friends
("Good-bye to all my friends, my beautiful and lovely friends"),
Stevie seemed to anticipate the anger some of them would feel.
Although one reviewer predicted that "none of her beautiful friends
will bear her malice", the Jacobsons did not again invite Stevie to their
home until after the Second World War, and when Suzannah re-read
the book in 1969 she felt as angry as ever about Stevie's references to
her husband and herself, as well as about those to Jews. Stevie had
agreed to dine with Suzannah on Christmas Day in 1970, but she was
by then seriously ill, and Suzannah never saw her again. She was
resolved, had Stevie come to dinner, to ask candidly why she had
treated some of her friends so unsympathetically in *Novel on Yellow
Paper*. When Suzannah received an invitation to a Memorial tea party
held for Stevie at the Ritz she declined, feeling it would be hypocritical
of her to attend.

The impression of anti-Semitism in Stevie's writing persists among
some readers and needs to be addressed. Naomi Replansky, an
American poet who had discovered Stevie with great delight in an
anthology around 1950, began a correspondence which continued for

two decades during which she twice visited Palmers Green. After reading *Novel on Yellow Paper*, Naomi wrote Stevie a mostly adulatory letter about it, but she did take up the novel's depiction of Jews:

> The little bits in the book about Jews however, that too startles my friends but in a different way. I am busy explaining the bits & defending—. There is too much of an emotional accrual to the word Jew already, it would seem there is no unique personal vision of it possible nowadays, it is sad but true. For example, when you talk about the practical commonsense of the Jews I too (Jewish) can't help but wince slightly, knowing so many impractical, nonsensical Jews, & disliking pigeonholes, so you see even what is meant as a compliment comes out all wrong. Oh of course I know all this kind of sensitivity is just as stupid as the thing that created it; & I know for [indecipherable] that Jew is in the eye of the beholder; I was not a Jew until my first discovery of anti-Semitism & then I was a Jew, malgré moi.

Armide Oppé, a close and long-time friend of Stevie, says she finds no anti-Semitism in *Novel on Yellow Paper*, "only a consciousness of Semitism   I say this as a half-Jew myself."

Stevie's "consciousness of Semitism", her interest in the role and fate of Jews in society, accounts no doubt for the inclusion among books she owned of a 1928 edition of Belloc's *The Jews*. By present-day lights this work is patently anti-Semitic, and even at the time of its publication it raised the charge of anti-Semitism against Belloc—a charge he denied. Although Stevie admired some of his books and at times imitated his verse, Belloc's belief that Jews constitute "an alien body within the society they inhabit", and that therefore a "Jewish problem" existed which needed solving, cannot necessarily be attributed to her. It is true, however, that before the Second World War, Jews in England and their non-Jewish countrymen did often feel alien to each other. (In this regard we do well to remember that, as Noel Annan points out, Britain was "renowned among Jews as being the least antisemitic European country.") Sir Stephen Spender, himself partly of Jewish origin, has written that in his youth, "although I was never anti-semitic, I despised some . . . qualities in myself which I thought of as Jewish, and my feelings for the English was at times almost like being in love with an alien race." Most of those who knew Stevie well think that her attitude toward Jews in *Novel on Yellow Paper* was the usual pre-war attitude of the English: not so much ethnic or religious hostility as a greater sense than people have today that Jews are "different" or "exotic".

It was not only Jewish friends who were offended by their portraits in the novel. Nina and Doreen Woodcock report that neighbours were "hurt" by Stevie's depiction of them, and this is confirmed by a letter written to Stevie during the Second World War which says: "Charles is grieved at your having put Prue [by then dead in an air raid] in your book, he didn't mind a bit before: I told him that you actually like P. very much, and thought a lot of her work . . . She didn't mind at all I know, and he knows too. Only it is a pity it's in print. I thought I'd better warn you."

At first Aunt was furious about the book but softened when she read so many favourable reviews. A cousin wrote from Hull and mentioned how affectionately Stevie depicted Aunt and that, too, helped. The two lovingly apologetic paragraphs in the printed book that follow Stevie's wildly funny description of Aunt costuming herself as a fan to attend a church masquerade are not present in the typescript.

> Darling Auntie Lion, I do so hope you will forgive what is written here. You are yourself like shining gold. When I think of what some women are like, I am full of humble gratitude and apprehension that I have you to live with.
>
> And why apprehension? Because the Pompeys who are so clever to poke fun at the noble Lion, with his golden coat, his elegant tail, his neat shape, and the precise footfalling of his padded gait, have but themselves to thank if Lion in royal huff and puff, go move his house.

However "an elderly cousin" who worked in Somerset House and until that time had managed their money affairs, "took grave exception" to Stevie's portrayal of him as "Uncle James" ("Alf" in the typescript) and they never again heard from him. His three-page, single-spaced letter, Stevie called "the most utter rubbish on the sex theme". She mentioned him in a letter to Rupert Hart-Davis and said his was the only "beastly" letter she'd had, though she added that Rosa, Herman, all the Larry-party crowd and Leonie refused to speak to her because they considered her an anti-Semite. Her friends, Josephine and Harriet, stuck by her she said, adding in her letter that "These are all characters in Pompey in case you forgot."

Stevie returned to the troubled subject of Jewishness in *Over the Frontier*. Pompey says that she will fight only with the people she likes, the people she can live with. In principle she supports freedom for the Irish but could not fight on behalf of a people she finds unsympathetic. She tries to dodge an inquiry about the Jews but her questioner persists. Pompey confesses her despair at the racial hatred she feels, says with self-indicting irony that she has dear friends who are Jews, but admits that whatever their cause and however right it be, she would not fight with them. Her real friends, those she deeply loves, happen not to be Jews. She resolves the discussion by making personal friendship the motive for moral action and proclaims the transcendence of friendship over policy. None the less, in this same novel, Stevie depicts cultivated Jewish society sympathetically in the person of old Aaronsen whom she finds "Very essentially civilized, urbane and international, in this sweet sweet cultured sentimental manner, that is so what Allemagne barbare detests". She adds these significant paragraphs:

> *Barbare* is always so foreign to me, you know, it rouses in me such a fury to destroy, to be so cruel, with more than battle cruelty, to be so cruel to tread upon the ecstatic face of this idealismus barbarus, that it takes me out of my rhythm. . . .
>
> But in permitting myself to become so destroyingly furious and so intolerant, is it not to find in myself a little of this very barbarismus I so much should like to wipe out for all time from all people? No, I do not think this in me, and in those who feel as I do, is barbarismus, intolerance yes, but it is the sharp spear of defence against infamy that must bear for a sharp point the uttermost sharp point of a righteous pure intolerance.

Pompey says that Aaronsen and she "have often discussed the perils and provocations of idealismus in the abstract", and reflects rather puzzlingly on "what he and his race have brought upon their heads in the practice of and flight from a mutually incongruous idealismus."

Stevie said in 1960, looking back, that "I didn't want to write novels. It's much harder work than poems. And I don't think I do it awfully well. You can play about with a poem when you are doing your other work."

Not for a moment had she abandoned her poems. By October of 1936, Cape had agreed to publish a collection and she wrote in that month, on yellow paper, to Rupert Hart-Davis there, enclosing "the poems with the beastlies pinned on to make it more fun for everyone". It was Hart-Davis who had suggested that the poems be illustrated although he feared it might be too expensive to do so. Stevie's response was to ask if she might do the drawings herself. He was delighted with them and arranged to print the drawings with the poems. Apparently Stevie had sent him a huge batch of poems because he told her "We shall obviously have to cut down the length of the MS by a good deal." Evidently Stevie had submitted many more poems to the *New Statesman* after her first were published by them, but in October she was complaining to Hart-Davis that "this time next year I have no doubt they will still have them, unless you care to take action." "A lot of the poems", she told him, "go to music—some of it my own and some other people's. For instance 'The nearly right and yet not quite' goes beautifully to (I think it is) the Coventry Carol."

On 10 October the *New Statesman* published Stevie's poem "Lord Barrenstock" which features one of the themes "E.B." alluded to in her report to Curtis Brown:

> Lord Barrenstock and Epicene,
> What's it to me that you have been
> In your pursuit of interdicted joys
> Seducer of a hundred little boys?

Stevie was amused that Raymond Mortimer had placed "Lord Barrenstock" on the same page as "two rather grand poems by E. Sackville-West and Binyon". (Sackville-West was known to be homosexual.) She surmised that Kingsley Martin, the editor of the *New Statesman* whom Stevie disliked, must have required a general anaesthetic when he saw this.

In November of 1936, only three months after it appeared, Rupert Hart-Davis wrote to Stevie to tell her that Jonathan Cape was reprinting *Novel on Yellow Paper*. "We still have something over 200 copies of the first impression left, but we want to make sure of not running out of stock around Christmas." In the same letter he enclosed a contract for "the book of poems" (*A Good Time Was Had By All*) and mentions that "you have already finished another novel". On 1 January, Hart-Davis wrote again to say "The sales of N.O.Y.P. have been keeping up very nicely. It has never sold less than forty in any one week since it was published (except once when it was thirty-three). The grand total is now just over thirteen hundred." Hart-Davis seems to have enjoyed his correspondence with Stevie and replies with such comments as "What a splendid letter. Both Hamish and I enjoyed it thoroughly."

However Stevie was vigilant in her own interests, demonstrating a practicality which probably she learned from Aunt and summoned in her careful and astute dealings with publishers, the BBC, and in a minor way with the stock market later on. It was a vigilance she could not afford to surrender. In a letter signed "Love, R", Hart-Davis began: "Stevie my dear, You write like an angel, but you argue like a fiend"; and later, "What a little tiger you are—not a shark, but a tiger, and since you have the attractiveness of that beast, we have to give in to you every time." The argument centred on royalties and, although Hart-Davis contended that the scale of royalties she suggested was "quite impossible . . . one cannot possibly pay 20% until at least 5,000 copies have been sold", he none the less agreed that Cape would "revise the royalty scale and will start you right away at 15% rising to 20% after 5000." Later in this letter he tells Stevie that "Believe it or not, *Novel on Yellow Paper* itself shows a loss of £3.14.0 on our balance sheet. This is simply because we spent a great deal on advertising it—much more than was economically practical—since we believed that we should get it back in the long run by making your name known."

One of the first reviews of *Novel on Yellow Paper* to appear and the one Raymond Mortimer mentioned in his letter to Stevie was David Garnett's in the *New Statesman*. Garnett began by saying that "This week a book has made me happy for two days", though warning that "ten to one you won't like this book at all" (the judgment Raymond Mortimer disputed). None the less, Garnett insisted *he* liked *Novel on Yellow Paper* although he found it difficult to say why. "It is written in a slangy, rather infantile jargon, with lots of Americanisms, German sentences and foreign mishandlings of English; with Gertrude Stein's repetitive mannerisms and tricks borrowed from Hemingway . . . I adore it because the result so completely expresses the author's character and gives pungency to her wit and her sense." Like so many of the critics, David Garnett theorized that "Miss Stevie Smith . . . isn't writing a novel at all", and added that she was "saying just what she feels about herself, her employer, her aunt, her lovers, her friends and the good people, or not-so-good people, with whom she stayed in Germany. So her foot-off-the-ground is just a device for telling the truth which couldn't be told otherwise." (Garnett's view was echoed in 1969 by the novelist Paul Bailey when *Novel on Yellow Paper* was reprinted: "Pompey Casmilus, the soliloquiser, is indistinguishable from Stevie Smith, the poet.") Garnett found he agreed with many of Stevie's assessments—about Germany, for instance. He calls the author "an extraordinarily sensible girl": "Whether she talks about Gilbert Murray's translations and Racine, about sex, or about being hopelessly and unsuitably in love, she always tells the truth and talks sense, and the slapdash pseudo-American slang acquires for me a strangely poetical quality." Quite enthusiastic praise this was from Bloomsbury, which in 1936 was still a significant literary presence in England.*

---

*Except for a few peripherally social ones, Stevie had few Bloomsbury links. In 1968, she applauded L. P. Hartley for a letter he published contra Leonard Woolf. Stevie was delighted that Hartley scored against "that awful Bloomsbury self-righteousness & self-deception". This she wrote to Anna Browne more than thirty years after she received Clive Bell's letter of praise, sent to her about *Novel on Yellow Paper* soon after its publication. Bell wrote:

David Garnett's review of *Novel on Yellow Paper* led to a meeting with Stevie and an invitation to his home near Cambridge.

> One summer's day, when we already were on comradely terms she suddenly appeared at Hilton Hall in the company of Miss Katherine Cox. . . . They were a magnificent contrast. Miss Cox was dressed as though for a party at Henley or some such. That is a long summer frock sweeping the lawn, a blouse up to her chin and a magnificent "picture" hat. I will not swear to white gloves but certainly a parasol. She was a tall woman and the effect was majestic.
>
> Stevie, on the other hand, was as nearly naked as was possible for a woman at that date. A skimpy top and shorts. I think the garments were pale green. Stevie was very brown herself. The sun shone. It was hot. I hope there were strawberries and white currants.
>
> My son Richard was about ten and William eight. They at once took to Stevie and soon a game began. You were chased and had to get home.
>
> In those days Hilton Hall was not the Stately Home that my daughter-in-law has made it with clipped and hoovered lawns. There was a patch of grass on which one could play stump cricket and badminton but it was encircled with stinging nettles. Running at full speed, Stevie tripped and rolled into a thicket of nettles. . . . what fun she was.

Perhaps this was the visit Stevie recalled when she wrote to a friend, "I stayed once years ago with David Garnett at Hilton Hall . . . and swam with several others, including Angelica [Bell Garnett], in a very very deep and dark stretch near a very sinister looking bridge."

"During the last eighteen months or so, there has been no more striking feature of English letters than the rise and sudden arrival of Miss Stevie Smith", wrote a reviewer in 1938. "With two books —*Novel on Yellow Paper* and *A Good Time Was Had By All*—and some scattered poems in periodicals, she has managed to become one of the principal subjects of discussion at literary tea-parties." Stevie once said that the artist wants attention, assessment, and employment. By 1936, with the publication of *Novel on Yellow Paper*, she had achieved all three and was indeed impressively launched.

---

footnote continued

I have only one fault to find with your enchanting book. Why do you say that you are a "foot-off-the-ground person"? I think, I hope, I know what you mean. But if I do, then assuredly I am a foot-on-the-ground person—both feet. And yet I find your book one of the best I have had for some time, and—as you might say—I read an awful lot of books—in several languages what's more.

7

Novel on Yellow Paper
Considered

The reviewers of *Novel on Yellow Paper*, many of whom are now prominently part of the literary history of the 1930s, agreed with the critical judgments David Garnett expressed in the *New Statesman*. A survey reveals that most found the novel to be original, sagacious, affecting, and witty. Even the few who faulted it mentioned redeeming qualities. The reviews convey with period flavour the kind of critical reception Stevie's novel received.

The *Times Literary Supplement* critic thought it "impossible to classify—an autobiography perhaps, but of the mind more than of a physical existence, and an autobiography with Shandyesque digressions." (Affinities between *Novel on Yellow Paper* and *Tristram Shandy* were observed by many reviewers. Stevie acquired her copy of Sterne's novel in 1921.) Noting such comparisons are dangerous, the *TLS* reviewer said "nevertheless, Sterne is the writer who springs to mind in the midst of Miss Stevie Smith's ramblings". "Curious, amusing, provocative and very serious" were adjectives used by the reviewer who continued:

> The book is a peculiar one, containing evidences of a cultivated and thoughtful mind, interrupted by and often expressed in the language of a devotee of lower-class American films. Miss Smith's odd and very individual method of setting out her ideas makes her style less effective when she is serious, or at least when she is being serious about her own personal affairs. This is not because she cannot express strong emotion, for violent feeling, especially of impersonal indignation against abuses, abounds in the book. But as soon as it becomes personal, in spite of her efforts at detachment, it

naturally loses the satirical lightness of touch which characterizes
the rest. So the apparently autobiographical love-story which rears
its head occasionally and takes up most of the end of the book is,
though deeply felt, less convincing than the rest.

The reviewer for *The Evening Standard* wrote:

One begins reading, a bit annoyed at what seems an affectedly
haphazard way of writing, and thinking, "Too much fluff." After
about 40 pages, one says, "This girl's no fool." Later on "She
knows a lot." Then "She's as good as gold," and finally one is in a
state to chime in with her designation of herself as "the cleverest
living Pompey," and to lap up her *obiter dicta* on Euripides and free
love and the German people, and wish there was more.

The poet Edwin Muir reviewed *Novel on Yellow Paper* for *The
Listener* and was less enthusiastic. He found it "compellingly exasperating" and said it "contains such a well-sustained picture of a dis-
illusioned flapper's mind [an opinion to which Stevie strenuously
objected] that it is probably worth reading . . . To keep up this sort of
thing without a break for over two hundred pages requires a considerable degree of literary skill. The result is excruciating; but it gives a
picture of a mind which is worth exploring if one has the patience."
The Spectator called *Novel on Yellow Paper* "young Joyce out of Anita
Loos" and got wrong Stevie's word "Minorelizabethanismus", as she
complained to a friend. Despite the reviewer's objections to *Novel on
Yellow Paper* he ends by saying "It is worth the effort."
Ian Parsons seized the chance to review the book, emphasizing its
originality in form and content. The word "novel", Parsons wrote,
"has been stretched to the uttermost to include it":

For it is not subject to the laws of writing but rather to the more
subtle laws which govern conversation: it has the pressing immediacy of word upon thought, the immense parentheses, the
sensitive obedience to the opening side-track which is worth following, the returns upon itself, the one-touch-and-leave-it-at-that,
the let's-really-go-into-this, all the ranging freedom of talk, whose
unity is solely (and sufficiently) achieved by consistent sincerity in
the speaker . . .
    It is impossible to give any impression by short quotations of the
exuberant vitality of her diction or the beauty of her curious
stammering and yet emphatic utterance. She—or Miss Smith, for
their relationship is clearly as close as that of Sterne and Tristram
—is a poet and a wit, equipped by nature to attack the powers of
dullness and folly with the two most powerful weapons there are.

Parsons must have felt some satisfaction in saying publicly what in private he had said unsuccessfully to his colleagues at Chatto and Windus.

Storm Jameson was in 1936 an immensely popular author. The dust jacket of *Novel on Yellow Paper* quoted her as calling it "a work of genius". Below this quotation were the words of the novelist, Rosamond Lehmann: "This complete, intense and immediate offering of the essence of a character for one's acceptance is something new in my experience." Two "high spots" for Miss Lehmann were Stevie's "apprehension of what is vulgarity", and "the part about suicide and death which all parents of sensitive children should digest".

Writing in *New Literature*, Marie Scott-James accounted for part of the appeal *Novel on Yellow Paper* had when it appeared. "In these days of Fascism and Communism, revolution and threats of war", she wrote, "how delicious it is to be amused"; and called Stevie a "new entertainer, original and acute, who is amusing in the most nourishing way". A Bristol fan corroborated this view when he wrote some years later to tell Stevie that *Novel on Yellow Paper* was "the sort of book I must have by me during the war".

The Irish playwright Denis Johnston put his praise of Stevie in a fan letter to which she replied in September 1936. In another letter to him that same month she included her press cuttings and delightedly said how very nice for the most part they were. As Stevie proudly tells her new friend, she has had a letter from Kenneth Clark, then director of the National Gallery. She rejects, though, Clark's cavil about the space she devotes in *Novel on Yellow Paper* to a plot précis of *The Bacchae* (he would have been far less happy with the typescript version in which Stevie gives a noticeably ampler account). "Dear Pompey," he had written,

> (if you will pardon the liberty, but I have tried dear Madam & dear Miss Smith, & do not like either form at all) I liked your book very much indeed. It is not at all like what Mr. Cape's readers imagined, not all a humorous work, but very poetical & often very profound. I think you have invented a beautiful style with the rhythms of speech made poetry, & the rhythms of poetry disguised in slang. You have an ear which allows you to do almost anything. I have only one criticism. I didn't like your summaries (or recreations, or whatever you call them) of the books & plays you were reading. I know that a book is apt to play a great part in one's thoughts & get muddled up with the emotions aroused by actual life: but even so, when it comes to writing out the plot of the Bacchae etc., it becomes rather a bore.
> However, don't bother about the second half of this letter.

Writers can't expect to please everyone. But Stevie must have felt vindicated when in 1958 Mary Renault, a novelist well known for her depictions of ancient Greece, wrote to her and said, "Recently I re-read [*Novel on Yellow Paper*] with as much delight as ever, especially the spine-chilling account of the Dionysus–Pentheus passage in the Bacchae."

In her twenties, Stevie had a very low opinion of *John O'London's Weekly*. She referred to it in her journal as a "tripey rag". And although subsequently she reviewed for it from 1940 to 1954, in some years with regularity, she wrote contemptuously of the magazine to John Hayward and others. Nevertheless it was in *John O'London's* that Richard Church wrote a rhapsodic review of *Novel on Yellow Paper*:

> I have discovered something original this week, and if it infects my prose I can't help it, the book is so infectious. . . . It has no chapters; it just runs on, *couramment*, like a fugue, sometimes bubbling over the Stein, with cunning little repetitions, oh so coy and ingenuous, you wouldn't believe how much cleverness and insight they carry along with them.

Church quoted Pompey saying she has many treasured friends outside of her office, and he prophesied, "this book will bring her a great many more dear friends out of the office".

Naomi Mitchison, the novelist, essayist and literary journalist, was one who encountered Stevie as the author of *Novel on Yellow Paper* and became a treasured friend. On "Christmas Day", probably in 1936, she wrote to "Dear Stevie Smith":

> After watching your things in *The New Statesman* which were so much better than all those young men they usually have, I bought your novel & have just finished it & am now in a pure fan-mail mood about it. It kept on being real & so inside one that one no more objected to your dillyings & dallyings than if they had been in one's own mind. It has also in parts great wisdom & it is, technically, either extremely crafty or a work of genius: I am not sure which you would rather it was, as the former is your own doing, the latter merely a breath of the Gods. I did so like separate things, the part about the Jews & about the Bacchae & about the kind of people who are dangerous & about children & death. I shall give it to my sons to read & also to a sculptress called Gertrude Hermes who appears to be very like you or the you your readers cannot help imagining you to be, for the whole thing is so magnificently unlike these bloody boring novels one keeps on coming across. I hope anyhow you are really a woman but that, I think, you must be. Perhaps, if you liked, some time we might meet?

Meet they did and eventually Stevie met Gertrude Hermes as well.

Another person who came into Stevie's life as a result of the publication of *Novel on Yellow Paper* was Sir Osbert Sitwell. Stevie had read his work in the 1920s and mentions in *Novel on Yellow Paper* "Les Sitwells", along with D. H. Lawrence and Aldous Huxley, as leading literary chaps. On 3 February 1938, Sir Osbert wrote to "Dear Miss Stevie Smith" from Hyères in the South of France to say:

> I am so much enjoying your witty, sensible, in-growing book, that I thought I'd write and tell you so (as I love receiving letters of praise, and hope you may, too?). I began *Novel on Yellow Paper* last week. . . . What a relief to find a novelist who can make everything interesting.

Stevie replied, but the whereabouts of her letter is unknown. Sir Osbert wrote back to thank her for a "fan letter". She obviously discussed with him *Over the Frontier* and her design for a third novel, and they continued to correspond. Stevie wrote enthusiastically about several of Sir Osbert's books in her years as a reviewer, although her interest in Edith Sitwell was slight. Sacheverell she referred to in an admiring way as "sinister".

How did Stevie manage to make *Novel on Yellow Paper* an alluring book for so many distinguished readers? E. M. Forster rightly insists that it is the company of compelling and lovable characters we seek in novels. And the poet Robert Nichols, puzzled as to the identity of "Stevie Smith", touched on this in a letter:

> I do not know who Miss Stevie Smith is. I never heard of her before. But she's evidently a born writer with a most remarkable ear for the beauty of speech rhythm and the personal idiom of speech. She is also able to create character. I don't judge a novel on whether the characters are endearing or no: I judge them on aliveness. Nevertheless I confess to delight when I discover characters that are endearing, and Miss Smith's "heroine"—though that would be the last thing "Pompey" would care to be called—is one of the most endearing personages I have met among modern novels. And what's so nice is that she doesn't set out to be. She's just herself, even as that remarkable personage her aunt "the Lion of Hull" is herself.

Virtually all the reactions to *Novel on Yellow Paper* emphasize the appeal of Pompey Casmilus. And indeed the appeal Pompey had for her readers seems to differ very little from the appeal Stevie's friends attribute to her.

Pompey had been baptized "Patience", a name rooted in the Latin verb "to suffer", one may recall, but later on when she grew up and

moved around London she acquired the name "Pompey" (paralleling
the shift from Peg to Stevie). Pompey emphasizes a decaying yet
elegant quality about herself, rather like a crumbling statue sculpted in
antiquity, she says, imagining herself to be one of the old Roman boys
down on their luck who visited around, cadging free meals from
friends who, in turn, found them entertaining and useful at parties.
Stevie describes Pompey's life as one of meeting with friends in town
and weekending in their country houses, not unlike her own life at this
time, and indeed resembling a social rhythm Stevie kept to long after
the English weekend was finished, in the acerbic view of her friend,
Elisabeth Lutyens. Next to wit and chic, Pompey lists "lovely to stay
with" among the qualities she most prizes in her women friends and
she feels fortunate to be the recipient of so much hospitality. Indeed
gratitude to friends is, in a poignant way, a motif throughout Stevie's
novels and points up her persistent feelings of loneliness. Stevie's
poem "Numbers" had by now been written and was soon to appear in
A Good Time Was Had By All.

> A thousand and fifty-one waves
> Two hundred and thirty-one seagulls
> A cliff of four hundred feet
> Three miles of ploughed fields
> One house
> Four windows look on the waves
> Four windows look on the ploughed fields
> One skylight looks on the sky
> In that skylight's sky is one seagull.

Of this poem Stevie later wrote to a friend:

> It is based on the same old bore about feeling lonely. The one poor
> creature in that house is alone & he has the feeling that instead of
> loving hearts there are only numbers around him, that is, indifferent
> Nature, that, in his despair, he feels is only numbers, or must be
> expressed as numbers, as Number is at the opposite pole from the
> feeling heart & (in *his* mood) to expect Nature in any other way, is to
> make a mockery of it, & of himself. Pretty glum, but I'm afraid at
> root a lot of the poems are.

The image of a decaying Roman statue is a controlling one in *Novel
on Yellow Paper*, revived continually by the associations evoked each
time Pompey's name is mentioned. The atmosphere it engenders is
that of the shabby gentility, most attractive to Stevie, that characterized her social background and was manifest in the Edwardian villa
where she continued to live. And almost all her life she had, as well,

the presence of Aunt—a Victorian by formation who must have stood, for all her familiarity with the Stock Exchange, in startling contrast to the spirit of the 1920s and the ensuing decades during which they lived out their long life together. *Novel on Yellow Paper* is in outline a tale of a girl, very exuberant and impulsive, who journeys to an office, meets friends, travels abroad, and falls in love, but who always, like Stevie, returns to the quiet, darkened Victorian household she found, finally, preferable. Pompey is clever, intelligent, brave, candid, loving, sometimes defeated—an appealing someone to whom women and men can feel close. Such fun to be with, one thinks of Pompey. And what a splendidly witty girl.

Wit is, of course, a captivating element in *Novel on Yellow Paper*, and Stevie used it consummately to entertain her audience and animate her narrative, as often she did in life. Ian Parsons remembered that when someone used the phrase "a cool thou", Stevie retorted, "What's a *hot* thou?" No wonder Sir Phoebus asks Pompey in *Over the Frontier* to return from her holidays with witticisms he can incorporate into his speeches. Inez Holden, a novelist who became Stevie's friend in 1930 or thereabouts, once described lunching in a restaurant with her. Behind Stevie sat a woman wearing a velvet hat. Above the woman, on the wall, hung a clock. "What time is it?" Stevie asked Inez. "Look behind you," she told Stevie, who glanced around, turned back and said, "Oh yes, I see: half-past plush." And about a boring woman they knew in common Stevie once told Inez: "I was just sitting in a small restaurant the other day when upturned our mutual." Inez, who suggested to Stevie the title of her poem "*Le Singe Qui Swing*", was a perfect audience for these jokes.

No doubt Stevie polished her original and natural wit in all the weekending she did in these years and indeed throughout her life. In *Novel on Yellow Paper* Pompey talks repeatedly about her pleasure in visiting friends.

> I don't mind saying I am a lucky girl and get entertained pretty freely one way and another. And a lot of my friends, now it is funny how it has all turned out, have moved away to the country. So now I go week-ending, and there you get new angles on life. And there's not one of them where I go that's not lovely to stay with, excepting some whom I won't put in just here to strike a minor key.

And later in the novel:

> I am a *toute entière* visitor. That is what I am being all the time. I visit and visit and visit, my darling friends, my less darling friends, my acquaintances. I am very grateful to them all. In visiting I find a very

great deal of comfort and satisfaction, and each least place where I visit I am so enchanted and so happy that it is another visit, and that at the end of the time I may say: Good-bye and thank you, good-bye. And perhaps as I have said they will stand and smile, and say: Good-bye Pompey, come again soon. That is the very highest pleasure to me, that it is a visit that comes to an end, that may recur, that may again come to an end and be renewed. The rhythm of visiting is in my blood.

Failing to amuse one's companions if invited to dinner or a weekend house party remains, of course, a cardinal sin in England. Stevie seems in matters conversational to have been sublimely a Star. A quatrain John Hayward called "Peevy" and sent to Stevie, gaily pays tribute to her allure as a conversationalist:

> Oh dear I feel ever so grievy—
> Have to lunch elsewhere and can't meet Stevie
> (No wonder they only provide a fork
> The cutting edge, of course, will be Stevie's tork!)

Her wit was "basically English", according to Veronica Wedgwood, "like Edward Lear in its flair for observing, though Stevie was essentially more profound than Lear." And Helen Fowler recalls that "She couldn't talk about anything from Satan to Brussels sprouts without saying something fascinating."

Stevie quickly seized the real point of any conversation and would go straight to the heart of a discussion either to demolish the point or bolster it. She had no illusions and never glossed over painful realities. Stevie could pick up a topic and elaborate it at great length. Her penchant for the odd topic is illustrated in a review she wrote of a saint's biography. She ends the piece by saying that the saint continued to preach when he had only one tooth left in his head, and from this infers that preachers in the Middle Ages were in fact able to give sermons, even though toothless: a point, Stevie adds, about which she had often wondered. She would notice, as she did with Lady Lawrence, a child by the seaside: "That's a difficult child," she would say, "she won't be very nice to her mother." "A sort of gale went through the house when she laughed," said Anna Browne, recalling the account Stevie gave of one day finding on her doorstep a rather middle-aged "Fairy Queen" got up with wings and a wand, accompanied by a young man. Everyone in the road, except Stevie, knew from watching television about this advertising ploy. The Queen asked, "I wonder if you've got Fairy Snow in the house?" Thinking she wanted to borrow some, Stevie went to see and returned to report she had none. "May I just ask you a simple question?" said the Fairy Queen a bit impatiently. (Everyone who watched television knew the answer.) "Why is it that Fairy Snow washes whiter than any other washing powder?" "I expect because it's got bleach in it," Stevie responded, feeling quite pleased with herself. Furious, the Fairy Queen drew herself up and declared, "Fairy Snow has *no* bleach products in it." Her male companion was doubled over with laughter.

Stevie confessed years later that she muddled words in *Novel on Yellow Paper* out of nervousness and fatigue though people mistakenly thought of these as witty neologisms. She gives the example of having said someone was brought up in "affable" circumstances rather than "affluent": people praised her cleverness when in fact she was self-confessedly "slightly off beam". Stevie's wit was "offhand", according to her friend Sally Chilver, "but she invented witticisms so complete that they laid one flat and gave the conversation, at the same time, an interesting turn." Her gift as a conversationalist was evidently rare and virtuoso. Norah Smallwood remembered Stevie's penetrating way of putting a question and her riveting flights of speculation. She went on to say that Stevie was affectionate but could make a story out of you. Stevie valued gossip. People love gossip, she wrote, "and gossip is what novels so splendidly are". A number of her friends recall that Stevie could spice up gossip by being "biting", "mordant", "*méchante*", "barbed"—but most of them agree she never said vile things about others and was, basically, kind.

Luckily for her readers, Stevie alternated the social expenditure of conversation and humour with quiet periods in Palmers Green. In an interview she said she felt fortunate not to have moved exclusively in a literary society where all her ideas would have been dissipated in "high-powered conversation", the fascination of which she obviously felt. Probably we have her novels and poems because she did, periodically, rusticate in Palmers Green with Aunt who considered Stevie's writing "unnecessary".

And yet, however brilliant her talk, Stevie was mercurial and slipped suddenly into moods of deepest melancholy, illustrating Virginia Woolf's view that "nothing thicker than a knife's blade separates happiness from melancholy". Nearly all her friends point to this melancholy into which her gaiety might suddenly and unexpectedly shift. She could be "a girl . . . for the glooms", and sympathized with Lord Balfour's saying that if offered another life on earth "he would say 'No' as the chance of being even *tolerably* unhappy was too remote".

Despite the *élan* with which Pompey sails into London, *Novel on Yellow Paper* exudes, all in all, a vast nostalgia for the "richly

her sadness
melancholy

compostly loamishly sad" Victorian days, the pictures and writings that reflect them, the menacing laurels, and even the "damp Victorian troubles": deaths, weepings, and everywhere decay and dissolution perceived in the gaslight. Pompey was partial, in her Mrs Humphrey Wardish way, "to sickly green gaslight". All of this was embodied for Pompey, as it was for Stevie, in the poems of Shelley and Alfred Lord Tennyson. With apparent pleasure Stevie described the ambience of George Sand's childhood: "Gothic castles, preferably in ruins, wild wet woods, galloping, maddened steeds, nameless sorrows and cascading, moonlit waterfalls were what people liked."

In *Novel on Yellow Paper* style ennobles a plot which in its essentials might have appeared in one of the many women's magazines which the firm Stevie worked for had published over the years and with which she daily dealt: in *Peg's Paper*, for instance. A new style, says the American critic Irving Howe, is "a new way of looking at the world". In pointing, as did nearly every reviewer of *Novel on Yellow Paper*, to the originality of its style, they indirectly refer to the obliquity and originality of perspective which all her friends remember in Stevie and which is there to experience in her writings.

In *Novel on Yellow Paper* Pompey praises a friend for exactness of form and says that form is something she can't seem to acquire. Although Mrs Woolf wrote that "when we are writing the life of a woman, we may, it is agreed, waive our demand for action, and substitute love instead", Stevie seems always to have felt in her fiction the lack of a story logically and compellingly told, and admired in the manner of an envious student the conventionally spun tales of Somerset Maugham. There is plot in *Novel on Yellow Paper*— Pompey falls in and out of love twice, for instance—but it is constructed in layers which emerge as one recollection leads Pompey to another, and not in a linear way.

Stevie seems to have turned to myth as compensation for lack of narrative drive in *Novel on Yellow Paper*. Enchanted as she was early on by Grimm's *Fairy Tales*, she seems to have had a high awareness of myth and its capacity to universalize common experiences which, without it, remain private or drearily day-to-day—or worse yet, descend to banality. Invoking various classical stories in *Novel on Yellow Paper* (and elsewhere in her fiction and poems)—the stories of the Bacchae and of Phaedra were favourites—Stevie touches the commonplace to the universal. Some readers find in her heroine a transcendent type (for Ian Parsons, Pompey was Don Quixote), yet she remains memorably and abidingly an original fictional character called Pompey Casmilus. Stevie was aware of contemporaries who had combined their characters with classical types: Joyce's mix of

Leopold Bloom and Odysseus, to cite the most famous instance. But not many writers linked the quotidian to the mythic and managed to come up with such a good read.

Casmilus is another name for the god Mercury—or Hermes as he was known to the Greeks. Pompey Casmilus places herself under Mercury as her "tutelary deity", and the aspects she ascribes to him are the same ones Stevie later attributed to the character she described in her riddle poem, "The Ambassador":

> Underneath the broad hat is the face of the Ambassador
> He rides on a white horse through hell looking two ways.
> Doors open before him and shut when he has passed.
> He is master of the mysteries and in the market place
> He is known . . .*

Stevie, who wore prominent hats and inherited her nickname from an equestrian, gave her fictional stand-in the name Casmilus half in mockery. Like the god she describes, Stevie (as a story teller) looked forward and back as she sped along. Foot-off-the-ground, she was a "master of the mysteries" who yet was remarkably at home in the market place (read the West End publishing world). By making something mythic of herself and of her experiences, she tried to widen the appeal of her autobiographical tale.

Stevie's other resource was the interior monologue through which she presents her character and story. As the reviewer in the *Daily Mail* indicated in praising the style of *Novel on Yellow Paper*, the work "seems to have been genuinely typed, direct from the mind, on to yellow paper". The American poet and publisher, Jonathan Williams, who was a friend of Stevie, is not a fan of her novels but describes her narrative technique usefully as "the Old Interior Monologue, come home to daydreaming on the Piccadilly tube-line to and fro from Wood Green Station, London, N22." It is in *Orlando*, a book whose tone and narrative style may have influenced Stevie (her copy is marked "Leipzig—1929"), that we are told the most important part of style is "the natural run of the voice in speaking". The poet Robert Nichols made a significant connection when he wrote to Virginia

---

*Stevie said of this poem: "I will tell you who he is, he is Mercury. . . . In his Phoenician form he is known as Casmilus. That is a most beautiful name. It is only to be found in the 1823 edition of Lemprière, a misprint, I think, for Camilus." Although the 1823 edition of Lemprière's *Classical Dictionary* does *not* contain the misprint Stevie mentions, the 1832 edition (which she owned) does. But it is Janus, not Mercury, who was associated with "looking two ways" and having "doors open before him". Mercury was ambassador of the gods, conductor of souls to Hades, and patron of merchants, thieves and poets. He is usually depicted wearing a winged cap.

Woolf on the appearance of *Novel on Yellow Paper* to say, "You are
Stevie Smith. No doubt of it. And *Yellow Paper* is far and away your
best book."

In addition to Proust, Woolf, and Joyce, Stevie had read Dorothy
Richardson, so that the experiments novelists then were making in
varieties of first-person narration were familiar to her. But she was
familiar as well with Fielding to whom she refers in *Novel on Yellow
Paper*, and the address that eighteenth-century novelist made to his
reader is one Stevie's narrator practises: "Reader, do you ever feel
sea-sad, loamishly-sad, like Tennyson, with that sadness too deep for
words?" Or, "But first, Reader, I will give you a word of warning."
The reader is repeatedly addressed in the novel as "you", in a manner
that is intimate yet playful.

Stevie observed other story-tellers at work in first-person narration.
About Sinclair Lewis's *The Man Who Knew Coolidge*, for instance, she
remarked that it is a wonder the book can be read, it being solely a
lengthy monologue. However, one can read it with ease, she continues, and she asks if anyone else had ever achieved such complete
self-revelation. Stevie once praised Edmund Wilson's use of the first
person in *I Thought of Daisy*. She said it gave his book, as it always did
in her opinion, the "discipline of the single focus". Her own way with
first-person narration is distinctive and stays identifiably in the ear.
Muriel Spark remembered it years later in a review of a writer whose
style she found "reminiscent of Stevie Smith's in her *Novel on Yellow
Paper*, the words coming direct, without let or hindrance, from the
author's personality." And Mrs Spark mentions in both an "unin-
hibited vivacity" that is gained.

Not all of those who influenced Stevie's style were belletrists.
Gertrude Wirth, the friend with whom she met for German conversation earlier in the decade, had eventually left London to live in Paris,
and she and Stevie corresponded with some regularity. By 1936 Mrs
Wirth and her husband had gone in flight from Hitler to Shanghai. It
was there that she received from Stevie a copy of *Novel on Yellow Paper*
along with a letter in which Stevie, according to Mrs Wirth, "apologized for having more or less fallen into the style of my letters from
Paris." (In *Novel on Yellow Paper* there is a Gertrude who writes to
Pompey in Germanicized English.) Long after Stevie had given up
writing fiction, a literary friend wrote encouraging her to resume: "I'd
been hoping . . . that all this enforced attention to the realistic novel
might compel you to write a new example of your own other sort:
because you, Cortés, if not so stout, have discovered another country
where no other writer has yet followed you."

Years later, when she had published many works, people unfairly referred to Stevie as someone who had written *one* book—*Novel on Yellow Paper*. Perhaps that provoked some of the ambivalence she felt about her first book. In 1969 she said that *Novel on Yellow Paper* "has hung like a millstone round my neck", and friends recall how heartily she disliked *Novel on Yellow Paper* later in life. On the other hand, Marghanita Laski found Stevie quite willing to discuss the novel, and it was the one of her books among several owned by Miss Laski that Stevie chose to sign. Writing in 1969 to her old friend Denis Johnston, Stevie praised his play *Moon in the Yellow River* and added it was a far better work than *Novel on Yellow Paper* which she didn't at all admire. Molly Everett, a friend of Stevie in her last months, said that when Stevie saw a new copy of *Novel on Yellow Paper* Mrs Everett bought in 1970, she said with horror, "What are you doing with *that*—Terrible, terrible." And then she started to laugh.

"I am already sick of my first book," says Pompey in *Over the Frontier*. What is lovely, she adds, is "quickly to write a poem". Two novels and a number of stories were to follow *Novel on Yellow Paper*, not to mention the many essays and hundreds of reviews Stevie wrote right up until her death. But she understood herself to be at truest pitch a poet, and even braided a number of poems into *Novel on Yellow Paper*, a work Inez Holden considered a "long comic-tragic poem". More and more as time passed, Stevie hunted "the beauty and subtlety of unuttered thought" in the guise of poet.

8

Poems, Politics, *Over the Frontier*

When Arthur Pearson Ltd joined with the larger publishing house of George Newnes, either late in 1936 or early in 1937, Stevie acquired a second employer, Sir Frank Newnes. He was, like Sir Neville Pearson, son of the founder of his firm. Although together these men owned controlling shares in Newnes and Pearson, they had in fact very little to do. Stevie told friends it had been concealed from the baronets that they weren't essential to the operations of the firm, which published trade journals, ladies' magazines, and a prestige periodical, *Country Life*. According to her own account, Stevie had a bell in her office with which she could summon Sir Neville and Sir Frank, although they had none to summon her.

Her office was on the first floor of Tower House, a building in Southampton Street, off the Strand, where the new firm was located after the merger. A small, dark, windowless room furnished with volumes of *Country Life* in leather bindings, it suited her well. She decorated the walls with a selection of her drawings, and once wrote a rum account of the office, littered in the absence of Sir Neville with vases, books, boots, holsters, underwear, fishing tackle, oak chests, and family portraits. It had as well a hot plate, radio and typewriters. Generally Stevie was good-humoured about the oddness of the set-up, went about her work, and was friendly to everyone. Dressed in neat but not fashionable clothes, she wore her hair straight, sometimes with a fringe. Although at times the baronets gave her "fidgets", they paid her twice what she was worth (she insisted), inquired after her health, and were long-suffering.

Sir Neville had been astonished to discover when *Novel on Yellow*

Paper came out that Stevie was an author but, she said, he forgave her for the depiction of him in the book and finally was amused by it. Stevie must once have complained about her job to V. S. Pritchett whom she met in these years and at whose home she attended a party. He wrote to her to say: "I'm sorry you work in a Gothic ruin. Has the thought of growing ivy over G. Newnes ever occurred to you?" He then enclosed verses for Stevie, commiserating with her on being interrupted by "such who entering her native bower / Molest her ancient, solitary brain." Of course, in reflecting on Stevie's travails as a secretary, some sympathized with the baronets who perhaps merited more attention to their few needs than Stevie was disposed to extend. All in all, though, she seems to have been an attentive and "faithful dog-o".

When Sir Neville attended meetings Stevie often spent hours writing stories, poems, and letters to friends instead of "reading bloody manuscripts or writing cheques or working out dividends or putting things away preliminary to taking them out again tomorrow." The poems Stevie assembled for her first collection, *A Good Time Was Had By All,* which was published by Cape in April of 1937, represented "the scourings of about ten years of illicit office scribbling". This Stevie said to Naomi Mitchison, who had written a review of the collection. Stevie seemed to her "like a bird, not an exotic bird but one of the plain-coloured English birds, restless hedge skirmishers, good survivors in any weather." "I have written a quite different review from the one I'd meant", Mrs Mitchison told Stevie, "partly based on what you yourself said about Blake", and she wondered "what the recognized modern poets are going to say" about the poetry.

When Stevie appeared on the literary scene, the Georgian poets were yielding to the Auden generation and to Dylan Thomas. "Stevie Smith . . . stands outside any tradition of the day", writes one critic, "and in so doing acts as a comment on what is happening elsewhere." She had virtually no social or artistic links with those poets and her work seems startlingly *sui generis* when viewed alongside theirs. "There are always certain people who are aware of their epoch but don't let themselves get done in by it", Mrs Mitchison wrote in her review of *A Good Time Was Had By All*. "Shelley was done in, but Blake was not. . . . Such people don't have to be 'we'; they can be 'I', proudly and bouncingly as Blake was. . . . Stevie Smith bounces with Blake."

Other reviewers pointed to Stevie's success with satire, citing as examples "Lord Barrenstock" and "Major Macroo" among other poems. George Stonier, in whose critical judgment Stevie had "absolute confidence", reviewed the book for the *New Statesman*, pointing to the same style he had found in *Novel on Yellow Paper* and that had allowed its author to be "brilliantly funny and intimate at the same time". It provided as well "passages of unexpected beauty". He praised "Eng." and "Bandol (Var)" as feminine counterparts of D. H. Lawrence's "Pansies". "*A Good Time Was Had By All* is doggerel", wrote Stonier, but "doggerel of a peculiarly attractive and personal sort, which contains its own poetry." In the *London Mercury* the quintessential Janus-like character of Stevie's poems was commented upon, the reviewer calling them "lyrical-sardonic", and referring to "shocks of pain and laughter". Like *Novel on Yellow Paper*, the poems too were "grimly entertaining" and had "an agreeable bite". The illustrations this reviewer found "delicious". The *TLS* also mentioned Stevie's illustrations: "Even poems which are apparently serious may suddenly be turned into satire by a mocking drawing", the reviewer said, but found that "such lightning changes of mood suggest some uncertainty of purpose at times".

Stevie frequently did manifest uncertainty about her writing, although fear of libel as well as self-doubt was often the cause. A few months before Stevie's first book of poems appeared, Rupert Hart-Davis pointed out to her that "when you sign your contract you warrant that there is nothing libellous in the book, and if this turns out not to be so I don't see how the poor publisher can be blamed." Stevie's lifelong concern about libel is understandable in a writer whose work is so autobiographical. She must once have complained about having to make deletions to a woman who corresponded with her about translating *Novel on Yellow Paper* into Swedish, for the woman responded: "We will find out about libel laws here; it would be excellent if the libel could be put back in the translation." In *A Good Time Was Had By All* Stevie perhaps feared displeasing Eric by the publication in book form of "Freddy". But her more urgent concern was the reappearance in England of Major Rowland Raven-Hart, OBE, brother of Hester Raven-Hart, her childhood friend at the Yarrow Convalescent Home.

"Major Raven-Hart has chosen this inauspicious moment to descend upon his motherland so if there is a slap up libel action over the poems you will keep undercover won't you", Stevie wrote in 1937 to Naomi Mitchison. "I am pretty sure that Raven-Hart was also Major Macroo", Mrs Mitchison speculated, referring to the character called Hawkaby Cole Macroo in the poem "Major Macroo". The poem, which Stevie published in *A Good Time Was Had By All*, is yet another of her animadversions on marriage and how awful it can be. Major Macroo wisely married "A patient Griselda of a wife with a heart of gold / That never beat for a soul but him / Himself and his slightest whim." He left her for long stretches and cared not "if she might be unhappy or bored." Also "He'd several boy friends / And she thought it was nice for him to have them." His selfishness and unconscious cruelty, qualities that Stevie may have remembered in her father, infuriated her: "Such men as these . . . Never make a mistake when it comes to choosing a woman / To cherish them and be neglected and not think it inhuman."

"I wonder if Major R. H. has recognised himself," Naomi Mitchison asked Stevie. "Damn it all, he ought to be proud to have inspired you." Several weeks after the publication of *A Good Time Was Had By All,* Stevie told Mrs Mitchison that "Major R. H. has pottered over to Dublin and I see his wife and she hasn't guessed so I trot her round and hope for the best. (You see with what fatal ease I drop into doggerel?)" Long after Stevie's death, Mrs Mitchison speculated that "Stevie had a rather peculiar flirtation—perhaps no more—with Major R. H. . . . [who] keeps coming into the poems in various disguises." No other evidence exists for this speculation which of course may be accurate. It is probable, in any case, that Stevie feared both Major Raven-Hart and his wife would recognize his lineaments in "Major Macroo" and retaliate in the law courts.★

No such recognition or, in any case, no such retaliation ever occurred. But in April and May of 1937, right after the publication of *A Good Time Was Had By All,* Stevie had a row with Cape over royalties, and for a time thought of changing publishers. She discussed this with Naomi Mitchison who noted: "you mightn't be so popular with Jews and all the successful publishers are Yiddisher boys, not that the Xtians are any better really." Stevie concluded, "No no,—no Yiddisher would accept me I fear", apparently because of lingering allegations by former friends who were Jewish that *Novel on Yellow Paper* had an anti-Semitic strain. However, Cape yielded to Stevie in the royalty dispute and she told Mrs Mitchison: "Rupert, Hamish and

---

★According to Stevie's poem, Major Macroo had "fads", the parallel of which were Major Raven-Hart's canoe trips on the Nile, Mississippi, and other major rivers. He was "everlastingly writing books about canoeing up the Irrawaddy etc." Stevie once wrote. The Major's military experience in Egypt and India during the First and Second World Wars, and his many canoe adventures, must have fascinated Stevie, who in her youth had wanted to become an explorer rather than a secretary. As for Major Macroo's "boy friends", some correspondence is suggested by Raven-Hart's *Canoe To Mandalay* (1939) and *Canoe Errant on the Mississippi* (1938), for example, which contain photos and reminiscences of the boys who were his companions on canoe trips. The close of *R.A.F.ing It* (1943), which Raven-Hart published under the pseudonym L.A.C. Errant, is an encomium to male companionship.

even Jonathan [Cape] I see now behaloed, couleur de rose, just this side
of sainthood."

Stevie had submitted her second novel to Cape early in 1937, and at
the end of March Rupert Hart-Davis responded "with unmixed
delight":

> I am not at all sure that I know what it is about, but that did not
> interfere with my pleasure at all. . . There is some magnificent
> writing in it, and I think it is a better book than N On Y P, since it
> has more shape and unity. There are still lovely long digressions,
> but the whole thing has a beginning and an end, whereas N On Y P
> just began and just finished, if you know what I mean. The
> description of the proof-correcting episode made me laugh till I
> cried. The night-riding is excellent. This obviously has some deep
> inner significance for you, c.f. many of your poems.

But Hart-Davis also raised in this letter the topic of libel. "I suppose
there is no possibility of LIBEL, sub-surface or other?" he asked.

Evidently there was. "I am sorry to have to tell you", Stevie wrote
to Denis Johnston in November 1937, "that my book *Over the Frontier*
does not now come out until January by reason of the machinations of
the proof readers, me and Rupert, and the awful business of steering a
clear course between libel and obscurity." Denis Johnston sympathized with Stevie in his reply: "As for *Over the Frontier*, really this
libel business is becoming such a racket that something will have to be
done about it. . . . But I still don't see why Cape should want you to
pay for resetting passages their lawyer wants out." Stevie seems
herself to have had lingering fears about libel in her second novel
because Hart-Davis wrote to her shortly before *Over the Frontier* was
printed, in reference to her letter "about General Ironside. I considered
the passage carefully again, and decided that it was O.K. to leave it,
particularly as Rubinstein [Cape's libel lawyer] had passed it." Earlier
that year Hart-Davis had written to say, "I have forgotten to tell you
of one of my chief reactions. It is that I can now see a line of
development for you as a writer, whereas after seeing simply N On Y
P and the poems one hoped, but could not tell."

In 1937 Stevie's literary output was remarkable. Spurred by the
success of *Novel on Yellow Paper*, she saw her first collection of poems
into print that spring, and in the ensuing months managed as well to
place poems in the *New Statesman*, *London Mercury*, *Night and Day* (that
short-lived but glittering magazine), *The Bystander*, and *Granta*. (Most
of these were reprinted the following year when Cape brought out a
second collection of Stevie's poems.) *Over the Frontier* was finally

published early in 1938. Meanwhile Stevie began a new novel which she called *Married To Death.*

Throughout 1937 Stevie made friends of a number of new people, worked full time at her secretarial job, and conducted her life with Aunt in Palmers Green. However exhilarating this was, it must as well have been fatiguing for Stevie who all her life succumbed so easily to exhaustion. Weekending helped to revive her energies but it posed another problem she mentioned in a letter that year. "I am not a very good girl to my Aunt, you know, I am hardly ever at home . . . she gets lonely, does the Lion . . . home ties are trying because of this guilty feeling if you do not heed them; well then I might see less of my so-dear friends and stop rushing off Fridays for weekends, but . . . I like to see them, so everything is just fixed to be an uproar of distraction. . . ." In June 1937, Stevie went for a holiday with Aunt to Hunstanton on the Norfolk coast she loved so much. She needed the holiday, she said, because she was very anaemic. She stayed at a hotel and encountered "two friends of mine . . . oh yes, of course they are in the next novel", and their two children. The holiday was a success, partly because Stevie enjoyed "vicariously the experience of a family life". She bathed, walked and rode horseback until she got bucked off and was so frightened, she said, by stories an Irish groom told her about riding accidents, that she resolved to sell her boots and stick to walking. She incorporated this incident in a piece she published two years later in which she says, "I never rode again from that day to this." "There are parts round the Norfolk coast that look like nobody had ever been there before just dunes and sand and sea and nobody at all. . . I hate being back [home]."

In Palmers Green that year, Stevie attended plays put on at the Intimate Theatre, a local repertory group situated a few streets away from Avondale Road. One evening she took Aunt there to see *Hamlet,* and was amused by the Lion's preparation:

> She has been reading this grand old play and she says: It is very sad. There are a lot of well known sayings in it; there are a lot of people killed. Yes of course she has read it before but as a young girl I think she has forgotten, but staunchly she reads it through in the most repulsive sort of family bible edition of Shakespeare dated about 1860 full of engravings of him all head leaning on pillars rather like the statue in Leicester Square except that there are no pigeons to misbehave themselves.

Along with her friend Narcissa Crowe-Wood, a magazine editor on whom she had based "Harriet" in *Novel on Yellow Paper,* Stevie joined the London Gate Theatre in 1937. This interest in drama, kindled in the play-acting she did as a child, was no doubt intensified by her friendship with Denis Johnston, to whom she wrote regularly for some years after his fan letter about her first novel. They finally met in June of 1937 at the Café Royal. Stevie had proposed the meeting-place, suggesting that she wait "just inside the door by the fireplace among . . . the azaleas . . . until I am (a) claimed or (b) thrown out as a bad advertisement for the cooking. . . . I have *no* fringe", she told Denis Johnston. "I had one for a week but under pressure from friends I brushed it back & now it has grown full length. . . ." She describes herself in the letter as "very skinny rather again the Russian famine touch", and draws a picture of herself with the notation "5 ft 2½ inches". Stevie wanted to take Johnston's play *Moon in the Yellow River*, which she very much admired, to a clever young man who ran the Intimate Theatre in Palmers Green, in hopes of persuading him to stage it there.

Stevie regretted not being able to visit Denis Johnston in Ireland, although that summer she considered accompanying Harriet (Stevie continued to refer to Narcissa Crowe-Wood by the fictional name she'd given her) on a Bank Holiday cruise there. Harriet had fallen for someone, Stevie said, who taught dancing on the ship, and Stevie's task was "to be in the way and be in the way". This mariner's daughter had second thoughts about the rough sea, noting, "I do not make a very good sailor." In the end she did not go.

In June, she attended a party at Naomi Mitchison's. Mrs Mitchison's brother was J. B. S. Haldane, the scientist, known too for his Communist affiliations and in England for the writing he did for the *Daily Worker*. Mrs Mitchison entertained many English socialists and at her parties Stevie met some of them. "You will have to come again & meet more Reds! Also Wyndham Lewis", she wrote after Stevie had thanked her for a previous party. Stevie, conservative in her sympathies (if indeed someone so non-political can be characterized in standard terms), complained to Denis Johnston, "I like Naomi but of course we are poles apart as far as World Problems goes, but if she thinks she's going to rope me into the Haldane-Communismus gang she is mistaken."*

---

*Stevie's antipathy to Communism continued into the 1950s, when the Cold War caused many English writers to mistrust their government's peaceful intentions. She had joined a writers' committee formed "to stop the drift to war", but had soon become uncomfortable with its public letters and manifestos. On 14 February 1951, she resigned from this committee, saying it appeared "to have a pro-Russian, anti-Government, anti-American bias, in other words to tread closely upon the Communist party line." Two weeks earlier she had made known her sentiments

Two years later Stevie published a piece which reflects her response
to members of the Left she met in 1937 and continued to see for a while
thereafter. In it she ponders the excesses of both Right and Left and
exposes both as foolish. But finally she judges the tale of "Righties" to
be "safer, much safer, for mankind than the tale of the Lefties . . . Yes,
I will say it, 'I want more security' is a better song than 'I want more
happiness'." She adds, in a comment on the Right's preoccupation
with dividends, that theirs is not "a wicked gospel" because "where
there is a maximum of security, that is to say as even a distribution of
wealth as possible, there is a maximum of individual liberty." This
was probably Stevie's version of attitudes she inherited from her Tory
Aunt, who was "well read in the political game". Certainly Stevie was
original in thinking of an even distribution of wealth as a goal of the
Right. Politics had, as well, a comic aspect which Stevie faintly
mocked in telling Denis Johnston about "more talkie from Naomi
Mitchison, and she's got world problems on the brain". Generalizations are often alien to the artist, and they sometimes were to Stevie, as
she suggested when she wrote "I can so much more be interested in
Helen on Lou [her lover] than in Helen on Chamberlain."

> "No, I am not interested to concentrate upon politics [Pompey
> tells Professor Dryasdust in *Over the Frontier*], fascism or commu-
> nism, or upon any groupismus whatever; I am not interested to
> centre my thoughts in anything so frivolous as these variations upon
> a theme that is so banal, so boring, so bed-bottom false, so suspect
> in its origin. *C'est la vie entière que c'est mon métier.*"

"All my rich friends are communists, they have the cosmic conscience, you know, they have it rather badly", Stevie wrote. She
probably included in this category Margaret Gardiner, another person Stevie met at this time, who was a friend of Auden, and of other
writers of the 1930s, and who thought of herself and her friends then
as left wing. Miss Gardiner had a long relationship with J. D. Bernal,
the scientist, by whom she had a son. Stevie knew Desmond Bernal
and presumably met him with Miss Gardiner. Stevie later based the
character "Tengal" on him in her third novel, in which she recounts
a story he told her of his education by the Jesuits as a schoolboy at
Stonyhurst.

---

about Communist-leaning writers in a letter to the secretary of PEN's English Centre:
"relying on the pledge of secrecy, I do think it might be a good thing to have a
president who is known to be an active non-fellow-traveller. . . . One has quite
enough F.T.s as it is, breathing holy brotherhood down one's neck!"

Miss Gardiner once yielded to the importunities of a male friend that she introduce him to Stevie. He expected the author of *Novel on Yellow Paper* to be a bewitching glamour girl and was immensely disappointed when Stevie entered the room. In Miss Gardiner's view, "Stevie transformed herself very brilliantly, though, from a scarecrow into a character." Once Miss Gardiner went off in the evening to a party, and returned home after 2 a.m. to find Stevie sitting forlornly in the kitchen:

> She had asked whether she could stay for a few days while her aunt . . . was away. "What's the matter, Stevie?" I asked. "Why haven't you gone to bed?" "There was nobody to warm my milk," she said plaintively. I was amazed. "But couldn't you do that for yourself?" "No." She shook her head. "Aunt always does it for me."

Everyone remembers that Stevie wanted to be cosseted. Miss Gardiner and her friends took her seriously as a "minor poet" but did not think of her in the same category as Auden who, in Miss Gardiner's recollection, made no mention of Stevie's work. She thought of Stevie, "who always called one 'dear' ", as an "oddity".

Stevie also made "Cambridge communist friends" after the publication of *Novel on Yellow Paper*. They were people of "very highly trained intelligences, their hearts are in the right place, but also their heads are very good". However, added the sensible Stevie, "They have the schoolmaster-mistress approach to reality. . . Behave yourself. . . . War? Oh, dear me, no, that will not be at all necessary." Exactly which of her friends qualified for this criticism is not clear, but by 1937 she knew many people in Cambridge including, of course, John Hayward. Surprisingly enough, one Cambridge friend with whom she corresponded in the spring of that year was Lydia Keynes, the wife of Maynard Keynes.

In earlier years as Lydia Lopokova, Mrs Keynes had been a leading ballerina of the Diaghilev Company. When she no longer danced, she acted and did some broadcasts on the BBC. This Russian-born artist who, Virginia Woolf wrote, "mauled" Shakespeare and "speaks English like a parakeet", may have invited Stevie to compose monologues for her, because in June 1937 she wrote and voiced some enthusiasm about one of the two Stevie had sent. It is a four-page monologue spoken by a woman living in Shanghai (where Stevie had recently written to her friend Gertrude Wirth) who writes in her head a letter to Charles, her husband, cruelly absent for two weeks. The speaker thinks of what she really wants to say ("Darling darling I love you so much") and the perfunctory remarks she will allow herself to write down ("it is very hot here, the dog has had a fight"). While waiting for Charles she does the trivial things that have to be done, and grows ferocious and cruel herself. With only the company of a cat for consolation, she stains with her tears the paper on which she writes.

The piece is reminiscent of passages in Stevie's first two novels, particularly those with strong echoes of Dorothy Parker. Mrs Keynes thought there were "possibilities in Charles, but I cannot concentrate on him at present, because I am a nurse to J.M.K. who is making slow progress. Perhaps in two or three weeks time I will be in London, and then I will be delighted if you come to Gordon Square. In the meantime do listen to me and criticise on June the 7th (Variety Programme)." Keynes had suffered a heart attack in the spring of 1937, and from then until his death in 1946 his wife devoted herself to nursing him. That may be why Mrs Keynes never did perform Stevie's monologue on the BBC.

"I was in Cambridge last weekend staying with Lyn Newman", Stevie had mentioned in a letter to Mrs Keynes, and it was perhaps through Lyn Newman, and her husband, the mathematician Max Newman, that Stevie met Rachel and Horace Marshall. In a letter dated "June 27", Mrs Marshall wrote to ask if Stevie remembered meeting "a family . . . at a garden party in Cambridge", and to invite her to stay at her home for a summer weekend. Horace Marshall's sister Ray was married to David Garnett, and Rachel noted in her letter that "the David Garnetts particularly want us to bring you out to see them at Hilton". (Perhaps the Marshalls were present on the day Stevie visited Hilton Hall and fell into the nettles.) Stevie became a lifelong friend of Horace and Rachel Marshall, often spent the weekend with them, and probably met at their home a new friend —the famous left-wing economist, Joan Robinson. A photo of Stevie in the Marshalls' garden shows her sitting on a bench, a cigarette in one hand and a book in the other, wearing a dress ending near the top of her thighs. She is barefoot.

In late June of 1937, Stevie received a less than encouraging letter from David Garnett. She had sent to him a draft of *Married To Death* about which he commented:

> This is a painful letter to write and probably to receive. I have done my best with *Married To Death*, but I can't read it, and what I have read leaves a confused impression on my mind. I cannot describe it better than I could describe the landscape of what I see when I have been swimming under water.
>
> One does not know who is who or what is what as you say yourself in one place.
>
> I think it is absolutely fatal for you to write about yourself any more. A book must have shape, bones, foundations. It ought to be

built like a house. This is liquid, a flowing stream of words. And while you write about yourself you will write in this pouring way . . . It is in fact day-dreaming, the changing flow of thoughts & words pouring like a millstream & working the typewriter keys like the . . . mill pounding away at night & frightening Don Quixote.

It is terrible—because *you are a writer*. Every page shows what a fine writer you are: & the pages add up into an impenetrable dossier of private day-dreams. You must think of writing as a form of architecture: as a matter of logical balance of forces—seeing that the beams tie the walls together before you put on the roof.

Write a story about people unlike Pompey & having no relation to her. Write about anybody but yourself. Somehow you must escape from this poisonous day-dream. . . . you've been writing for yourself & not for us. Yours ever & with faith in your future.*

Much of what David Garnett told Stevie about architectural weakness in her manuscript was said as well by Desmond Shawe-Taylor in his review of *Over the Frontier*: "I rather think she may have to drop Pompey", was his reluctant conclusion.

Despite the assurances Rupert Hart-Davis gave about her future, Stevie was uncertain where to turn, as she indicates in verse and a drawing she pasted into a copy of *A Good Time Was Had By All* and sent to Hamish Miles (the editor at Cape with whom she proofread *Novel on Yellow Paper*). She captioned her drawing:

> Children who paddle where the ocean bed shelves steeply
> Must take great care they do not paddle too deeply.

The brink of a precipice where children play is labelled, "NOYP, AGTWHBA, OTF" and below these acronyms for the books she had completed by 1937 is a great question mark. Stevie adds these verses:

> So you see Hamish there is one thing more
> Remains to be said that has not been said before
> I am grateful to you yes but there is the fear too
> That always the next time I may disappoint you
> This thought turns my verse and alas ends my rhyme for me
> Goodbye goodbye goodbye.   Excuse
>   Stevie.

---

* *Married To Death* appears to have been a novel Stevie wrote and then abandoned. Twelve years after she received Garnett's letter about it, she had her just-published third novel, *The Holiday*, sent to him because, she said, "you were so very kind to me when I first began to write and I know how very disappointed you were in the manuscript novel I sent you after *Over the Frontier* was published and it was certainly very bad. I do hope you will not be disappointed in this one."

"Look out for choppy weather in January", wrote Stevie to Denis Johnston, referring to the publication of *Over the Frontier* in 1938. On several counts its appearance filled Stevie with alarm. Nearly a year before *Over the Frontier* appeared, Stevie had confessed to Naomi Mitchison her misgivings about the reaction Sir Neville Pearson would have to her second novel. "Phoebus is getting hot-cattish I mean cat on hot brickish about 'all this yellow paper' and I'm beginning to wonder if he will stand for the new novel. . . . You are lucky not to have a Phoebus to square." A few months before the book appeared, Sir Neville warned Stevie to be careful. "I think it is an amusing situation, rather a little bit dangerous too, for instance they made me alter lavatory to washplace, that shows you how refined we are in Southampton Street." In November Stevie complained to Denis Johnston:

> I am no martyr, I cannot feel really cross with these directors, indeed I see their point, or occasionally I get a glimmer of it. Happy is the secretary who has no history, that sort of thing I take it, and as Sir P. said: Caesar's wife . . . But I am Caesar's wife, I said, meaning of course that in that way I was above reproach, and not wishing in any way to make trouble with Calpurnia but it's no good. . . . I am sailing all the time a bit near the wind, but if it gets too strong, I shall take a reef and tie some grannies, I do not intend to be a martyr to art for art or any hysteria of that sort; it is entirely a question of cunning and not of principle. How much can I get away with? it's a gamble.

In this second tale of Pompey Casmilus's adventures, Stevie is occupied with an image that abounds in writings of the 1930s and early 1940s—the frontier. Auden and Isherwood are two whose works prominently feature it. (In 1937 the Cambridge-based Group Theatre produced the last Auden–Isherwood play, entitled *On the Frontier*.) The frontier was for obvious reasons increasingly present in the consciousness of Europeans as the 1930s advanced, and by the end of the decade frontiers preoccupied the world. The second half of the novel, and its rather surrealistic account of an espionage mission Pompey undertakes, is remarkably and deliberately contrasted to the first half where Pompey agonizes over her failing affair with Freddy, attends to her job in a publishing house, weekends with friends in the country, and conducts her cosy, familiar life with the Lion Aunt in her London suburb. "Don't go running down the safe and ordinary. Hang on to it", Pompey advises in *Novel on Yellow Paper*. But midway through the second tale, and while convalescing at a *Schloss* on the Baltic (from flu induced by her bust-up with Freddy), Pompey disregards her own advice.

"Out of desperation over her lonely 'Life in Death'," wrote one critic, "but also at the cost of her freedom, [Pompey] places herself in the service of a political and warlike power, opposed to a democratic ideal." She sets off on a mission that gradually leads her over the frontier into the midst of a war and, in the end, to self-knowledge:

I may say I was shanghaied into this adventure, forced into a uniform I intuitively hated. But if there had been nothing in me of it . . . should I not now be playing, in perhaps some boredom, but safely and sanely enough, with those who seem to me now beyond the frontier of a separate life?

Years later, reviewing an author whose point was that "wars, oppressions and cruelties are first in our own hearts", Stevie said, "it is a truth that simply cannot be too often repeated."

The novel is not finally a success, in part because of the jarring shift mid-book from realism to surrealism. "There is nothing worse than the not perhaps quite funny enough", says Pompey—inadvertently pointing to another flaw in this novel. Stevie said in later life, "It is horrible, I am so ashamed of it", and told John Hayward it "was nothing to be proud of". At the outset, though, most critics liked it. One reviewer found it, although "not always rationally comprehensible", a "considerable" advance over *Novel on Yellow Paper*, "for the first lacked form". The *Times Literary Supplement* liked the novel too, although the reviewer denied it that designation: "'Over the Frontier' is not . . . a novel. It has no plot and it contains no integrated characters. It is rather a . . . record of Pompey's spasmodic thoughts and emotions and *obiter dicta* . . . on whatever comes into her bewildered heart and analytical head." Malcolm Muggeridge, who also reviewed the book, said: "Every variety of groupismus addict would be able to find something to object to in it, and therein lies its virtue." Once again Stevie drew the attention of notable reviewers. The novelist Frank Swinnerton reviewed *Over the Frontier* in a batch of new books by H. G. Wells, Sinclair Lewis, and G. E. Trevelyan, among others. He found Stevie to be "deliciously inventive", though he thought that some readers might be "fatally perturbed by Miss Smith's pidgin English". He confessed that he "did not understand what 'frontier' it was that Pompey crossed in the book, and I shall not here grope after hidden meanings." Nearly everyone had difficulty with the second half of the novel where the plot grows puzzlingly fantastical.

Rosamond Lehmann praised "the feeling of strong sensitive intelligence" in the first half but "not the fancy surface", and ultimately got no thanks from Stevie. Years later Miss Lehmann remembered that

Stevie "took criticism very badly (like most authors!) . . . [and] I got a rather huffy letter from her—for which long after, she apologized very winningly." The relationship between Stevie and "die schöne Rosamund" as Stevie rather wickedly referred to her—always misspelling her name—had begun in a brighter key.

They met presumably at the lunch party to which John Hayward invited both writers after he read *Novel on Yellow Paper.* In the spring of 1937 Stevie sent Rosamond Lehmann a copy of *A Good Time Was Had By All,* and the day after receiving it Miss Lehmann wrote back to say how "very funny and moving too" she found the poems and how "sometimes really brilliant" the drawings seemed. "Eng." was her favourite poem:

> What has happened to the young men of Eng.?
> Why are they so lovey-dovey so sad and so domesticated
> So sad and so philoprogenitive
> So sad and without sensuality?
> They love with a ci-devant feminine affection
> They see in their dreams a little home
> And *kiddies*
> Ah the *kiddies*
> They would not mind *having* babies:
> It is unkind
> Of Nature to lag behind.

Perhaps it was a poem directed at surburban Freddy, but its theme would have been congenial to the legendarily beautiful and romantic Miss Lehmann. Her husband was doing ambulance work at that time in Spain. "He can't come back till he knows fascism will be defeated", she told Stevie. "Reading your poems has given me a nostalgia to see you again", and she invited Stevie to come to her home in Ipsden, near Oxford, for a quiet weekend on 15 May, just after the coronation of George VI. "John Hayward will tell you it's very unfrightening, staying here."

> She came down and spent the day with my sister Beatrix and myself. . . . She was *extremely* talkative and amusing . . . after I'd taken her to Goring to catch the train back to London, Bea and I agreed that her plainness was quite startling. Then Bea said, "But *did* you see her *legs*? Very pretty legs." (True!) We didn't meet again.

When *Over the Frontier* was published in January 1938, Cape sent Rosamond Lehmann a copy and she wrote shortly after to Stevie to say, "I revelled in the first part & found it as moving, as funny, as fascinating, as intelligent as ever—in fact, even more so, I think—".

She went on to give criticism of the second part that today seems more cogent than much of the adulatory comment tossed off by reviewers in 1938.

> I felt it has worried *you*, you hadn't been able quite to bring it off. I feel so uncertain about the clues—as if they were too private; and I lose the scan of Pompey offering up her incandescence for all to profit by . . . I am your true admirer. I know what it's like to be told one's second book has failed—when one *knows* really it's a step forward. . . . I *know* your third book will be the best yet. Pompey is one of the most adult and enlightened women I've ever known. But sometimes I feel she lets herself be caught in her illness & weakness & little-girlness—then I am disappointed.

Rosamond Lehmann later guessed that her saying the novel "didn't quite come off" infuriated Stevie and that "the tremendous success of *Novel on Yellow Paper* had perhaps gone to her head". "Little-girlness" probably hit home. As her style of dress and behaviour (waiting for someone to warm her milk) show, Stevie did often play the "little girl", although she would no doubt have resented advertence to this by others, particularly when made in a critical way by such a woman of the world as Rosamond Lehmann. Certainly she resented Miss Lehmann's comments and became "cross and snappish" in the phrase of the latter, who felt impelled to explain:

> About the baby talk. I suppose it's true, it hits a tender spot and "lets the side down." I feel embarrassed at having my own infantilism laid bare . . . childishness in grown-ups seems to me utterly different, and bound to be different, from being a child. I don't *quite* understand about the child being the direct heir to the kingdom the power and the glory. You seem to be saying that to be a child is the best thing, and *I* think to be grown-up (in the way I mean it) is the best thing. Child-likeness *is* a part of this, of course, To remain at the kind of anarchistic or disintegrated stage which (I think) you describe . . . seems to me to be destructive, and as if it *must* end in fantasy and daydreaming about life. Of course I don't think of people like your baronet & V. Brittain as adult in my sense. I agree with you about them. They are examples of a "frightful maturity."

Derek Verschoyle, then literary editor of the *Spectator*, apparently fuelled the controversy by telling Stevie that several of Miss Lehmann's letters were malicious in intent. Stevie did sort out much of their dispute in a letter she sent to Rosamond Lehmann later in 1938. "When I got your subsequent letters I saw at once what had happened. You thought I had asked Cape to send you my book; and thinking that had a perfect right to criticise it." She dispatched a copy of her second book of poems, *Tender Only to One*, published in November of 1938, and invited Rosamond Lehmann to send her reactions. But this represented less a peace treaty between them than a fragile truce.

Edwin Muir gave a more enthusiastic review to *Over the Frontier* than he had given to *Novel on Yellow Paper*. He found it "brilliantly sustained, a wonderful piece of fantastic character writing", and noticed something central about the book that applies in a larger way to Stevie's work: "once you put on a foolish mask you have no need to fear seeming foolish. You can say what you like." Another critic mentioned Stevie's attitude of "brilliant clowning", pointing to the storyteller's clown mask she so often adopted. This mask links Stevie to other writers and visual artists in the period who found that the persona of the sad clown allowed them a viable and comfortable voice somewhere between the comic and tragic.

Stevie had read Kafka's *The Castle* several years earlier and found it brilliant. The repeated references to the *Schloss* (castle) in *Over the Frontier* naturally bring Kafka to mind and it is not surprising that Muir, who together with his wife Willa translated Kafka so success- fully, was reminded by Stevie's novel "curiously of Kafka, though the style and the quality of imagination are so different." Marie Scott- James noticed the Kafka connection as well in what is, all in all, the best review of the novel to have appeared in 1938. She spoke of the novel's "shadowy world, part Kafka, part Carroll". A critic with whom Stevie eventually corresponded and whose literary assessments she approved, wrote that Pompey felt "most comfortable as a guest who always is glad to come and just as glad to leave again." He called *Over the Frontier* a novel about "the danger of compulsion and being held prisoner which threatens those people who are born to be guests"—an interpretation that also links Stevie's writings to Kafka's.

Storm Jameson could not "understand the last third of this book. It is baffling"—although she thought it the tale "of a mind peculiarly sensitive to something which is happening in the world, which perhaps only a poet can feel." Probably one can agree with the reviewer who wrote that "the author tries to land too big a fish". But what in retrospect we can now see is that in 1938 the frontier was a frightening fact and simultaneously a timely and arresting metaphor, and that Stevie exploited both dimensions. When *Over the Frontier* was reprinted in 1980 one reviewer emphasized the metaphorical aspect of the frontier. "There is another frontier faced in the book . . . [Stevie Smith's] much apostrophized 'dear Reader' understands, however, that it is the wilderness of her own *animus* that she is exploring." The terrible experience of Pompey's long night ride in *Over the Frontier* turns out, rather as in Auden's poems of the 1930s, to have the nature of a Quest, and Pompey is rewarded for carrying it off so honourably with a redeeming truth: "power and cruelty are the strength of our life, and in its weakness only is there the sweetness of love."

In February of 1938 Stevie told Denis Johnston how glad she was he had liked *Over the Frontier*. She had had "a mixed bag of brick-bats and bouquets, and now it's more or less all over, and I suppose I ought to be getting along with the next one." A short time later the *New Statesman* offered prizes for a poem in the style of "Hopkins, Shelley, W. H. Auden, Oscar Wilde, Stevie Smith, Edward Lear, Hilaire Belloc, Wordsworth". Second prize went to "More About Pompey":

Alfred Lord Tenyson (Alfie to me,
At least when we were in the bath together in my flat at
    Cockfosters, chaps),
I so-o liked to have him sitzend auf meinem knie . . .

That others were already parodying Stevie's style may have made her self-conscious, but it is a measure of the fame she had attained by then.

9

No Private Peace

"To keep out or not to keep out of war, that was at that time the question," Stevie wrote as 1938 closed and a new and ominous year commenced. *Over the Frontier* is a novel about war in which Stevie cast Pompey as something of an insouciant and *stürmisch* Maid of Orleans, the military woman disguised in a man's uniform, reviving however unconsciously "a favourite character throughout the nearly four centuries of British broadside ballad tradition". This literary figure of the woman soldier was soon to have real embodiment in England among the many women who either served in the armed forces or took their turn as fire-watchers or air-raid wardens in the war effort at home.

Stevie went to first-aid lectures which she detested but, as she wrote to Rosamond Lehmann in 1938, "now and then things get so frightful in Germany I think one ought to make sacrifices, so now we are having gas lectures on top of the first aid ones . . . my doctor said the other day that it would be better to keep away from them and that a rather bad first aider was a burden . . . one doesn't know what to do."

Despite the threat of war and other hardships, these were years when new friendships brightened Stevie's life. "Friendship . . . is the stuff of life", she wrote, but characteristically included as well "the revolt from friendship". "I am so grateful to my darling friends, to all my darling friends", said Stevie. (But in a poem she wrote at that time, "I am glad . . . that my friends don't know what I think.")

One of the friends Stevie saw quite often in the 'thirties was Inez Holden, who brought out her first book in 1932. Stevie described her to Denis Johnston as "rather off the handle quite nice, and really funny"; and to Osbert Sitwell as "so amusing, and not groupy at all, she always makes me laugh so much . . . the whole time." During the war Inez's merriment no doubt helped cheer up Stevie. She spoke of Inez as one of her "academic gairls" in the old-fashioned pronunciation which she used fondly but in a spirit of mild fun too.

Stevie's third novel, *The Holiday*, has a character called Lopez based on Inez, and another character by the same name, also based on her, appears in one of Stevie's short stories. Inez reciprocated: in a fictionalized diary published as *It Was Different At the Time*, Inez Holden describes a character she calls Felicity with whom she went about London in April 1938. Felicity had "a forehead fringe and when she talks her eyes move restlessly from side to side", as Stevie's piercing dark eyes did. "It was as if I had a girl-Eddie Cantor galloping round my room", Inez wrote, in a description echoing Pompey in *Novel on Yellow Paper*, who has a "fringe awry that is looking always like Eddie Cantor".

Inez was a friend of Barbara and Alan Clutton-Brock whom Stevie met, perhaps through her, in 1938. At the time, the Clutton-Brocks lived in Essex and Stevie visited them there. Later on Alan Clutton-Brock inherited Chastleton, a beautiful Jacobean house in Gloucestershire, where Stevie loved to stay, although she could be wry about the bevy of cats that wandered about. On one visit Barbara Clutton-Brock came upon Stevie sitting in the garden, scribbling a note. She rose at Barbara's approach and the sheet of paper fell to the ground. Her hostess, who bent to retreive it, couldn't help notice it was a thank-you note Stevie had written before she'd even started for home. Mrs Clutton-Brock viewed this as one of Stevie's "economies". Stevie and Alan Clutton-Brock shared a strong dislike of Roman Catholicism which was, his widow judged, the basis of their friendship; but Mr Clutton-Brock, a professional art critic and painter, did not admire Stevie's drawings and thought she should publish her poems without them. Stevie was annoyed by his advice and said she didn't intend them as works of art but rather as pointers to meanings in the poems.

The Clutton-Brocks were another of the families Stevie adopted. They had two daughters, one of whom adored Stevie, the other who disliked her intensely. Their mother remembers a rivalry for attention Stevie demonstrated when she visited and the children were about. The Clutton-Brocks found her surprisingly easy as a guest, although she was given to announcing during a visit, for instance, that a friend had turned up in the neighbourhood and might she be asked to dinner? She liked to take walks in the countryside and must have been, Mrs Clutton-Brock thought, the only person ever to have encountered a snake in Gloucestershire. She became quite hysterical over that experience, her friend remembers: "I never saw her like that before." When Stevie asked Alan Clutton-Brock for a painting and he invited her to choose one from many that were stored upstairs, she went off and returned with two in hand, one by him and the other an amateur work painted by his wife. Stevie liked them equally well. Barbara Clutton-Brock said Stevie had no critical judgment about pictures and very little interest in any of the arts apart from writing.

This is curious because Stevie went to galleries with some frequency and in 1938 published not only her essay on George Grosz's pictures as part of her second novel, but also an essay in the *New Statesman* called "Private Views" about the summer shows held annually at the Royal Academy. She does comment on the pictures, although gallery going appears to have been as much a social experience as an aesthetic one for Stevie. A friend who was a serious collector and sometimes visited exhibitions with her recalled that Stevie was quickly wearied by walking about in galleries, but had a remarkable way of sizing up a collection and "making her own" the few items that interested her. Once they went to a private view at the museum in Norwich and Stevie soon dropped on to a sofa in the centre of one of the rooms. Another woman came over and flopped down alongside her, inquiring if Stevie were tired. "My friend is *still* going round the exhibition. She's been here half an hour already," Stevie told her. The other woman replied, "I too have a *very strong* friend."

Ivor Brown, an old acquaintance of Stevie and subsequently editor of *The Observer*, reacted to her essay "Private Views" with an article the following week in the *Manchester Guardian* in which he criticized Stevie's style for its "jerky jabs at the attention". Stevie was just then corresponding with Osbert Sitwell who told her Brown was an "idiot" and aptly added, "if he is out to find a modern author *without* a sense of rhythm, why pick on you who are the essence of rhythm?"

An anxiety Stevie suffered in these troubled pre-war years, and for a long time after, arose from lack of money. "The pressure of money is . . . not to be laid down, I guess, this side of the grave", she wrote. Reviewing was one way to add to her modest secretary's salary, and it was in 1938 that she began in a slight way what burgeoned for her into a lifelong career as one of London's most entertaining and intelligent book reviewers. Her earliest reviews were written for *The London Mercury* whose editor, R. A. Scott-James, admired her writings and published some of her poems and drawings in 1937. His assistant there was Armide Oppé, who became a close friend after they met in connection with the work Stevie published in *The London Mercury*. Scott-James may have been the editor who invited Stevie to submit trial reviews in 1937. The reviews met with approval and Stevie was offered the job. To Rupert Hart-Davis she confessed:

> The problem now is how can I say what I think in a vigorous and highly offensive manner and avoid the consequences. I tried the synopsis method but the editor said: Put more of yourself into it. So how can I do that if I do not like the book, and not be offensive . . . you see I am not so certain of my judgement, but you see also I have a quick reaction to these books and if I do not write then it is getting dull and not what the editor wants. Perhaps it is not a good thing for authors to review books.

Stevie was no doubt thinking of her own history when she wrote to answer a man who asked her advice about reviewing:

> Reviewing is a difficult ring to break into unless you have already written something yourself or can catch the eye of an editor with some specimen reviews which he may like the look of. Rightly or wrongly editors like names . . . Why not also send them a poem or two? I published a good many poems in this way before I produced the books, and as you are also interested in the financial aspect, I need hardly point out that if you publish first in periodicals and then in a book, you are selling twice over—to say nothing of the anthologies.

The first of her reviews for *The London Mercury* was of Osbert Sitwell's *Those Were the Days*. She describes the contents clearly but also ruminates in a fascinating way about issues raised by the author —a speculation, for instance, that the period leading up to World War I "may prove too much a shadow of 1938 . . . Is the human race then never to be trusted round the corner where the hobby-horses of ideology neigh?" The advice to "put more of yourself into it" was probably the soundest an editor ever gave to Stevie. The review pleased Osbert Sitwell and he wrote to tell her so.

Less responsive in 1938 was Kingsley Martin who, as editor of the *New Statesman*, overruled his literary editor and rejected a review Stevie had written of both Jan Masaryk's *Modern Man and Religion* and Father Gerald Vann's *Morals Makyth Man*. Stevie had praised the clarity of Thomistic thought in the Dominican's book and faulted Masaryk for his "misty and fear-ridden Calvinism". Kingsley Martin was identified with anti-fascist groups although, Stevie said, "he only disliked *other* dictators". She described to Osbert Sitwell a rally she attended in June 1938 of one such group at which the poet Cecil Day Lewis spoke while Kingsley Martin sat on the platform. Day Lewis quoted Göring, saying, "When I hear the word culture I reach for my gun." (Often attributed to Göring or Goebbels, the line was actually from *Schlageter*, by the Nazi playwright Hanns Johst.) Stevie told Sitwell: "When I hear the word Kingsleymartin, I reach for my gun. And when I hear the word Daylewis I reach for my gun." She sarcastically described in this letter "the *mouse* of those *mountains* of oratory . . . a telegram to Freud [in flight from Hitler], groupily wishing him a happy stay in England."

Nina Condron, a journalist whom Stevie saw often in these years, was also a friend of the American writer Kay Boyle and sometimes spent holidays with her in Austria or the South of France. For a time in 1938 Stevie was planning a holiday jaunt to join them near Gervais-les-Bains but was put off by what she inferred from Nina Condron's conversation would be a discomforting bohemianism in the Kay Boyle *ménage*: "Why put cups on the table, why not on the floor, the English are so conventional, if you want to throw somebody out of the window, why not do it, the English are so conventional (this will become a poem if I am not careful)." She tried to arrange for Malcolm Muggeridge and Inez Holden to go along, as well as Denis Johnston, and proposed that they stay at the pub instead of *chez* Boyle. In those years Stevie had only a fortnight in July for her holiday, and as the others were not free then, she dropped this plan. In the end she spent the two weeks in Scotland—frightened away to the west coast, she joked, by "the wraith of Edith [Sitwell] haunting the Northumberlandshire uplands".

Despite Stevie's efforts to discourage him, Denis Johnston went to Austria that summer and Stevie wrote in September to ask how he found it. She was attending first-aid lectures on bones and bandages given at Tower House by a Colonel whose name, Thurlow-Potts, she very much enjoyed. But the talks convinced Stevie she would be hopeless at first aid, so she decided if war came she would try to get a job at the BBC to replace men there who would be called up. It was her hunch that the paper shortage would lead to the suspension of many publications and the shrinking of reviewing jobs; she thought to take advantage of her "fine port winey voice" in radio broadcasting, and asked Denis Johnston for introductions at the BBC. In the same letter, Stevie declares her support for Eden, Churchill and Attlee. "Horrible times" is her envoi in the letter. The next day she wrote again to Johnston to tell him Sir Neville Pearson had discouraged her BBC aspirations, saying that "my voice is awful, I mean not clear and with a lisp (!!) and also that if war comes I shall be for once in my life indispensable here."

"My poems will be out later on this autumn and I think I really am not going to write any more", Stevie told Denis Johnston in 1938.

Rosamond Lehmann was awaiting her copy of Stevie's new book of poems, *Tender Only to One*. Miss Lehmann was anxious about war and described herself to Stevie as having "a perpetual hollow in the diaphragm, and occasionally everything seems no use". She appears to be questioning a contention of Stevie when she asked, "How do we know that the suicide rate declines in wartime? All the young men are saved the trouble of killing themselves by having it done for them." She was sick, she said, of all public activities. "I'm not even attending gas lectures . . . I want to retire and write reminiscences of my childhood." Rosamond Lehmann told Stevie she was longing to read her poems: "they are exactly what I feel like reading", and she asked Stevie again to "forget the misunderstandings".

In August of 1938 Stevie had written to Rupert Hart-Davis at Cape, asking that he delete five poems from *Tender Only to One*, including what was to become its title poem. He obviously refused to delete that, and also "Little Boy Sick". She was "not certain" about "Souvenir de Monsieur Poop" but decided to retain the poem as it balanced off so many short ones. But she rejected "Father Damien Doshing" (with its echoes of Belloc), about a priest who took in washing:

> This conduct in a man of God
> His flock thought very very odd
> And to his priestly honour detrimental
> They told the Bishop who though transcendental
> In his views was seriously shocked.

"Lulu" is another poem Stevie deleted from her second book of poems:

> I do not care for Nature
> She does not care for me;
> You can be alone with a person,
> You can't be alone with a tree.

A number of other poems which Stevie had published in magazines in 1937 were not included in *Tender Only to One*, probably owing to limitations of space or uncertainty on her part about their suitability.

The book appeared on 18 November and Miss Lehmann wrote a few days later to say, "I think this selection is the best thing you've done . . . Your special qualities come out stronger than ever; your talent for pointing the deathly joke and causing the wry smile. . . . Some of the lines haunt me, and I kept on repeating 'the world inherits wormliness' ["Reversionary"]—which is what I feel since the Munich agreement . . . The illustrations are wonderful." She went on to praise "The Photograph" and "Infelice"—a poem that contained, according to one review, "all the romantic, self-deluding women in the world". Rosamond Lehmann agreed with an assessment Stevie once made of George Stonier that "when it comes to poetry . . . he does know", and it was Stonier who reviewed this new collection of Stevie's poems for the *New Statesman*. He wrote that the title poem, illustrative of her infatuation with death, "exploits a distraught humour".

> She writes alternately sharp and sentimental verse . . . to nursery tunes; she is in love with Death . . . and with the scenes of childhood; her writing has the air of an odd only, lonely child . . . What is less easy to suggest is the mixture of nostalgia and parody, of poetry and jingle.

Stonier did not refer in his review to "Souvenir de Monsieur Poop", which Stevie put in the collection and then showed to him too late to withdraw when he expressed his dislike for it. Stevie told Rosamond Lehmann that she couldn't sleep for thinking of it. "How I let it get in I can't imagine, I get so sick of my poems I cannot really bother to read them in proof and am always chopping and changing." She wished there were a litmus test for poems, "blue for bad and pink for good", to help her make better judgments. Monsieur Poop is an arbiter of poetry who stands for all the stodgy editors and critics Stevie faced in the late 1930s, especially Ivor Brown. (Brown's piece in the *Manchester Guardian*, in which he criticized Stevie's prose, made a point of Stevie's being young, praised Housman, and disingenuously referred to his fogeyism. Stevie's Poop is prejudiced against the young, likes Housman, and disingenuously refers to himself as an old fogey.) But her assault on Monsieur Poop is uncharacteristically leaden.

In those days, just before World War II, the "happy dogs of England" about whom Stevie wrote a "political" poem included in this collection, were still free to bark, but increasingly Stevie and her countrymen were aware that "If you lived anywhere else / You would not be so gay." At a time when old stabilities seemed threatened by extremes of Right and Left, the Church of England was one symbol of stability to her, and in *Tender Only to One* Stevie, who was "well up in Church history" and "could correct anyone on Augustine & Pelagius", defended it in her poem "The Bishops of the Church of England":

> I admire the Bishops of the Church of England
> No man can be a Bishop of the Church of England
> And a fool.
> A man can be a Bishop of the Church of England

And a knave.
But
Fortunately
Few if any of the Bishops of the Church of England
Are men of ill will
They do their best
To resolve wisely
To govern effectively.
They are the butt of the ignoramus,
Of the sentimentalist,
Of the man who makes
Of his own bad temper and incompetency
A Movement for the Amelioration of the Sufferings
Of the Oppressed Members of the Lower Middle Classes.

A few months later Stevie said in *Eve's Journal*, a Newnes and
Pearson magazine for which she wrote a monthly column called
"Mosaic":

> "Why doesn't the Church say do something?" This is another cliché
> of the Lefty good-hearts. The Church? There's a very special dirty
> look for the Church, and as for the Bishops . . .! The printers won't
> pass it, so we fall back on dots. But the Church has always said:
> "Man is Sinful." And Man has never denied himself the pleasure of
> proving the Church right. What can the Church do or say but stick
> to its mandate, keep the churches open and pray for a change of
> heart?

In *Eve's Journal*, as in her fiction, Stevie wrote autobiographical
narratives, although she deployed such customary thin disguises as
Bottle Green for Palmers Green. Some small incident served as a
slender plot on which Stevie draped her views about quotidian
concerns or depicted unusual characters. In one brief essay, for
instance, she described drives she took into the country with a friend.
They go to tea at the home of Lord Bubble. With his pigeons dyed
green and blue and yellow he is obviously Lord Berners, and Stevie
gave a brief but entertaining account of this sybarite's life at Faring-
don. Her caution emerged in this essay, however, when she came to
mention Lord Bubble's young lover and heir whom she identified
only as "Mr._____." It was difficult to find material for her column,
Stevie said, "and I don't know that I am yet absolutely OK for libel".
Many of the poems in *A Good Time Was Had By All* are occupied
with Eros: the dictionary of slang lists a sexual meaning among those it
gives for the phrase "a good time" and it is this meaning T. S. Eliot

exploits when using the phrase in "The Waste Land". In *Tender Only to One*, though, the subject shifts to Thanatos: in the title poem the name of the "One" is Death, and the titles of other poems tell the story: "Death's Ostracism", "The Deathly Child", "Proud Death with Swelling Port", "Upon a Grave", "To a Dead Vole", "Siesta". Others, too, such as "The Boat", "Silence and Tears", "The Doctor", and "Arabella", are meditations on death, its inevitability and its allure. As Stevie wrote in "Come, Death (1)":

> Who would not rather die
> And quiet lie
> Beneath the sod
> With or without a god?

To what extent the imminence of war intensified Stevie's disposition from childhood to see death as a friend is difficult to say. But as the refrains in "Upon a Grave" indicate, "In Death is sorrow shed" and "In life is sorrow known" is her Sophoclean thought in this volume and recurrently afterwards.

Stevie's embrace of death seems, finally, less related to World War II following in rapid succession an earlier and devastating war than it does to something more personal. In "Fuite d'Enfance", the penultimate poem in *Tender Only to One*, Childhood—the speaker—refers to two loves ("One is my father / And one my Divine"), and asks, "Which shall I follow . . . / And following die?" Stevie fought hard throughout her life to preserve what was valuably the child in her. Deprived at an early age of her terrestrial father and mother and born into a century that witnessed the Death of God, this heir of Blake walked about as a child not of Innocence but solely of Experience.

"We are not innocent, yet innocence is what we would wish for", says Celia in *The Holiday*. And again, "I wish for innocence more than anything, but I am conscious only of corruption." In "The Lads of the Village", with its echo of Blake's "London" in the line "let man have his self-forged chain and hug every fetter", Stevie says it is upon the fields of experience that "pain makes patterns the poet slanders". And one is reminded of Blake's "The Tyger" in the opening lines of Stevie's "Little Boy Sick":

> I am not God's little lamb
> I am God's sick tiger.
> And I prowl about at night
> And what most I love I bite,
> And upon the jungle grass I slink
> Snuff the aroma of my mental stink, . . .

> My tail my beautiful, my lovely tail,
> Is warped.
> My stripes are matted and my coat once sleek
> Hangs rough and undistinguished on my bones . . .
> Consider, Lord, a tiger's melancholy
> And heed a minished tiger's muted moan, . . .

Stevie immortalized her Aunt as the Lion of Hull, but—as in "Little Boy Sick"—she very often at this time presented herself in the image of a tiger. At the end of *Novel on Yellow Paper* she symbolizes the sorrow of Pompey in "the tigress Flo" (Florence was Stevie's first name, of course) and this just two paragraphs after quoting a couplet from Blake:

> There was pity and incongruity in the death of the tigress Flo. Falling backwards in her pool at Whipsnade she lay there in a fit. The pool was drained and Flo, that mighty and unhappy creature, captured in what jungle darkness for what dishonourable destiny, was subjected to the indignity of artificial respiration. Yes, chaps, they worked Flo's legs backwards and forwards and sat on Flo's chest, and sooner them than me, you'll say, and sooner me than Flo, that couldn't understand and wasn't raised for these high jinks. Back came Flo's fled spirit and set her on uncertain pads. She looked, she lurched, and sensing some last, unnameable, not wholly apprehended, final outrage, she fell, she whimpered, clawed in vain, and died.

Over the Frontier is only a few pages under way before the reader is asked, "What of the heart of Pompey that lay down to die with the tigress Flo, to decline upon her paws, to give up a life that was so hateful." In *Tender Only to One* Stevie points in "The Photograph" most directly to the tigerish nature which was all her life the source of so much suffering, but of glory too.

> They photographed me young upon a tiger skin
> And now I do not care at all for kith and kin,
> For oh the tiger nature works within.
>
> Parents of England, not in smug
> Fashion fancy set on a rug
> Of animal fur the darling you would hug,
>
> For lately born is not too young
> To scent the savage he sits upon,
> And tiger-possessed abandon all things human.

It was this tigerish nature that made Stevie think of herself as "a desperate character".

Right after the publication of *Tender Only to One* Stevie felt "very depressed and deathly" about her writing, and was convinced it annoyed people who presumed her motive for writing was in fact to annoy. She continued to watch with horror Germany's moves on the world scene. In January 1939 she joined PEN after Storm Jameson convinced her that the writer's club would extend aid to refugees. (Ironically Stevie came down with German measles in February 1939.)

Aunt, who would not have a radio in the house in those years, disliking the "noise" so very much, kept up with the news in the papers. Listening to a neighbour's radio, Stevie heard the news of Chamberlain's visit to Berchtesgaden and speculated about the outcome of this journey in an article she published in March of 1939. She took issue with isolationists in "Bottle Green" and elsewhere who asked with some condescension, "Where the hell is Czecho-Slovakia?" They will find out all right, wrote Stevie, "if Germany goes on her *Drang nach Kolonien*". While a neighbour absurdly insists that nothing can happen to her sons because the family has signed the Peace Pledge, Stevie has a vision of "darkness and a great wind blowing over dead battlefields, and the stench of death without honour, and the ridiculous sad cry: We never knew." Stevie speculated in her essay about what posture America would take in the growing conflict. As for England, she concluded: "If there is no

Stevie's drawing of a child on a tiger skin which she placed by her poem "The Photograph"

possibility of two opposed ideologies existing side by side, then the choice must be made, even the choice of war." Nor can those who sign peace pledges evade their responsibilities in her view: "there is no existence of a private peace; you fight for your country or, refusing to fight, you yet fight, and directly for the enemy. That is perhaps the ultimate most horrible demand of war; the State must have your conscience."

Soon after war did break out, on 3 September 1939, and Stevie at the age of thirty-seven made what efforts she could for her country. She became a fire-watcher and in that capacity met Norah Smallwood, the publisher, patrolling a West End district to which they'd been assigned. (Fire-watchers were usually assigned to a building and stayed there overnight with a stirrup pump. If an incendiary bomb fell, the fire-watcher attempted to extinguish it or, if unsuccessful, summoned the fire brigade.)

The war was scarcely begun before Stevie heard from another friend who had suddenly been whisked away to Rumania. This was the novelist, Olivia Manning, at whose recent wedding Stevie had been bridesmaid. They had met in the early 1930s when both were aspiring writers and Olivia was renting a bedsitter in Chelsea. During those years Olivia enjoyed visiting Stevie's home in Palmers Green on weekends. According to a mutual friend Olivia found there, with Stevie and her aunt, "an atmosphere of security and comfort which must have made her room in Oakley Street seem even chillier and more threadbare." Soon after the success of *Novel on Yellow Paper*, and on Stevie's recommendation, she took her first novel to Cape, and they published it.

Olivia wrote in September 1939 to tell "my dearest Stevie" that she and her husband, Reggie Smith, were in Bucharest where he had been sent by the British Council. They had packed and left London hurriedly. Olivia asked Stevie to find out what had become of their flat and, if possible, to move her books to Stevie's house at Palmers Green. She inquired after mutual acquaintances such as the poet Louis Mac-Neice, the novelist and critic Walter Allen, and a writer called John Mair who was later killed in action (as was Olivia's only brother, for whom she grieved throughout her life). "You know how much I want to see you again", wrote Olivia, although this sentiment was not to prevail in their troubled relationship.

Olivia and Stevie both loved gossip and intrigues and this penchant may have imperilled their friendship. In October Olivia wrote again to Stevie, one of "the people I like best". She was alarmed to have discovered that Margaret Gardiner had found out "Everything" although neither Olivia nor Reggie told her (this probably referred to a miscarriage Olivia suffered). "I do hope you told her to keep quiet . . . I am not letting everyone know my secrets and I am quite serious when I say I do not want people to know", she wrote, suggesting that Stevie was very much a confidante of Olivia's at the time. Olivia filled her letter with whatever gossip there was and asked Stevie to send the latest news from London, adding that she longed especially for "the horrors of Bloomsbury".

In 1939 Stevie sent some short stories to John Lehmann for publication in the magazine, *New Writing*, he was then editing, but he rejected them. She succeeded, though, in publishing two stories that year —one, a prose-poem in the *New Statesman* called "Surrounded By Children", which depicts a summer's day in Kensington Gardens and Hyde Park, and the children, some of them spoiled and unpleasant, who play there. Suddenly "a famously ugly old girl", a caricature of the author with her "wisps of grey hair carelessly dyed that is rioting out from under her queer hat", enters. She spies a luxurious perambulator and, attracted by the security it offers, tears off her clothes and climbs into it. In a gesture that is partly prophetic she stabs herself with a hatpin and draws blood. Nightmare might define the genre of this story in which Stevie seems to depict her struggle to climb back into the condition of infancy or, as once she described it in a poem, to "Storm back through the gates of Birth." In "Childhood and Interruption" Stevie depicts the attractiveness of the pram:

> And underneath the pram cover lies my brother Jake
> He is not old enough yet to be properly awake . . .
> For a little while yet, it is as if he had not been born
> Rest in infancy, brother Jake; childhood and interruption come
>   swiftly on.

And in a poem written much later, "To Carry the Child", Stevie says that "To carry the child into adult life / Is to be handicapped."

This is a poem central to an understanding of Stevie and the perils she risked in her effort to preserve such qualities of childhood as the originality, freshness and directness she deemed essential for a poet. But carrying the child throughout life often entails the perpetuation of attitudes and behaviour that are painful, isolating, and paralysing. Many of Stevie's friends remember times when they were surprised by her childish behaviour: Racy Buxton, for example, who was pushed aside by Stevie when walking with her on a path to the sea. "I don't want to see your back," Stevie had said, "I want to see the sea." And Lady Lawrence, who recalls "the impishness which at one party led [Stevie] gleefully to pick out all the smoked salmon in the sandwiches, leaving the bread and butter for the rest." Elisabeth

Lutyens wrote: "She would *demand* company—someone to talk *at*—insist on being driven round and about and having her creature comforts catered for. Given these—and she had her own way of childishly screaming for attention—she could reward you by being very, very funny, with a devastating gift of observation and awareness of the ridiculous in our human predicament."

In "To Carry the Child" the "man in the poem is meant to have the child in him and to feel the nuisance it can be," Stevie wrote, "but then to see it is a nuisance for the child, too, and that where the two of them exist together, then each has a right to exist, and some value." She wrote the poem in her sixties, when she must have given thought again, in a calmer way, to Rosamond Lehmann's disapproval of Pompey's "little-girlness", and her view that failing to grow up keeps one at a kind of anarchistic and destructive level. Her last stanza even incorporates some of Rosamond Lehmann's diction:

> But oh the poor child, the poor child, what can he do,
> Trapped in a grown-up carapace,
> But peer outside of his prison room
> With the eye of an anarchist?

And in a stanza that seems poignantly autobiographical, Stevie wrote

> The child in adult life is defenceless
> And if he is grown-up, knows it,
> And the grown-up looks at the childish part
> And despises it.

"The Herriots", Stevie's short story about Peg Lawless and her desolate married life in Bottle Green, appeared in the December 1939 number of *Life & Letters To-day*. A few months earlier this magazine printed Stevie's review of Llewelyn Powys's *Love and Death*, which she followed with two more reviews in that journal in 1940. Stevie began as well to review regularly for *John O'London's Weekly*.

In July of 1940 Stevie sent Naomi Mitchison heartfelt condolences on her friend's loss of a baby daughter. She attempted to console Mrs Mitchison by saying how well out of the present world the infant was:

> I think it will be rather a horrible new world and a pretty hungry one too, if we carry on the war, as I have no doubt we shall; one gets the habit and coming late to full strength, as we always do, must of course put that strength to the test . . . Hitler's war is practically over now; if it wasn't for us, and what an "if." Ours is just beginning. The timing seems wonky to me.

Stevie goes on to describe to Naomi Mitchison one of her early assignments for *John O'London's*—an essay on H. G. Wells. "It is like being at school again, doing these set jobs", she complains, praising Wells for exciting atmosphere and good characters but faulting him for having "no conception of spiritual strife or of the importance of spiritual things at all. As if machines could be either good or bad without minds to drive them and make of them either buses or tanks."

From July to December of 1940, the time of the Battle of Britain and then the Blitz, 23,000 English civilians were killed in air raids, more casualties than the Armed Forces suffered during the same period. Stevie managed to leave London in August for a visit to Cambridge where she saw the Marshalls, John Hayward, Max Newman, and other old friends. In London she continued her work as a fire-watcher, patrolling an area of the West End near her office. One night she and Norah Smallwood, on duty as a pair, wandered in through the open stage door of the London Coliseum. Mrs Smallwood recalled that:

> Stevie began to have a look round. She loved looking and nosing into things. And in no time at all we found ourselves in one of the dressing rooms, and on the dressing tables were these white boot boxes, again which she lost no time in looking into, and inside one of them were these false noses and moustaches, which Stevie fell on with cries of delight and joy and started putting on and wearing and miming, you know, twizzling and twirling, as she used to do. And of course we ended in a hilarious heap. Fortunately nobody around.

In portions of the typescript of her third novel, *The Holiday*, deleted in the published edition, Stevie described her protagonist awakened in her home by sirens and anti-aircraft guns. This character says that the best sleep she gets is when she's on fire-watch because only then can she have a rubber hot-water bottle from the first-aid cupboard. Stevie wrote to tell Rachel Marshall in Cambridge about the father of a friend who came back from Wales to stay overnight at his home at the edge of Palmers Green. "He had never *been in a raid* and had only heard aeroplanes 'going over'; sounds like peace doesn't it? The poor man picked the night of our big blitz for his stay, so I guess by the morning he'd got an angle on it. That was a blitz." Stevie then gave an eloquent and harrowing description of how knocked about London looked:

> Yesterday morning we paddled to our offices through piles of broken glass, in the sunshine the streets sparkled like diamonds with the stuff, smashed so small, looking like diamonds and frost. There is a large bomb crater in the road just outside what is left of St. James's, Piccadilly, it's deep, all London clay and with fountains of water cascading down into it from broken mains; smells of gas and

burning everywhere, but all that has been dealt with by now, I mean the gas, water and burning. It's bloody silly knocking each other's cities to bits like this, I'm sure these air terror tactics are futile, they only work when there is no air defence, no near-parity, and the whole thing is followed up by mechanised infantry. The effect it has on us seems to be fury and the intention of holding out these next few months at all costs until we creep up to parity and plus.

Inez Holden had rented the mews house at the back of H. G. Wells's London home and Stevie, writing to Rachel Marshall in 1941, told her friend, "I often have tea with Inez at weekends and H. G. Wells comes across from the big house for twenty minutes or so for a nice bit of toast. He thinks we shall win in the end because we can't afford to lose, but it may take another two years or so."

10

The Devil of the Middle

"How long will the war last . . . shall we win the war, how does it
go?" queries one of the characters in Stevie's fiction. In a passage
deleted from *The Holiday*, the novel Stevie published after the war, she
wrote that "The war works upon us. It is that middle period . . . the
devil of the middle has hold of us, we are exasperated, we feel that we
are doing nothing at all, so we have the guilt feeling too. We work
long hours . . . Every war has this middle period."

Writing to Stevie shortly before the war, Osbert Sitwell had said he
shared the sentiments of a song about war that the painter Walter
Sickert (whose pictures Stevie thought "stunning") had learned as a
boy, "If you'll excuse me, I've had some." So had Stevie, but she
seems not to have insulated herself in any way from the sufferings war
brought to her and to those around her. This is poignantly manifest in
a correspondence she had with John Gabriel, a second lieutenant in the
Royal Berkshire Regiment, who first wrote to Stevie on New Year's
Eve in 1940 from The Old Vicarage in Grantchester, where Rupert
Brooke had lived. "I've known and loved your poems for some time,"
he said, "but yesterday I happily exchanged a Christmas book token
for 'Over the Frontier' and ever since a particularly boring phase of
Army life has been brightened delightfully. . . . You remind me of a
laughing butterfly flickering over the head of James Joyce. As an
antidote to war you are to be highly recommended."

In one of her letters to Gabriel, Stevie mentions that she likes Orwell
"very much", and praises the magazine *Horizon*, although she says, "I
can't get [Cyril] Connolly to take any of my poems. . . . That
grave-yard note of mine is jolly difficult to place." Gabriel's wife of

eight months replied, thanking Stevie for her letter and saying her
husband had been reported "missing believed drowned". Stevie was
terribly upset and wrote to tell Clothilde Gabriel how much her
husband's fan letters meant to someone who was "not a very popular
author". She lamented "the way that war always takes the best, those
who . . . are most needed when peace comes." That same month
Stevie reviewed for *Aeronautics* a posthumously published collection
of stories by a British airman, whose wife's words about her husband
in the book's preface are similar to those John Gabriel's widow wrote
to Stevie.

In 1941 Stevie was regretting her lot as "a penurious author" and
missing many of her friends who had fled London. "I wouldn't really
be out of London but I do wish some of the others were here." Inez
Holden was a companion she could count on, and in September they
attended the PEN World Congress lunch and sat with Arthur Koestler
and Cyril Connolly, among others. Stevie admired very much the
stoicism with which the English endured bombardment and proved
that "cities and races cannot be wiped out", and she found fault with
Vera Brittain's depiction of England under siege:

> A stranger reading Miss Brittain's observations of England's civ-
> ilian population under fire might get the impression that not one
> stone was left standing upon another, that not one square foot of our
> island was left unbombed, that its heroic population, in the throes of
> total death, could look for some measure of race survival only to
> those few British children who had got away in time to the Colonies
> and States . . . She notes also that terror bombing may brutalise the
> bombed, but not that their growing indifference to it may blunt the
> edge of a weapon dearly prized by the Germans. . . it is unfortunate
> that never once does she catch the authentic voice of England, as
> little hysterical as the growl of her guns.

During the war years, the patriotic Stevie none the less kept a
balanced view and even could commend a book about Germany "that
is not steeped in blood and horrors. So many writers give the
impression that Germany is a cross-section of the Wagnerian dwarf-
land. That cannot be the whole truth." She reiterated her pride in "the
British lion getting really off the ground as wars approach their second
and third years." "Our times have been upon the rack of war", Stevie
once wrote, and agreed with a commentator who found not the least
of war's torment to be the drive to undo the victim. Years later she said
that though the Nazis were hell, their cruelty lasted only thirteen years
whereas the Inquisition persisted for seven hundred. Throughout the
war Stevie's unwavering patriotism was accompanied by historical

perspective and a refusal to yield either to hyperbole or sentimentality. Replying in the pages of a magazine to an essay by a man she describes as "muddle-headed", Stevie told him to

> keep your anger for yourself—and your pity. History is mainly about wars and the consequences of wars. The history of our own country is the history of success in war . . . that is a hard lesson for you, but it is true . . . And what do we learn of war from history, Jack? We learn that war decides. Is cruel. That Fred dies—and Pete—and Vera—and Joan" [childhood friends the man had mourned in his essay].

Lucky are "those in close immediate touch with an older calmer undriven generation . . . a Lion of Hull, to draw from them at need upon their strength. Oh I do love my Aunt and so much I admire her. She is so reasonable, so balanced, so sound, and yet so kind and practical." At home, the comforting presence of the family Lion continued to steady Stevie. Christmas 1941 was "a nice quiet family Christmas" with Molly, home for the holidays, down with a bad cold. Stevie was busy with housework and wrote on New Year's Day to tell Rachel Marshall "if I had it always to do I don't think I should ever do anything else." In the same letter she said how cheered up she was by her friend's praise of her poems: "I am never very certain about them, the inspiration or whatever it is comes in such a vague and muddled way, and I am not sure that I don't sometimes get the wrong poems into print."

In the early 1940s Stevie was still writing poems, but had difficulty placing them:

> the *New Statesman* has been wonderful, but of course they have to get their politics in—and that doesn't leave much room. Apart from that I am not doing much writing, except some reviewing, *John O'London, Country Life, Modern Woman*—and (don't laugh) *Aeronautics!* There's a nice collection for you, all our own [Newnes, Pearson] papers of course. Only *Aeronautics*, which has a peach of an editor, pays me, so don't you go thinking honey I'm making my fortune, I guess I'll never make that.

As early as March 1940 Stevie seems to have proposed to her publisher a new book of poems, but Rupert Hart-Davis had written to tell her "there isn't an earthly chance of our publishing another volume of your poems at the moment." By 1941 Hart-Davis was in the armed forces, but Jonathan Cape consulted with him about Stevie's insistence that either they do a third collection of her poems or she would move to another publishing house. "We should be sorry to see work of yours appearing under another imprint, but we can understand that if we are disinclined to put out another volume at the present time you would want to try elsewhere", wrote Jonathan Cape to Stevie in February. And he added, "Will you get your poems together and let me see them?" Sales figures for the earlier books of poems were discouraging: only 400 copies of *Tender Only to One* were used, and many of them went to reviewers. *A Good Time Was Had By All* ran to 780 copies, but again a number of them went to reviewers.

Late in 1941 Cape agreed to publish a new book of Stevie's poems, along with drawings about which she was more than ever enthusiastic. The plan was to publish in paper covers in the hope of finding "a larger public" for her. "There are so many drawings, which I think are so much better than they used to be, and I can't get poems to tie up to them", Stevie wrote on New Year's Day. Her little office was wrongly referred to as the art department, she joked. Stevie had been asked by Cape to choose 80 out of 200 poems for her next book and, as usual, could not decide which to keep and which to discard. Nor could she determine which of her drawings to fix with particular poems. In a letter to John Hayward she exclaimed, "I wish I wasn't cursed with indecision or that I had not got this love for my drawings and hatred for my poems." C. V. Wedgwood, who had replaced Hart-Davis at Cape, made a provisional selection, then returned them to Stevie who was so inspired by her appreciation that she wrote six more poems and did twenty-four new drawings, compounding her problem. "The office looks like a paper chase", she complained. One of the poems about which Stevie couldn't decide was her "obscene poem", called "Goodnight" (see p. 65). Veronica Wedgwood favoured its inclusion but John Hayward advised against, and in the end Stevie discarded it, partly from fear of libel. As she told Hayward, "I take your opinion and will not publish it, though even if I wanted to I expect Cape would buck at the last moment. One has to be careful with these transcripts from life, I suppose."

Assembling the copy for her new book kept Stevie busy in the spring of 1942, an unusually lovely spring, although cruelly the air raids went on. "London is looking perfect just now with all the trees looking green and the flowers out", Stevie told John Hayward. She went to see the new Tate acquisitions at the National Gallery and enjoyed the Sickerts and Steers, though not the pre-Raphaelites. "I can't stand 'em you know. . . . A lot of highclass people with irritating voices . . . looking very Rex Whistler." (Stevie herself certainly looked very un-Rex Whistler.) Lord Rothschild, at whose Cambridge home John Hayward was living, organized a sale on behalf of the Red Cross, and Stevie generously offered to donate the manuscript of *Novel on Yellow Paper*. Hayward dissuaded her from doing so, but accepted two other "choice pieces" for which Lord Rothschild thanked her. During the summer Rothschild exhibited manuscripts that distinguished writers had sent him, and Stevie went to a private view of the exhibition where she saw the Queen and "was very impressed by the sherry".

In July Stevie was in Cornwall and wrote to Aunt from the Lizard telling her of an outing at Kynance Cove where Aunt and she had picnicked during an earlier visit. Stevie mentioned that she had "painted several cows". Back in London she lunched with chums like Inez Holden and discussed politics with Victor Gollancz. She also started a new novel and by mid–August had written 25,000 words of it: "it pours out with never a point or a point virgule. . . . Everybody is in it and everybody's conversations. I do hope I see you again soon," she told John Hayward, "so that I can put some of yours in, heighho." August was also the month that Stevie's story "In the Beginning of the War" was published in *Life & Letters To-Day*. An examination of the typescript of the new novel Stevie was writing, which was published several years later as *The Holiday*, reveals that "In the Beginning of the War" was originally a part of it. In the short story she expresses her patriotic pride: "The Germans have inflicted little pain on England because England is so strong . . . The Germans are asking for it, now they will get it." The girl in the story composes "a jingo poem": "For every blow they inflict on Jewry, And other victims of their fury, They ask for death on bended knee, And we will give them death and we, Will give them death to three times three."

Stevie heard again from Olivia Manning in 1942. She described to John Hayward how Olivia and her husband Reggie Smith "both Got Out in Time [from Bucharest] and after wandering through Greece and Asia Minor and possibly the Holy Land (*vide* missionary journeys of S. Paul) fetched up in Cairo, where they now are." Stevie sent Hayward one of Olivia Manning's stories, hoping he could steer it into print. He perhaps sowed seeds of doubt in Stevie about Olivia's loyalty by finding her "patronising". At the time Stevie defended her, saying: "I am absolutely certain Olivia doesn't mean to be patronising, she was always very interested and generous about my writing . . . (and it wasn't half as mutual as it oughteravbin)."

"Stevie survived on her friends", one of them said, and a poem in the book Cape was soon to publish, "Dirge", seems to imply this was true:

> From a friend's friend I taste friendship,
> From a friend's friend love,

> My spirit in confusion,
> Long years I strove,
> But now I know that never
> Nearer I shall move,
> Than a friend's friend to friendship,
> To love than a friend's love.

But her relations with friends were sometimes troubled, and darkly so, as the poem goes on to suggest:

> Into the dark night
> Resignedly I go,
> I am not so afraid of the dark night
> As the friends I do not know,
> I do not fear the night above,
> As I fear the friends below.

In October of 1942 Stevie found herself writing in anger to her friend, George Orwell: "you are the most persistent liar and these fibs are always coming back to me from other people." During the war Orwell was at Bush House with the Overseas Programme, and Stevie probably met him first with Inez Holden, although they had other links. (Stevie's friend Rosemary Cooper, for example, briefly sublet her West End flat to him, and Alan Clutton-Brock had been at Eton with Orwell, who often visited the Clutton-Brocks when they lived in Essex.) In later life Stevie claimed that she used to fire-watch with Orwell.

Stevie's branding him a liar came about because of a reading of her own poems she had hoped to do on the BBC, although she had "never broadcast before or had a rehearsal". Orwell claimed that his secretary had reminded her a few days before the broadcast of the particulars but that the messages may have been misdirected. Stevie did not accept Orwell's explanation. "You never gave me a date for the bloody broadcast or breathed one word about my reading my own poem," she fired back. "I'm bored to death by the lies." Over a decade later Stevie read *Sunset and Evening Star*, a book in which Sean O'Casey took the view that Orwell wanted the world to die with him, and when he saw it wouldn't, turned people into beasts in *Animal Farm*, then destroyed the world and people in *Nineteen Eighty-Four*. Stevie said O'Casey was "dead right" about Orwell's

> sick-man fancy of a pool of self-abasement for all the world to dip in, and his sick man's lust for extreme future cruelty. And will he not be a disappointed ghost if 1984 when it comes, comes with the Bank Rate at four per cent and Mr. Priestley's successors still

whining cheerfully about nothing worse than currency restrictions
and passports? Was it contempt based on ignorance of what makes
people tick, and the British people especially, with their long
tragi-comic history of being tyrants on the right hand while blasting
tyranny with the left, that put such a gloom of Petainismus into his
books?

Despite the low opinion Stevie came in the 'fifties to have of George
Orwell, and despite her anger at him in 1942 over her missed
opportunity to read her poems on the BBC, during the war years she
evidently was fond of him. Some of her friends at that time even recall
her attitude as romantic, and a few go further than that. Norah
Smallwood insisted that Stevie confessed to her "an intimate relationship" with George Orwell, and quoted her as saying, "I was
living with George Orwell and it wasn't easy." As Stevie never left her
home in Palmers Green, Mrs Smallwood presumed she referred to
lunch-time trysts and the like.

Bernard Crick, Orwell's recent biographer, repeats an anecdote
both Anthony Powell and Malcolm Muggeridge report, in which
Orwell told of having had sex with a woman in a London park as they
had nowhere else to go—a circumstance he incorporated in his
fictional tale *Keep the Aspidistra Flying*. As Crick sceptically indicates,
"The name of Stevie Smith has . . . been persistently linked with this
tale." Kay Dick, who knew them both, derides such speculation
and Sally Chilver calls the park story "one of George Orwell's leg-
pulls". Whether or not the poem "Conviction (iv)" refers to this
incident is not known, but it was written in the years Stevie knew
Orwell:

> I like to get off with people,
> I like to lie in their arms,
> I like to be held and tightly kissed,
> Safe from all alarms.
>
> I like to laugh and be happy
> With a beautiful beautiful kiss,
> I tell you, in all the world
> There is no bliss like this.

The drawing Stevie placed next to the poem (see page 198) in the
collection which, in 1942, she was readying for publication, depicts a
man and woman making love out of doors, with an animal looking
on.

Stevie did, by her own admission, portray Orwell as Basil in the
novel *The Holiday*. After Orwell's death she wrote that she "knew

George and his first wife, Eileen, quite well and saw a good deal of
them during the latter part of the war and for a few years after." In
correcting a misinterpretation of some words Orwell had addressed to
her about women (girls do not "play the game"), Stevie wrote "all this
comes into *The Holiday*, in various fairly lengthy conversations
between the writer, (it is a first person novel) and two characters who
divide between them many of George's opinions and characteristics as
I saw them. I seem to remember I had the idea at the time that splitting
George into two might lessen the danger of libel, not much of a danger
really." Basil Tait is the character in *The Holiday* to whom Stevie
attributes the remark Orwell made to her. Celia, the autobiographical
narrator, says he "is rather a fool. . . . You would say: He is promising, he is like a fourteen-year-old boy you know, he thinks girls can't
play." He enters the novel earlier, along with Tom Fox, and these
are the two into whom Stevie divided Orwell. Stevie depicts in a
wickedly skilful way what she took to be Orwell's narcissism:

> Basil knelt down and put a match to the fire. There was this icy
> feeling coming out of Tom, but Basil inclined to conversation. He
> said that very soon the population would be only forty million. He
> said that the cruelty of the Germans was nothing to what the cruelty
> of the English would be if the English were really up against it in the
> matter of losing their property, that is their goods and their money
> and a chance for the kids.
>
> But all the time that Basil was speaking he was watching Tom. I
> thought: Basil loves only Tom; Tom is his dear one, his mate, his
> friend, he is his D. H. Lawrence of a David-and-Jonathan situation;
> and I felt a contempt for this, and for this hysteria of a masculine
> *agape* that runs through our English literature and through our life
> too . . . It is not entirely a homosexual thing (and not entirely not);
> it is innocent so far as it goes, but innocent?—but juvenile . . .
>
> Basil loves Tom, he does not care for women, he loves only Tom.
> He thinks that women are biologically necessary and resents the
> necessity, he is like a twelve-year-old boy, he thinks "girls are no
> good." But he loves this cousin of mine, this Tom, Basil loves him
> (oh, in the way of friendship, in the way of friendship, he is not
> getting off with the fellow, oh, no offence in the world).

Stevie, who in later years referred to "that rich nursery of English
talent, the lower middle classes", defined perhaps the profoundest
difference between her views and those of Orwell when she wrote:

> Basil and [his] chums are still in this violent revolt from the virtues
> of the middle-classes; it must be a sort of snobism, it has to be that.

But perhaps it is only partly that, only a little snobism but chiefly a restless and temperamental dislike of that type of person. If it was not for Basil and Lopez and you and me, and perhaps Tiny, the middle-classes would be unbearable. But also without the middle-classes we should be unbearable. There must be that variety of virtue as of experience. Basil is strong, simple, with a fine writer's mind and the reasonable faith of an intelligent revolutionary. These people see the middle-classes as obstructionist box-dwellers, whose only thought is for themselves and for their families, as if that was not the common thought of the greater part of mankind. The free-blowing revolutionaries, the classless artists, these are the salt of the earth, for they have the power to see a thing while it is yet a long way off. But you cannot make a diet of salt, and it is through the use and practice of the middle-classes that the vision is made actual. I have not found the middle-classes against the new ideas, so much as anxious how they may be applied; but of course I am speaking of the less wealthy sort of middle-class person, such as we have at home.

A review Stevie published in the *Tribune* in 1943 praised exactly the middle-class virtue and respectability that, according to Stevie, Orwell scorned.

An actress friend of Stevie, who lived for a time in Palmers Green and at whose cottage Stevie stayed years later when she read at the Aldeburgh Festival, remembers Stevie saying Orwell once pursued her at night down a hall of Bush House—naked. But the actress friend hadn't understood which of them was naked, and never dared ask. Stevie remembered or imagined some such passionate encounter when she wrote, in *The Holiday*, of a rendezvous Celia and Tom have in a studio at "The Ministry"—a fictional equivalent of the BBC. Tom becomes mute and transfixed as if afflicted by madness as he takes Celia's head in his hands and strokes her throat before collapsing in tears.

One of the people who remember that for a time Stevie's fondness for Orwell "went very deep" is Veronica Wedgwood. Soon after Veronica Wedgwood joined Cape in 1942, Stevie was referring to her as "a friend of mine". Late in September 1942, she sent Stevie proofs of her new book of poems and said, "I have read the poems from end to end and laughed myself sick and been profoundly moved, turn and turn about." The book was scheduled to appear for the Christmas trade. Stevie wrote the blurb for *Mother, What Is Man?*, whose title comes from Francis Thompson's *An Anthem of Earth*:

Ay, Mother! Mother!
What is this man, thy darling kissed and cuffed,
Thou lustingly engender'st,
To sweat, and make his brag, and rot
Crowned with all honour and all shamefulness?

Stevie's blurb read:

> In 72 new drawings and poems Stevie Smith gives an answer that is
> tender sane hilarious and cautionary. She thinks that fear of life is
> more common than fear of death . . . She thinks that people have
> never before been so unhappy as they are now and that this
> unhappiness may lead to death. But it may also, if people are strong,
> drive them to throw off the tricky habits of thought which make
> them sad . . . they are full of the most buyoant [*sic*] *hope*. For if we
> are like this, we must surely wish to be something different. And in
> this way the poems may be truly said to embody the spirit of
> *Christmas love* and hope *for the New Year*.

One cannot help but feel that against a background of war and cruelty
Stevie is examining at close range in her third book of poems terrifying
questions about evil and the human heart, and indicating throughout
that nothing of this is alien to her. Humankind is described as darkened
by a kind of night in her poem "In the Night":

I longed for companionship rather,
But my companions I always wished farther.
And now in the desolate night
I think only of the people I should like to bite.

E. C. Bentley spoke of Stevie marrying, in *Mother, What Is Man?*,
nonsense with sadness and beauty as "a feat of modern letters" and
many years later this judgment still rings true.

As another critic noticed, "the jungle theme" appears and reappears
in this volume. "The Face" is about "a human face that hides / A
monkey soul within":

Sometimes the monkey soul will sprawl
Athwart the human eyes,
And peering forth, will flesh its pads,
And utter social lies.

There is in "The Smile" the "ancient girl . . . garbed in spite / [Who]
turns to rend, and lives to bite." One response to the question asked in
the book's title is Stevie's poem "A Man I Am".

I was consumed by so much hate
I did not feel that I could wait,
I could not wait for long at anyrate.
I ran into the forest wild,
I seized a little new born child,
I tore his throat, I licked my fang,
Just like a wolf. A wolf I am.

I ran wild for centuries
Beneath the immemorial trees,
Sometimes I thought my heart would freeze,
And never know a moment's ease.
But presently the spring broke in
Upon the pastures of my sin,
My poor heart bled like anything.
The drops fell down, I knew remorse,
I tasted that primordial curse,
And falling ill, I soon grew worse.
Until at last I cried on Him,
Before whom angel faces dim,
To take the burden of my sin
And break my head beneath his wing.

Upon the silt of death I swam
And as I wept my joy began
Just like a man. A man I am.

A number of reviewers gave particular praise to Stevie's drawings. George Stonier said in the *New Statesman*, "The chief advance on her earlier books seems to me in the illustrations which often diverge from the text. The drawings for 'Autumn', 'Where are You Going?', 'Study to Deserve Death', 'If I Lie Down', and 'Ah, Will the Saviour . . . ?' are a perfect blend of grace and incongruity." "Her drawings require a special word of praise", wrote another reviewer:

> they are an integral part of the verse, their informality and incision giving them the same validity as the gesture has in dramatic poetry. Indeed, their affinity to gesture is so close that Beatrice Lillie springs instantly to mind. One of these days, perhaps, if English cabaret reaches an intelligent level, we shall hear that mistress of the wryly mocking gesture reciting Stevie Smith's verses, and rejoice in that mating of original talents.

Stevie continued to make new acquaintances during the war, some of whom were to become lifelong friends. It was the period, for

example, when she met Sally Chilver. Working then in the Civil
Service, Sally went to a party at the home of the economist Margot
Naylor and met Stevie. Speaking of the early days of their friendship,
Sally Chilver said, "We started off as gossips, if you like." Stevie's
"rather stunning appearance, her deep and characteristic voice, her
gaiety, and also the fact she could talk about everything and was
interested in how people set up their lives and what they read" all
attracted Sally to her, and before long she was calling on Stevie in
Palmers Green with some frequency. When Stevie came to stay with
her she would ring up and say, "You live a very boring distance away,
and if I'm going to come and see you I require clean sheets, a bed for
the night, and you know what I can't eat." Soon Sally Chilver
and Stevie were boarding trains for Hertfordshire, where they
would walk in the countryside before returning on foot to Palmers
Green.

Helen Fowler thinks it was in 1942, or soon after, that she met Stevie
at the Marshalls' home in Cambridge. And Elisabeth Lutyens, England's most illustrious woman composer in the period, recalled meeting Stevie in the war years and thought they met at The George, a
famous pub next to the BBC which seemed to some of its *habitués*
rather like a London equivalent of the Deux Magots. Constant
Lambert, Dylan Thomas, Louis MacNeice and his wife Hedli
Anderson were among those who lunched there.

Another friend Stevie met during the war was Kay Dick, who had
joined the editorial staff of *John O'London's Weekly*. Although Stevie
worked as a secretary, she was in Kay's view "a terrific celebrity" who
had published two novels and several books of poems which made her
not only fashionable but daring to the young. Soon after meeting they
began to lunch together: at Rule's when they had reason to celebrate a
publication, or the Arts Theatre Club where, when money was scarce,
they read the weeklies that were available without charge. Sometimes
they sat in the inner courtyard of Inigo Jones's Covent Garden church
and Stevie would read new poems to Kay. "We nattered shamelessly,
and with, I hope, some wit, about mutual friends battling their way to
fame in literary circles: 'If there's one thing I do love it's a common
friend,' Stevie said." On occasion she went after work to Kay's home
in Hampstead and sometimes, because of air raids, she stayed the
night.

Stevie often was lonely and did crave companionship. As she told
John Hayward, she wrote less for fame than to find company, and
recently the writing she'd been doing was not admired: "It isn't just
that it has been turned down," Stevie said of her new novel, "but that I
really think it is the wrong sort of writing and really something to be

ashamed of, so I am really afraid to put pen to paper now." Veronica Wedgwood liked it but thought "it had tremendous libel dangers", and when George Stonier, the critic Stevie very much respected, was asked for an opinion, he read the novel and thought it "awful". "It is this terrible personal stuff," Stevie admitted to Hayward, "written in a jig of private feelings and secretarial odd jobs. Of course that is the way the others were written, but I know that *Over the Frontier* was nothing to be proud of, and *Novel on Y.P.* just happened to come off." Stevie set herself a course of reading Somerset Maugham whose writing she liked because it was "so controlled and cool, he has learnt what to do with private feelings". She told John Hayward she felt "so tired about writing and so nervous, and yet I do so hate this feeling that I am shut back into secretarial odd jobs."

Hayward replied immediately, "disturbed by the dismay you must inevitably be feeling at having, as it were, let yourself down." He urged her not to "suppose for one minute that it's all up with you, that you're done for & won't write another good thing all your life", and thought a course of Maugham inadvisable. "You have far too much original talent (more than he has) to need to cramp your style by following his methods—or, indeed, anyone's." Some years later Stevie viewed Maugham less admiringly. Although finding him a writer of excellence, she criticized his lack of heart which she attributed to "Maugham's fault of pride, the observer technique that sets the writer above his creatures".

Hayward speculated that Stevie might have exhausted the satiric vein, and said he always thought the danger for her was over-indulging her sophistication and consequently losing her innocence. "There was a hint of radiant, pubescent, insouciant, quizzical, tender quality about your novel on yellow paper which I felt, alas, was fading in its sequel", he wrote. He advised Stevie to leave *la bêtise humaine* alone for a while and forget: "(1) Stevie (2) that Man is Vile." He assured Stevie that the failure of her novel was only incidental and wondered if she might not profitably develop prose characters from persons who appeared in her poems. About most of them, he said, he'd always felt there was much still to be said. "*Don't* despair", were his final words to Stevie.

Stevie wrote to say how grateful she was. "I feel an absolute 'case', you know, and pretty hopeless." She liked his suggestion about amplifying characters who appeared in the poems, although what remained to be said was "pretty sad stuff . . . because most of them . . . seem to be suffering either from not being loved enough, in the warm cozy affectionate way that we all like rather, or from being loved in this way by the wrong people."

Stevie's letter to John Hayward is filled with lights and shadows. She had lunch with Malcolm Muggeridge—"marvellous good company, I really love him"—though she finds him "a tart piece . . . most malicious-catty-tonic". Stevie complains about her "so-interrupting baronets", but "they are fairly merciful employers", she adds, "an opinion I have always held in print and otherwise: 'What is the matter, Pompey?' 'Nothing, only I so much want to go away.' "

Late in the summer of 1943 Molly underwent "a perfectly awful operation . . . in an Ipswich Nursing Home", where Aunt and Stevie visited her. Molly had taught at Ipswich High School but came to dislike it very much. "My sister was not always so happy in her schools", says Celia in *The Holiday*. In her account of the horrors that made one of them, housed in a condemned building, seem "like a madhouse", there is presumably some exaggeration of conditions that did exist during the war at Ipswich High School: a vacillating head-master who thought it better to do the wrong thing than nothing; a sub-head who screamed and cried and made everyone nervous, forcing a sick child into the rainy cold playground. Children were whipped and spent days in damp air-raid shelters. One of the children taught by Pearl (in *The Holiday*, Stevie's fictional name for Molly) died of meningitis.

By 1943 Molly had joined the staff of Westonbirt School, an institution that offered a liberal education to girls from age eleven to eighteen. In wartime the school had been evacuated to Bowood and Corsham Court, and that is where Molly began her association with it, remaining on the staff until 1946, when she left to become County Drama Organizer for Buckinghamshire. Stevie called Westonbirt "Redesdale St. Mary" in *The Holiday*. After a disturbing start, Pearl is depicted as

> so happy at her school now, she has suddenly this warm feeling for everybody and for her staff-mates. . . . The headmistress says she is bringing a breath of fresh air and something besides gossip . . . when they were all leaving and the headmistress was there, all the girls ran up crying, Miss Phoze, may I sit beside you, may I sit on your side of the bus? So there is now this warm feeling of shoals safe passed.

Sisterly pride and relief are evident in this account.

In February of 1944 Stevie had been in Cambridge, staying with the Marshalls. She was so delighted when, after a poetry reading, she was presented with a bay wreath, that she brought it back with her to

war spirit

London and set it on a vase in her office. Both baronets congratulated Stevie, which led her to muse about why it is "that just slightly (ahem) ineffectual people are often so nice? It makes things very complicated." It was in February, too, that Stevie received a novel by Friedl Benedict (the writer, Anna Sebastian) from Veronica Wedgwood, who hoped she would review it in *Time and Tide*—a weekly Stevie wrote for at the time. The friendship Stevie formed with that writer was not a long one, for Friedl Benedict died in her thirties, but they were a hilarious duo, according to Kay Dick (who had been introduced to Benedict by Stevie), "like two slaphappy clowns". When they were together each "brought out the more ghoulish side" of the other's humour.

The D-Day invasion took place on 6 June 1944, and it was followed by the V1s and V2s—rockets which the Germans in a last-ditch effort rained on London. Stevie seemed proud of the English—particularly of the women who remained undaunted. In a magazine piece that appeared shortly after D-Day, she writes of an American friend called "Helen" who comes to London after an absence of five years and finds that English women "look well, a bit tired maybe, but healthy". Stevie is sympathetic towards the homesick American soldiers who crowd London and, through Helen, she comments on the increased friendliness among English people. "They talked more easily, and they didn't 'talk tired', or as if it was the fifth year of the war. In fact they still seemed so plum self-confident." No doubt Stevie enjoyed this mateyness and it enabled her to marshal more effectively her own forces for "the war within".

In "The Failed Spirit", another poem published in *Mother, What Is Man?*, Stevie "condemned" her feelings of loneliness:

> To those who are isolate
> War comes, promising respite,
> Making what seems to be up to the moment the most
>   successful endeavour
> Against the fort of the failed spirit that is alone for ever.
> Spurious failed spirit, adamantine wasture,
> Crop, spirit, crop thy stony pasture!

The drawing Stevie placed with this poem shows, she once explained, the failed spirit standing on a "disused railway track", and on the platform to the rear, in contrast to the spirit's isolation, "a she-ass licks the nose of her young foal".

As Stevie said in the blurb she wrote for *Mother, What Is Man?*, the poems were, among other things, "cautionary". "Plum self-confident" many English may have seemed to her in the fifth year of the war, but self-confidence was always in short supply for Stevie, and especially so in 1944.

I I

Under Cover of Friendship

On 2 September 1945, World War II ended: it had lasted one day short of six years. Stevie's life, like the lives of other Londoners, had returned to more tranquil and traditional routines. "I suppose", wrote one friend, "that now there's no more fire-watching you don't visit for breakfast." Peace arrived but was accompanied by post-war anxieties. According to another friend, Stevie suspected she might be suffering from cancer. Whatever the truth of this, she had consulted a physician about digestive troubles that summer. He prescribed a powder and tablets, and provided a form entitling her to "two pints of milk a day". Another physician took X-rays, checking for organic lesions, but saw no reason for her to cancel a holiday she was planning. So Stevie went off to Scotland for a few weeks with her childhood friend, Doreen Diamant, and husband. In addition to concern about her health, at war's end Stevie was nearing her mid-forties and had not published a book of poems for three years (five more were to pass before the next would appear). Also, her efforts to publish her third novel, *The Holiday*, were unavailing, and even her reviewing just then was sporadic and for the most part uninspired.

An uncongenial job was in part no doubt to blame. Did she still hold out for herself the possibility of an ideal job? "This ideal of happiness in work may sound optimistic", she wrote later on, "but it is very important because women are more affected by boredom than men, and at the same time less able to free themselves from boring situations. Their health is affected, they mope and grow miserable and very soon they are too weak to do anything about it—they just hang on. The capacity women have for just hanging on is depressing to contemplate."

Stevie aired this belief in the *Tribune*. T. R. Fyvel succeeded George Orwell as its literary editor in 1945 and often met Stevie for lunch at a nearby snack bar. He would discuss with her the latest books that had arrived and who might be a good choice to review them. He found her advice sound and her manner "businesslike". He found her as well "one of the rare women with whom one could have a regular friendship without any hint of sex entering into it, and this even though she was, basically, quite flirtatious." He wondered if Stevie's job at Newnes, Pearson was "good enough for her. Was it a convenient cave to hide in?" Stevie would probably have replied in the words of her poem "The Actress": "I can't say I enjoyed it, but the pay was good."

In an undated letter to Kay Dick, which was probably written in 1945, Stevie says she has finished a new novel. This must have been *The Holiday*, although it wasn't to be published until four years later, and Stevie by then had shifted the wartime setting to the "post-war". She says in her letter that many things remain for her to put into the novel and she hopes to place them "somewhere in the middle". She also points out some of the self-portraits she had already published in verse. These include the subject in "Death's Ostracism" who

> falling . . . will call the waves to friend,
> Come cover over all and make an end.
> No use, they will not do it, they swing aside.
> Death's ostracism in a dream he must abide.

Another self-portrait is the child in "One of Many" who grew murderously wild, driven to that pitch by being told

> You are only one of many
> And of small account if any,
> You think about yourself too much.

Also the speaker in "Love Me!" who fears "my Love me [will] never be heard" is a portrait of herself, Stevie told Kay Dick. Particularly poignant is her admission that "Lot's Wife" is, again, self-depiction:

> here in the streets of the living,
> Where my footsteps run to and fro,
> Though my smile be never so friendly,
> I offend wherever I go.

Of course, unlike Stevie, Lot's wife was married:

> Yes, here in the land of the living,
> Though a marriage be fairly sprung,
> And the heart be loving and giving,
> In the end it is sure to go wrong.

Like Lot's wife, Stevie in these years looked back. What she saw was the unsuccessful effort of her parents to make a life for themselves, and she found their failure the tale of many.

Stevie seemed in the second half of the 1940s to be divided by two conflicting drives: an effort to form deep and abiding friendships, and an impulse to risk libel even in depicting the lives of friends. These conflicting impulses were enacted in fiction—the literary form which occupied her almost to the exclusion of poetry between 1945 and 1950. "Blake was full of contradictions", Stevie noted in a review written during the period, noticing a further resemblance between herself and a poet to whom she'd often been linked. Contradiction, or at least ambivalence, is perhaps a condition of all poets; certainly it was of Stevie who, for example, described herself in doctrinal matters as "an unbeliever with a religious temperament".

One of Stevie's stories entitled "Life and Letters" never did reach print, although it was accepted in May 1945 by the editor of *Modern Reading*. It concerned a friend whom she depicted "squirming on his belly through the grass" to reach her. By June the editor was writing worriedly to Stevie, advising her that "if there is the slightest chance of your friend . . . really going to law (and who knows?) then for God's sake forget about Life and Letters." In July he returned the story "for surgical treatment", adding that "if the story does not yield to treatment you will have to write me another, with no male characters." Stevie did revise "Life and Letters" and resubmitted it in the same month. Two years later the editor, with elaborate explanations for its never having been printed, apologetically returned the story.

In 1946 Stevie was more successful at seeing her stories into print. The autumn number of the miscellany *Orion* printed "A Very Pleasant Evening"—a story which describes a few Oxford and Cambridge friends gathered during wartime at a London home for dinner. The highbrow conversation is interrupted by a doodle bomb which falls nearby and shatters the conservatory windows. Their host foresees the post-war years as ones in which "England will be of no importance whatever, there will be only America and Russia, but we shall have our famous character, of course." The story is rendered surprisingly effective by the surge of sexual passion that suddenly surfaces as the tale concludes. "A Very Pleasant Evening" prompted Desmond

MacCarthy to write of Stevie's nugget of genius "for seeing every-thing with a child's ruthless absence of sentimentality and with a child's illuminating irrelevancies of attention." Stevie did appear in these years at parties like the one in her story, to which came scientists, civil servants and literary types. A friend remembers her at a wartime party Margaret Gardiner gave at which a tweedy man said depress-ingly, "We're discussing the splitting of the atom."

In 1946, too, Stevie published in *The New Savoy* a work called "The Story of a Story". In it Helen is accused by Roland in a dream: "You go into houses under cover of friendship and steal away the words that are spoken." Here Stevie is giving the other side its due. Her own feelings are expressed by Helen who says the law of libel "is every-thing that there is of tyranny and prevention." Margery Hemming, in her copy of "The Story of a Story", wrote a key to the identity of its characters: Stevie, of course, is the writer Helen; Margery Hemming and her husband Francis (eminent civil servant and lepidopterist) are the couple, Roland and Bella; Ba is the art historian, Phoebe Pool; and Lopez is the writer, Inez Holden. The story Stevie was writing about in "The Story of a Story" is "Sunday at Home". One clue is the telltale goldfish that appear in both.

"Sunday at Home" was originally titled "Enemy Action"—a witty reference to the marital discord of its characters Ivor and Glory, and its macrocosmic parallel in the fall of doodle bombs. It was accepted in early 1946 by the editor of an anthology called *The Holiday Book*. A month later he returned it, after hearing from Stevie that the story was based on real people. Rejections and the threat of a libel suit did not deter Stevie, however, and during the next two years she kept trying to place the story. She wrote to Kay Dick in February 1946, describing the changes she had made in "Sunday at Home" so that Ivor and Glory could not be equated with the Hemmings. She changed her own name to "Greta", but Greta's conversation will be familiar to readers of such poems as "At School" and "From the Coptic". The central detail of Ivor sitting in a cupboard while the doodle bombs fall, talking to his wife through the door, she would not change. "I do not think this can be laboured into an imputation of cowardice", Stevie told Kay Dick, "because . . . the doodles reminded him of the 'experiment' in which he had been wounded." Francis Hemming was not involved in bomb experiments, but he was wounded in France during the First World War, and he did sensibly repair to cupboards when the bombs fell during the Second World War. "I think one could write a wonderful tale about a group of coterie friends writing about each other", Stevie went on in her letter, "and the zig-zagging antics they would get up to to foil the litigious." To Margery Hemming, Stevie became "a plain simple symbol of betrayal and treachery", and a door was shut tight against her.

Stevie thought that the stir about her story was "a storm in a tea-cup and that 'X' [Francis Hemming]—not having seen the ms.—imagines that all sorts of private matters concerning himself are in it." Stevie thought her friends difficult when they objected to her presentation of them in her writings and provoked a "personal war . . . so trivial and so deadly". Francis Hemming apparently threatened not only to drop Stevie but to sue for libel. The wife of the character on whom he is based tells the Stevie-based narrator that as regards her fondness for her friends and her willingness to make stories of them, "You want it both ways", and perhaps Stevie (like many writers) did. One cannot help being impressed, though, by her resolve to publish the stories at whatever personal cost. In doing so she sided with her Muse whose concern was to make "a strong communication" and who "never has any kindness at all".

Other literary projects that occupied Stevie in 1946 were a script on Thomas Hood broadcast on the BBC Eastern Service, and a study of Maurice Baring in a series called "The English Novelists". Stevie never completed this study of a writer she admired throughout her life, and years later apologetically returned the contract to the publishers.

In July Stevie had a holiday at Inez Holden's cottage near Andover where, two years earlier, she had met Antoinette Watney, an acquaintance of Inez. Mrs Watney and Elsa Barker Mill were London hostesses who invited Stevie to large and fashionable parties where she enjoyed chatting with Kenneth Tynan and other prominent guests and would on occasion stay the night. But usually a guest would drive her, sometimes reluctantly, the considerable distance north to Palmers Green. Mrs Watney once considered writing a piece to be called "Taking Stevie Home", as that was so often a project for her. "Stevie didn't really want to see me . . . what she wanted was the use of my car", complained Elisabeth Lutyens, and even such a close friend as Helen Fowler sometimes resented Stevie's efforts to cadge lifts.

In 1947 Helen Fowler was back in touch with Stevie after a two-year hiatus. Her letter was prompted by Flora Robson's reading on the BBC in July of Stevie's essay "Syler's Green: a return journey". This memoir of Stevie's childhood was a great success, rebroadcast several times in ensuing months. (Many years later Dame Flora was impressed by Glenda Jackson's portrayal of Stevie and said with little, if any, grasp of Stevie's achievement, "I understood only too well the dried-up wasted lives of the women of my age.")

Stevie wrote to tell Helen how glad she was that the broadcast, "which the BBC has had, in its usual waffling way, for about two years", had put them in touch again. "It's Palmers Green, of course", said Stevie, "or perhaps not of course to you non-suburban types, and I did not return to it for the reason that I never left it." Stevie thought "Flora . . . read it fairly well—awful vanity of the author—but a shade too honily."

Apart from a few reviews, Stevie's only other publication in 1947 was a story called "Is There a Life Beyond the Gravy?" which appeared in an anthology of *New Curious Stories* edited by Kay Dick. Twenty-four authors were represented in the book, but the *TLS* reviewer found only Stevie's story admirable, praising her "high and successful note of determined craziness". The story had been accepted for publication in 1945 and Stevie told Kay Dick that "I should never have written it, if it hadn't been for you." In addition to the title the story has other of Stevie's favourite puns, according to Sally Chilver, such as "kingdom kong" and the "pillar of asphalt". As Mrs Chilver points out, the story "is chock-full of Stevie's images and mythology and could well be illustrated from her drawings of monstrous man-high flowers, harpists, small girls in sailor-suits, and people running past."

The heroine bears the name Celia, which Stevie gave to her autobiographical protagonist in *The Holiday*, and Uncle Heber also appears in both tales. There is a Cas, too—not Casmilus, though, but Casivalaunus, "an old spelling for this shadowy British king", according to Sally Chilver, who calls him a figure from the realms of John Cowper Powys. Celia is employed at the Ministry where she is aide to Sir Sefton Choate, but also writes for the *Tribune*—all rather autobiographical detail. (Celia thinks of Sir Sefton: "Had he not written in his famous monograph on the Children of Israel, 'It is probable, in this rich oil-bearing district, that Lot's wife was turned into a pillar of asphalt, not salt'?" This statement, credited to Sir William Whitebait, Member of the Institution of Mining Engineers, forms the epigraph to Stevie's poem "Lot's Wife".) Although the characters carry on through most of the story in rather a this-worldly way, there are foreshadowings that suggest we are in the land of Shades. When at the end they chorus "We're all dead, we've been dead *for ages*", a pleasurable *frisson* overtakes the reader.

However strong Stevie's penchant for the other-worldly was, she did of course admire and imitate the practical ways of Aunt. Money continued to be a concern. In a review she wrote for *John O'London's* that year she queried with regard to a work that won an award for young British writers: "and those of riper years, is there no prize for them?"

Early in January 1948 Stevie wrote to Sally Chilver to thank her for a favour. Stevie told her she'd spent a very pleasant after-Christmas lull inking in the tunes Sally had thought up to accompany some of Stevie's poems and then copied out for her:

> This was a fascinating exercise as you know I do not understand scoring. . . . I played the tunes through and got my sister to do so too. The point is though that I am still haunted by the missed-shot tunes I seem to hear and cannot always reproduce even by singing, for instance there is real beauty to "The Lads of the village, we read in the lay . . ." but now I cannot get it right, it is exasperating, I really must do what E. Lutyens suggested and get them recorded, even if it costs me a fortune . . . What I shall do with your tunes is to get them so into my mind that some time a poem will fit itself to them, that is the way it does happen.

In public performances of her poetry, Stevie began more and more, as the years passed, to chant her poems in off-pitch versions of English hymn-tunes or folk songs. One friend found her performance reminiscent of Yeats (who was seldom on pitch either). Stevie later asserted that her reason for putting her poems to music was to ensure that "other people will get the right rhythm . . . I like, when I have written my poem, to make sure that if it ever is read by somebody else aloud it will be read properly . . . the way that another person may stress the poem, it might be better than my stress, but naturally one wants one's own way in these things." Friends did not otherwise think of Stevie as musical. One friend remembers Stevie's annoyance with the BBC Third Programme because it neglected books in favour of music. And to another, an opera buff, Stevie once rather scornfully said that music doesn't interest intellectuals or poets. Neither Helen Fowler nor Suzannah Jacobson thinks that Stevie had any interest in music. Another friend speculated that "Stevie wasn't interested in the other arts. She wasn't interested in literature, really. She just picked things up."

None the less some composers were attracted to Stevie's poems and created musical settings for them. Stevie claimed never to like the readings professional actresses did of her poems, and often she felt the same about the tunes invented by professional composers. Early in 1949 she told her editor at Chapman and Hall, who were to publish *The Holiday* that year, "Last night some of [my poems] were sung by Hedli Anderson to music by Elisabeth Lutyens (my tunes, some of them, I firmly state)." The audience responded enthusiastically, which may have persuaded Stevie to incorporate more chanting as time passed and she took to performing her own work. It also made her ask: "Does this suggest that there might be a wider public for them than the highbrow sort?" In all, Elisabeth Lutyens set ten of Stevie's poems, some with the intention of persuading Beatrice Lillie to perform them. But, according to the composer, Stevie was not pleased with her settings.

It was in 1948 that Stevie went to Poole, in Dorset, within easy reach of Studland Beach, for an August holiday with Betty Miller, a writer friend, and her two children, Jonathan (a physician and now the well-known theatre director) and Sally. Their father, Emmanuel Miller, a psychologist of repute, joined them when work allowed. Elisabeth Lutyens visited the holiday-makers there and remembered Stevie, aged forty-five, bickering with Sally, aged twelve, "in a childish way". Stevie would say: "You don't want that cream cake, I'll have it and you take the chocolate one instead." Almost immediately after this holiday Stevie sat down and wrote a fictionalized account of it, changing the names of the Millers.

Jonathan Miller learned in his twenties that he had been portrayed by Stevie, and not admiringly, when his mother told him that after reading the story she never again spoke to Stevie. She felt that for Stevie to have put her family into a story so directly and in such an unpleasant way was an unforgivable breach of friendship. From his early youth, Jonathan Miller recalled, Stevie visited his home near Regent's Park. His mother regularly wrote in the morning and after lunch went for a long walk in the Park. Stevie sometimes joined her on these walks. He recalls Stevie, Inez Holden, Naomi Lewis, and Cicely Mackworth as among the literary intimates of his mother, and Kay Dick and Marghanita Laski as less intimate friends who often visited. These women constituted a kind of Hampstead Set. He remembers Stevie, particularly, as being difficult and competitive with children. He believes she "probably hated children—she wanted to get on with quiet, civilized conversation, and children were in some way an obstacle whom she resented very much." One sharp memory Jonathan Miller has of Stevie is her being convulsed with laughter when she read in a Dorset paper the misprint: "Is there a life beyond the gravy?"

Stevie's depiction of her 1948 holiday with the Millers is called "Beside the Seaside". Subtitled "A Holiday with Children", it was published in a 1949 collection called *Holidays and Happy Days*.

> Children who paddle where the ocean bed shelves steeply
> Must take great care they do not
> Paddle too deeply.

These lines from a poem that appeared in *Mother, What Is Man?* are a cautionary epigraph as the story starts. Stevie must often have remembered the stories of her two paternal uncles who drowned, and these memories may have intensified the fears she had as a child for her father's safety at sea. Though an enthusiastic swimmer herself, she recurrently wrote of human perils in the metaphor of drowning. In the story the Miller children are called Hughie ("who was going to be a doctor when he grew up") and Anna, and their ages reduced to ten and eight. Helen, Stevie's stand-in, is fond of the children's parents, and of Anna. But from the start she is critical of Hughie's behaviour. Henry, the father, is tense and his wife finds him "more locked up in being a Jew than it seems possible". She tells Helen: 'You cannot know quite what it is like; it is a feeling of profound uncertainty, especially if you have children. There is a strong growing anti-Jewish feeling in England, and when they get a little older, will they also be in a concentration camp here in England?' Helen makes rather a shocking reply:

> 'One sometimes thinks that is what they want,' said Helen flippantly, getting rather cross, 'they behave so extremely. Well, that is rather an extreme remark of yours, is it not, about the concentration camps, eh, *here*? If there is an anti-Jewish feeling in England at the moment it is because of Palestine.' Helen paused and went on again more seriously, 'I do not hold with the theory that the Jewish people is an appeasing, accommodating people, knowing, as some say, on which side their bread is buttered, and prepared to make accommodations with conscience for their own advantage. No, I think that they are an obstinate and unreasonable people, short-sighted about their true interests, fanatical. They have not the virtues of a slave, you see, but also they have not the virtues of a wise person.'

Perhaps Betty Miller's anger came in part from her reaction to this passage. Margaret asks Helen not to speak like this to Henry and Helen agrees and then abandons her impulse to pursue the topic with Margaret when she looks closely at her gentle face. (Sally Chilver remembers a conversation about Zionism in which Stevie said— "'Oh dear, can good *really* come of it?' It was almost as if she had run a fast set of films through her mind and detected both the grandeur and incongruities present in the conception, in one go." Stevie's scepticism continued and in 1956, reviewing a collection of Israeli fiction, she wrote: "the people in these stories are loving and brave and hopeful and strange, but what hope have they?")

Henry returns to town and the two women are left with the children. The technical felicities of the story are many, among them the nature scenes that Stevie describes so affectingly. But unlike

Margaret, who enjoys the picturesque in the conventional adult way, Helen is "childish about it" and imagines herself three inches high and exploring strange land from a different perspective. The explorer's impulse, born early in Stevie, seems to have persisted. So Helen enjoys her adventures despite the grumblings of Hughie who tells his mother that he'd rather be back in the Edgware Road. "Helen sighed. She thought there must be something in Margaret's gentleness that drove her son mad. Hughie must wish to see her round on him, to make her angry, to make her cry?"

The climax of "Beside the Seaside" comes when the women escape from the men ("from Hughie . . . the restless son", thought Helen) for an outing. The women enjoy the day immensely but eventually Margaret has forebodings about Hughie and the trouble he will make. Helen's psychology seems wise. She tells Margaret, "It won't do Hughie any harm; in fact, it will do him good . . . take a long view, Margaret, the long peace is not always the peace of the moment. Hughie must learn." On their return, Hughie meets them with speechless fury. He is "pale with passion" and won't listen to his mother's apologetic explanations. When he does speak he calls Helen and the others "low, disgusting women . . . You are liars. I curse the day I was born." He steps up his fury, telling his mother that she puts his sister ahead of him and is crippling his life and endangering his sanity. He tells her: "people will point to you in the street and they will say, 'There is the woman who drove her son mad.'" Helen picks up a bundle of magazines and begins to hit Hughie, telling him to shut up.

Just then the father returns from a day in the clinic, making generalizations about badly brought up children and what their fate will be. Ironically he seems not to notice the storm of emotion that has engulfed his own family. Helen flees but reflects on how much, among other events of the day, she had enjoyed hitting Hughie. The story ends with a prose poem that lyrically renders the attractions of solitariness (with its portent of death) and of society, too. As Stevie wrote in *The Holiday*, "when I am with people, I do not feel so cold."

Stevie sent "Beside the Seaside" to Betty Miller who wrote to say that she enjoyed it very much as a story but was disturbed by Stevie's having put in it an episode that Betty Miller had told her in confidence.

The mother–son thread in "Beside the Seaside", and indeed the characterization of Hughie generally, is fraught with the same theme as a poem Stevie wrote around this time (and published in her fourth volume of poetry which came out in 1950), "A Mother's Hearse".

Once, after Betty Miller and Olivia Manning left a party, Stevie was asked to read some poems. Among her choices was "A Mother's Hearse":

The love of a mother for her child
Is not necessarily a beautiful thing
It can be compounded of pride and show
And exalt the self above every thing.

Oh why is that child so spoilt and horrible?
His mother has never neglected the trouble
Of giving him his will at every turn
And that is why his eyes do burn.

His eyes do burn with a hungry fire
His fingers clutch at the air and do not tire
He is a persecuting force
And as he grows older he grows worse.

And for his sake the friends are put down
And the happy people do not come round,
In pride and hostility against the world
This family upon itself is now curled.

Oh wretched they and wretched the friend
And this will continue without end
And all for a mother's love it was,
I say it were better a mother's hearse.

The child in the poem, according to all reports, was Jonathan Miller. Next day Olivia Manning rang Francis Wyndham to ask what Stevie read. He told her and then Olivia (maliciously, in his view) passed the news on to Betty Miller who was furious. Inez Holden rang up Francis Wyndham to warn him of the maze into which he was being drawn.

Francis Wyndham remembers a rivalry between Stevie and Olivia, and many friends recalled their relationship as a "spiky" one. Apparently their difficulties rooted in an affair which Olivia had had when she first came to London in the 'thirties and about which she wrote a piece in the late 1970s for *The Sunday Times* series called "First Love". The series was scrapped when a new editor took over the magazine, and the whereabouts of the essay is now unknown. But Francis Wyndham, who inaugurated the series, recalls that Olivia told of her love affair having been "messed up" by someone she called Scorpio in her essay, but identified to him as Stevie. Olivia Manning thought she would probably be heavily criticized for portraying Stevie just then in such an unsympathetic role, particularly as the play "Stevie" was a hit in London. This lover of Olivia was presumably Hamish Miles, Stevie's editor at Jonathan Cape. Olivia had told Kathleen Farrell once about an affair that went wrong and identified the man as "Hamish". ("I am uncertain of his surname but I think it was Miles", writes Kathleen, "I think Hamish worked at Cape's.") However Olivia did not in recounting this to Miss Farrell implicate Stevie.

Olivia had a great interest in the psychic. As she once told the novelist Francis King, "I have an absolute loathing of death. I really love life . . . and I do want to . . . believe that, even if it were far worse than any existence here on earth, there was the promise of an existence elsewhere." This certainly was a position that differed from Stevie's, and on a topic that mattered urgently to both of them. And Olivia spoke of this to Kay Dick when she said, "I have none of Stevie's sentimental feelings about death . . . I've no desire at all to be dead."

Throughout the 1940s Stevie seemed on her part to do what she could to be useful to Olivia. In 1947 she sent some of Olivia's poems to Kay Dick in the hope she would publish them in *Windmill*, a periodical published by Heinemann which Kay Dick, under the pseudonym Edward Lane, co-edited with Reginald Moore. And on into the mid-1950s she reviewed a number of Olivia's novels favourably. In July 1953, she wrote to tell Kay Dick that "Dear Olivia says she will come out & see me. That girl is an angel. . . . She has asked me to her party on the 23rd & I hope to be able to stagger forth." Kathleen Farrell recalls that "fundamentally, I am sure they loved each other. But Olivia, although she was such a good writer, was occasionally very unsure of herself . . . if she had a long and praising review and then there was one half-sentence which she did not like, it was that sentence which she would concentrate upon. I believe Stevie wrote a review of one of Olivia's novels which upset Olivia."

In 1955 Stevie published a review of *The Doves of Venus* in *The Observer* and, according to friends, Olivia was deeply offended by it. At the time there was gossip that one of the characters in the book was based on Stevie and, in an introduction to a new edition of the novel, June Braybrooke identifies the character Nancy Claypole as having "much of Stevie in her". Mrs Braybrooke acknowledges that there are "obvious attempts at disguise", but finds Stevie's spirit in lines spoken by Nancy, such as her response, hearing girls described as the doves of Venus: "It's true doves are sacred to Venus, but so are lots of other birds. Sparrows, for instance." Perhaps as revealing is Nancy's description of her father as "the world's most excruciating bore". Nancy is a plain girl who astonishes Ellie (the protagonist) by telling her she has gone "the whole hog" with men many times. Stevie may have been annoyed at some of Olivia's depiction of Nancy. In any event, her review was mixed, praising the novel as "beautiful" and one written "with a poet's care for words", but criticizing it for "moral naïvety" and the absence of balanced thought.

Probably this review brought to a head the grievances which Olivia had totted up against Stevie over the years. From then on friends thought of them as participants in "an odd love/hate relationship".

> Once at a Hampstead party everyone was twittering about Stevie and Olivia having both been asked when they were not on speaking terms, and so on—they rushed at each other with small cries of joy and sat together on a sofa the whole evening talking hard, disregarding everyone else.

Francis King did not know Stevie as well as Olivia, and he did not come to "hate" her, as Olivia did. But he agreed with Olivia "that Stevie was tricky and malicious". "I *did not trust her*," he writes. "Her tongue was as inaccurate as Olivia's but far more skilful—so that, whereas people often dismissed Olivia's gossip, Stevie's tended to be accepted. Far more damaging, therefore. Stevie was a great character . . . but she was not, emphatically, not a nice character, as so many people seem to believe."

Francis Wyndham remembers that after Stevie's death, Olivia was writing a memoir that included her, and indeed Olivia described the strange history of this in a letter to Kathleen Farrell, written soon after Stevie's death:

> A funny thing happened while I was reading the Stevie pieces. I thought Kay had been too kind to that sly puss so I started writing, on smallish pieces of paper, my own memories of some of Stevie's arch-bitchiness towards me, most of it dating back a very long time. In fact to the '30s, when we were very friendly and she was, as you know, bridesmaid at my wedding. As I wrote, I was lying on the sofa and I put each finished page on top of the books at my elbow. When I gathered them together I found that the first three pages were missing. I looked behind the books, under the sofa, every-where—no sign of them. . . . I can only think that Stevie had been playing a trick on me. Very much what she would do. The trouble is I cannot now remember what I wrote.

Olivia persisted in her animus against Stevie. In 1977 she wrote to Kathleen Farrell:

> No, I have not seen "Stevie" [Whitemore's play] and I am not sure that I want to. . . . Hugh Whitemore got Jean MacGibbon to ring me up and try to get information about Stevie. He particularly wanted to know *why* Stevie slashed her wrists. Fortunately I could

say that I had been asked by the *Times* to write a piece about Stevie so I did not wish to discuss her. I do intend to write it but am worried when I think about Stevie—should I reveal what she was really like? She had the most venomous tongue of any person I have ever known. I have letters from her in which she is viciously critical of the Millers—Jonathan in particular—on whom she was battening at the time. I think she became slightly milder towards the end when her aunt was dead and Stevie was dependent on the outside world. But the truth is the important thing and something (perhaps Stevie herself) seems to be stopping me from writing it. I did once start to write about her. I had about ten pages torn from a notebook and left them on the back of the sofa. When I came back they had disappeared. There was no one else in the house, yet the pages about Stevie had gone. I never found them and have never been able to write them again.

As Kay Dick has said, Olivia's animosities were legendary and a bit of a joke among her friends, and this should be taken into account in assessing her impressions of Stevie. The attention she felt Stevie in the afterlife paid to her literary efforts has its comic side. But perhaps Stevie was having her way with one of the friends she "feared below".

12

1949

Stevie's love for her cosy and familiar suburb continued to compel her, and she wrote affectionately of Palmers Green in "A London Suburb", an essay published in 1949. Coupling the strange with the domestic, she imagined her way back into the ancient past when the suburb was "a great woodland country-side . . . and across its wooded acres tore the wild boar and the red deer, and after them came the mounted nobility." Most memorably Stevie celebrated in this essay Grovelands Park (to which she gave the name Scapelands) which was so much a source of inspiration for her and where she often did escape from social pressures.

> The most beautiful place in the suburb is Scapelands Park, especially when the weather is wild and there is nobody about except the anglers. When the wind blows east and ruffles the water of the lake, driving the rain before it, the Egyptian geese rise with a squawk, and the rhododendron trees, shaken by the gusts, drip the raindrops from the blades of their green-black leaves. The empty park, in the winter rain, has a staunch and inviolate melancholy that is refreshing. For are not sometimes the brightness and busyness of suburbs, the common life and the chatter, the kiddy-cars on the pavements and the dogs, intolerable?

Stevie is realistic in her account. Not all goes well in the suburbs. In one home a husband, retired now for many years from the Merchant Navy, spends his days pacing an upstairs room. A trace of autobiography in this, no doubt: Stevie remembering that not all had gone well with the family at 1 Avondale Road.

On 20 February 1949, her own father died in the District General Hospital, Kidderminster, Worcestershire. He had lived in Worcestershire with his second wife, who predeceased him. After her death Mr Smith turned up in Avondale Road asking desperately "Why did this terrible thing have to happen to me?" Molly was unwell at the time and Mr Smith, receiving little sympathy from Stevie, returned home. Both sisters were named administrators of his will. (He left £470 1s.)

Stevie did not attend her father's funeral. She was preparing to read her story "Sunday at Home" (originally entitled "Enemy Action") on the BBC radio, and was disinclined to jeopardize this opportunity by taking time to pay her respects to a father she did not respect. Two years earlier she had begun her efforts to sell a story to the BBC. She sent a producer there "The Story of a Story" but he returned it saying he had read it in *The New Savoy* and thought it more suited to the eye than to the ear. The next year Stevie sent "Enemy Action" to the same producer, but he rejected that story for the identical reason. Then she sent Roy Campbell both stories. Campbell found things to admire but judged that they were "a bit above the heads of this audience". However, one of the producers wondered if Stevie might care to enter her stories in a competition then being conducted. The purpose of the competition was "to encourage writers . . . to interest themselves in the opportunities for story telling offered by Radio, with particular emphasis on the discovery of an author/broadcaster, i.e., someone who can express himself best when given the opportunity of using his personality to put over the particular atmosphere of the story he has written."

By early March 1949 Stevie had been round to the BBC to read her story in an audition, but the chief producer of the Talks Department, responsible for the short story competition, "came to the conclusion that it was hopeless. . . . I told her the story was impossibly difficult for her voice and that I should not recommend its inclusion. It was not in fact written for telling as her 'Return Journey' was. I am asking Miss Kallin to record all or part of the story to play back to Stevie Smith. She would like to hear her voice and how she tackles the script."

How these words must have dashed the hopes of Stevie who from her early years relished dramatic performance and appearances on stage. Now it seemed that just as the opportunity to broadcast her work to a vast radio audience had been within reach, she was to be denied it. Of course linked to her liking for theatrical performance was her firm belief (and a valid one) that her writing depended to an uncommon degree on voice and nuance. Hence her distrust of professional actors who could not be counted on to render accurately the author's orchestration of voices.

Over the years, with the help of Rachel Marshall, Stevie had been diligent in cultivating her voice. She must have worked hard in the days that intervened between her original reading at the BBC and mid-March when a producer at the Talks Department wrote, "allow me to say how *enormously* your reading has improved over the first time, and how much we enjoyed hearing it." The story was included among the runners-up in the competition and arrangements made for two rehearsals. The producer thought that "in view of the fact that you have improved 100% I am convinced that with more practice you would be able to assimilate and practise more points of production." When Stevie appeared for an early rehearsal, Anna Kallin found her "even more nervous and shy than I remembered her. . . . I rehearsed her for an hour, thought that she was rather good, recorded the whole story. . . . To me she sounded not at all bad . . . and I thought that with more time for . . . putting her at ease one could do something very good with her." So it was that on 20 April 1949, Stevie's voice went out on the air. She was paid twenty guineas for the story, and an additional five guineas for reading it.

"I think radio may, in the end, intrigue and amuse you", wrote Miss Kallin to Stevie. She was right: Stevie began with that broadcast a career as a reader of her own work that eventually brought her considerable income and fame. So successful did she become that years later Glenda Jackson reported herself reluctant to follow Stevie on to the stage when they both took part in a benefit. Douglas Cleverdon, who eventually headed the Third Programme, has said, "One can argue interminably, of course, whether a poet is the best reader of his own poems. But as far as the reading of Stevie Smith's poems is concerned, no one could touch her. She lived every poem that she spoke—the tone, the timing, the characterisation are all impeccable." The same was no doubt true of her story.

Armide Oppé, who often lunched with Stevie and was at this time sharing a house with Michael and Anna Browne in Regent's Park, gave a party with Mrs Browne to which she invited Stevie. Anna and Stevie met and spent much of the time talking together about the Norfolk coast which Stevie loved and where the Brownes spent summer holidays. After that they saw a lot of each other and Anna recalls that she and Stevie some days spent hours walking all over central London, "talking so hard that we were completely oblivious of our surroundings".

George and Olga Lawrence were new friends Stevie met about this time at the home of the Brownes. David Gascoyne, the surrealist poet, was a tenant of the Lawrences, and Stevie thought his work superior to hers. Twenty years later, the Lawrences estimate, Stevie plainly thought of herself as the better poet. (Of course one is startled to learn that Stevie knew Gascoyne's poetry as she rarely read the work of fellow poets, and wrote to Jonathan Cape in those years to say "I have never myself bought a volume of contemporary poetry.") Stevie's friendship with the George Lawrences took on, and soon she was visiting them at Brockham End, their country house outside Bath. Later Stevie was introduced by the Brownes to George Lawrence's brother and his wife, Sir John and Lady Lawrence, and they too became friends.

In June 1949, after years of trying to find a publisher, Stevie saw her third novel, *The Holiday*, reach print. Early the year before she had been negotiating with Peter Nevill Ltd for publication of the book. There were several delays and postponements of plans for publishing. But these were only the latest in a series of struggles she had in her effort to publish *The Holiday*. "I wrote it actually during the war", she told Kay Dick years later, "and I couldn't get it published—my word, nobody would take it. I kept on putting in things that were after the war, like the troubles in Palestine, and it was in such a muddle that I had to call it the after-the-war period." In October 1948 Stevie abruptly informed Peter Nevill Ltd that she had signed a contract with Chapman & Hall to bring out *The Holiday*. "I have not been happy about the long delays and changes and . . . I have no feeling of confidence that further delays and troubles might not have cropped up with them", she wrote to the man who had acted as her contact with the firm of Peter Nevill.

The Holiday is in part a novel inspired by the sadness of war and of "the post-war", and very much encapsulates its author's moods and thoughts during this time which "cannot be said that it is war, it cannot be said that it is peace, it can be said that it is post-war." (Throughout the typescript one can see that Stevie systematically inserted "post" before "war".) Says Celia's beloved cousin, Caz, "I do not know . . . that we can bear not to be at war." Celia, Stevie's autobiographical protagonist, insists that "crisis and exile are our lot". She is, in fact, a modern version of Persephone. Once when Stevie was sent a questionnaire about influences on her writing, she replied that she was "drawn toward the Grecian very strongly" and pointed both to the Bacchae passage in *Novel on Yellow Paper* and "the Persephone sub-theme in *The Holiday*". Celia does enjoy the lovely summer days but like her classical counterpart, she "knew how brief that beauty was; fruits, flowers, leaves, all the fair growth of earth must end with the coming of the cold and pass like herself into the power of death." Stevie wrote in the late 'forties a poem called "Persephone" in which she identifies other qualities shared by Celia and Persephone—loss of innocence, for instance:

> I am that Persephone
> Who played with her darlings in Sicily
> Against a background of social security.
>
> Oh what a glorious time we had
> Or had we not? They said it was sad
> I had been good, grown bad.

Here is Stevie's comment on "the classic story of Persephone":

She was stolen away by Pluto, the god of the underworld. He fell in love with her when he saw her playing with her maidens in a meadow in Sicily. In the legend, she was allowed to come back to sunny earth for six months every year, at the plea of her mother Ceres. The story of Persephone is a story of Winter and Summer, but in my story I have made Persephone a girl who loves winter, and snow, and the curious light you get when there is snow on the ground and you look up, if you are in the house, and the ceilings are bright with the reflection of snow. In my poem, Persephone even likes the dark places in her kingdom, which can be frightening. She might not like it so much if she had to stay there all the time. There is another thread in this poem—it is what she feels about her mother. She loves her, but at the same time she does not want to be sought for all the time, and wept for, and begged to come home. She wants to be herself, and free to stretch out and take her time.

None the less, the stanzas about the sorrowing mother cannot be read without remembering how Stevie and her mother, like Ceres and Persephone, lost each other:

> My mother, my darling mother,
> I loved you more than any other,
> Ah mother, mother, your tears smother.

In *The Holiday*, Celia is employed at the Ministry (which combines features of the BBC and of Newnes, Pearson) and she lives in a northern suburb with an elderly aunt. When her holidays come round, she leaves London and the Ministry and travels to her Uncle Heber's in Lincolnshire with Tiny, a favourite office chum, brother of her friend Lopez, and Caz, her cousin who, Byronic family rumours suggest, could be her half-brother. Celia is hopelessly in love with Caz.

Armide Oppé remains surprised that so far no one has looked

carefully at the motif of incest which runs through *The Holiday*. Incest is certainly suggested in the novel, but perhaps it was Stevie's personal metaphor for the duality which made her appear so paradoxical. In *The Holiday* Celia's beloved cousin is called Casmilus, the name Stevie gave to her autobiographical protagonist in the two earlier novels. In this novel where Stevie has divided George Orwell into two characters, she seems similarly to have divided herself into Celia and Caz. The difference is that the latter are male and female which allows Stevie to consider herself as the classic androgyne, sundered and kept so by some dark family history—separated from the "other" without whom she can never be content.

More broadly sounded throughout the book is the theme Hermione Lee has noted. Pointing to parallels between Stevie Smith's writings and those of Victorians such as Carlyle, Tennyson and Arnold, she says that "Stevie Smith's Victorianism is more than a penchant for Gothic families and Lincolnshire fields. Her religious anxieties seem markedly of the 19th century. In her novels and poems she was perpetually engaged in agonised conflict with the established Church." Perhaps the parallel is even more vivid when one considers *Robert Elsmere* and others of the novels of Mrs Humphry Ward, a writer of lesser rank who had for Stevie some special appeal. "There's something of Mrs Humphry Ward in me", says Pompey. Hilary Spurling helps us to see that when she refers to "the conflicts commonly endured by Victorian heroes and heroines in the throes of religious doubt, a struggle . . . at its most hectic in Mrs Humphry Ward."

Stevie certainly admired the Victorians. "Oh what a pity it is", she wrote, "that our strength has gone from us in the wars, that we cannot now write so strongly as the great Victorians. But it is not the wars, of course (great pictures were painted in Alva's Netherlands) it is the times, our strength is drawn out of us in a thousand directions at once and broken into pieces." Like Hardy, and other late Victorians, Stevie was born into a comforting religious vision of human existence dimmed if not extinguished by Darwinism. She refers to this in "Touch and Go", a poem published in several places, including *The Holiday*, before it was printed in her fourth volume of poetry.

> Man is coming out of the mountains
> But his tail is caught in the pass.
> . . .
> Do not be impatient with him
> He is bowed with passion and fret
> He is not out of the mountains
> He is not half out yet.

And so it is that in this novel humans are incomplete and the ape looms large. Celia thinks to herself at one point:

> we have come further than the apes and know what we are and do not like it and want to be something else. And I thought: We have come some way from the apes, but we think we have not come fast enough or far enough, and so fall into abuse and despair. But looking at the violent dark creature that is one step down from us, I cried out . . . I should go. . . .
>
> The idea is to go, go, go. Go from the apes . . . from the friends.

From her early years Stevie was attracted by the possibility of escape, and nearing the age of fifty she found that the idea still had a strong hold on her. Celia, in fact, does try to drown herself in this novel, led to it by despair that is deep: "I was oppressed by such a sense of melancholy sweet sadness, of a tragedy of huge dimension but uncertain outline, of wrongs forgotten whose pain alone remains, that I cried out in fear and threw myself upon the ground. There is not one thing in life, I cried, to make it bearable."

The novel, though, has other registers. As Hermione Lee further says, *The Holiday*

> gives a clear, bitter picture of middle class intellectual life just after the war, and is remarkably good at political conversations—about India, about the effects of German propaganda on the English temperament, about the social impotence of writers. It's also a very funny novel with a caustic eye for office life and suburban comforts [Sally Chilver had some reservations about verisimilitude in the depiction of the Ministry] . . . It's energetic in praise of independence, friendship (particularly between women), and professional camaraderie. But it is, principally, a document of suffering, a painfully close account . . . of Celia/Stevie's battles with despair.

Celia proudly tells us, "I am a middle-class girl, conditioned by middle-class thoughts, when I think of England, my dear country, I think with pride, aggression and complacency. I tie up my own pride and advantage with England's, I have no integrity, no honesty, no generous idea of a better way of life than that way which gives cream to England." These are sentiments similar to those in the poem "Voices against England in the Night" which Stevie incorporated into the collage which is *The Holiday*, and published as well in *Mother, What Is Man?*

In P. H. Newby's review of *The Holiday*, which Stevie greatly admired, he confessed, "I least know how to deal with [it] . . . The most obvious thing about Miss Smith is that she is not to the slightest degree interested in the telling of a story. . . . Yet I know of no novel which has caught so much of our post-war confusion of mind." (Perhaps the timelessness of what Stevie depicted is proved by the fact that the state of mind she reveals was essentially caught by her years earlier, during the war.) "It is the extraordinarily bold way she lays all her cards on the table that takes the breath away. I suspect that she has issued a challenge which hardly any other practising novelist would dare take up."

Newby was correct in saying that Stevie was not interested in plot, or at least not an adept at it: Joan Robinson once sent Stevie, on request, a detailed sketch that she might turn into a novel, although nothing apparently came of it. What Celia says in *The Holiday* applied to Stevie: she has "the pick-rag mind that was full of tags of the church, and the classics, but that was well enough as far as it went, and had served me well enough." In the country, at Uncle Heber's, Celia indulges in fascinating speculations about many subjects which flow freely during those country days which remind her of "the long summer days of childhood that are such a pleasure, nothing is ever such a pleasure again, tired and hot and happy the hours come to teatime."

Celia makes an observation revealing of Stevie when she touches on the middle years which she defines as those from twenty to fifty. She views them as a time when "intellectuality" dominates humans. But it is "instinctuality", associated with childhood and old age, "that brings with it so much glee, so much pleasure that cannot be told . . . and as I am by nature of this type of person, it is perhaps because I now run in these *middle* years that I am not enjoying it but must cast ever backwards to my childhood and forwards to my old age." Celia concludes with the sanguine assurance that "the feeling of full enjoyment will flood in again, we must get through these middle years".

Inez Holden wrote a long composite review of novels by Ivy Compton-Burnett, Betty Miller, Elizabeth Bowen and Cicely Mackworth, and included *The Holiday*. She praised it for its poetry and criticized it for excessive loyalty to "Mother Empire and England my England", and for "too many tears". Inez makes only one mention of Lopez, the character fashioned after her, but one wonders if she may have resented some of what Stevie wrote about Lopez and her boyfriend, Winsome Elliot. Celia concludes that they are driven by "intellectuality". Although no real rift occurred in their relationship, Inez and Stevie drifted apart, it seems, after the appearance of *The Holiday*. Or was it that Stevie was spending more time with friends who had families and with whom she could weekend and feel herself *en famille*? She said she was "glad" when, a decade later, the wife of a friend decided to translate one of Inez's stories into German. But, Stevie added, "God knows why I should be as the old thing has been very odd indeed for years now & will hardly address a word to me."

For Celia, who tells her uncle she is "nervy, bold and grim", the Russian-Marxist philosophy is no solution to post-war problems. Tengal (based, as we noted earlier, on J. D. Bernal) and "all his Elks" are linked with Torquemada, the villain of the Inquisition: "it might be the Communist Party itself". Celia writes a poem called "A humane Materialist at the Burning of a Heretic", which Stevie published in her next volume of poetry:

> When shall that fuel fed fire grown fatter
> Burn to consumption and a pitter patter
> Of soft ash falling in a formless scatter
> Telling Mind's death in a dump of Matter.

She glosses the poem in *The Holiday*:

> This humane person, this old-fashioned Trotskyite piece, is standing in the front row, he can smell the burning flesh of the heretic. This heretic has been very much tortured because you see his deviationist opinions are not very clear cut, so you see, poor fellow, he has not been able to frame a confession that will fit with what the torturers want, always what he says is not quite, as the editors say, "what we had in mind".

So the man is burned.

Celia then addresses a sort of sermon "to every writer that has sensibility and a desire to establish himself". Years later Stevie congratulated a friend who identified the source of Celia's sermon: "It's wickedly clever of you to have quoted that long bit from Dostoievsky." The sermon is verbatim from *Notes from the Underground*, a passage Stevie had copied into her reading notebook in the 1920s:

> And how persistent are your sallies, and at the same time what a scare you are in. You talk nonsense and are pleased with it; you say impudent things and are in continual alarm and apologizing for them. You declare that you are fond of nothing, and at the same time try to ingratiate yourself in our good opinion. You declare that you are gnashing your teeth, and at the same time you try to be witty so as to amuse us. You know that your witticisms are not witty, but you are evidently well satisfied with their literary value. You may perhaps have really suffered, but you have no respect for your own suffering. You may have sincerity, but no modesty; out

of the pettiest vanity you expose your sincerity to publicity and ignominy. You doubtless mean to say something, but hide your last word through fear, because you have not the resolution to utter it, and only a cowardly impudence. You boast of consciousness, but you are not sure of your ground, for though your mind works, yet your heart is darkened and corrupt, and you cannot have a full genuine consciousness without a pure heart. And how intrusive you are, how you insist and grimace. Lies, lies, lies.

Celia concludes that she has found "a masterpiece of astringency to dust away the russo-bourgeois-pre-revolutionary cobwebs".

But if Communism is not the solution for Celia, neither is Christianity. She believes that if men were less positive they would be more kind. "I detest your Christianity that will be so positive", she tells her dear Uncle Heber. "Unknown, unknown, unknown, let that be the life to come and the world that lies beyond." And she praises Confucius who "says that human beings are our concern, and not heaven or hell or the life to come, and indeed, with human beings, yes, just with this and no more."

Reactions to Stevie's novel gratified her. R. A. Scott-James, whose protégée in part she'd been in the 1930s, said in his review:

Some of us had begun to think that Stevie Smith . . . was something that happened just before the war, and would not happen again. . . Was Stevie Smith a pre-war product, gone like so many good things with the coming of war and post-war?
   But it has happened again. It has returned, not with quite so unrestrained a light-heartedness, but still with a light-heartedness which carries with it a terrifying burden of serious import, of apprehension, of increased awareness of the tragic problem of the individual and the corresponding problem of the human race in its present perplexities.

Richard Church called *The Holiday* Stevie's "most mature work. It brings to full flower those characteristics of temperament and technique which make this writer distinctive." And Howard Spring found it "a novel extraordinarily of our time: of its dissolutions, its closeness to anarchy; its balancing of the problem: Is the hope of to-morrow strong enough to avail against the frustrations of to-day, strong enough to pull us through the swirl of the river to the other side?"

Stevie seemed to say in the novel that the English would endure:

We are vacillating lazy and slow, but we have never in all our history for one moment entertained the idea that it might be a good thing to lose a fight. We are not a sophistical people and are saved the dangers that run with sophism; and our education has not yet succeeded in taking away from us the weapons of our strength—insularity, pride, xenophobia and good humour.

Despite all the critical praise for *The Holiday*, it was not published in America until many years later. William Morrow & Co., where the editors found its concerns "no longer quite current, such as the news that England is quitting India", declined it. "But we are sad to say no, because of the quirks and turns of Miss Smith's mind and because of her wit and wisdom." Morrow thereby terminated the option they'd had as a result of publishing *Novel on Yellow Paper*. Stevie sent the book on in September to Blanche Knopf who liked "parts of the book immensely" but thought it wouldn't sell. She apologized to Stevie for delaying her decision and attributed this to a liking for the book. "I want most sincerely, to see, to know about, your next book and I would hope to be able to publish it."

In England the novel did well. By October 1949, Stevie was writing to tell Helen Fowler that "it is sold out and they are reprinting hurrah. I never thought it/they would." *The Holiday* remained Stevie's favourite among the novels she wrote, and shortly before her death she told Kay Dick, "I like *The Holiday* very much better than *Novel on Yellow Paper* . . . it hasn't got those mannerisms, and also it's a period when I was older . . . I gave *The Holiday* to my present publisher [John Guest], and he couldn't make head or tail of it. He said, 'This is a terribly difficult book to read, Stevie.'"

What Stevie thought Palmers Green might make of it we do not know, but she was evidently concerned to keep her neighbours in the dark about the book. A Southgate bookseller wrote to her in June 1949 to say, "We shall most certainly respect your wish not to publicise your book locally . . . It is so true 'a prophet is not without honour save in his own country'."

"The train of death that you are waiting for is an excursion train, yes that is what it is: All aboard for a day in the country." For Stevie, the ultimate holiday was always to be death. But meanwhile she embraced terrestrial beauty more ardently than most have done and especially treasured the visits she began to make to Norfolk. In the fall of 1949, after seeing *The Holiday* through the press, collecting reviews, and dealing with the business attendant on publication, Stevie had her own holiday, "three heavenly weeks at Wiveton which is between Blakeney and Cley-on-the-Sea".

all windmills and salt flats and—at Blakeney Point—wonderful silver sand dunes and the most glorious bathing I think I have ever had . . . The social life was terrific much more than in sobre [*sic*] old

London and as long as the petrol lasted we pretty well combed up Norfolk. . . . I was staying with some friends called Anna and Michael Browne who had rented half of Wiveton Hall from some friends of theirs called Buxton.

Years earlier, when Stevie was perhaps eighteen, she copied this line from *Marius the Epicurean* into her reading notes: "Running counter to his almost morbid religious idealism was a healthful love of open air life." It is nice to think of Stevie in those weeks, happy, delighting in the countryside, and so neglecting her muse. As she wrote,

When I am happy I live and despise writing.

13

Not Waving

During the late 'forties and early 'fifties, when she was still working as a secretary for Newnes, Pearson, Stevie would sometimes visit the American writer Mary Lee Settle, who was living then in the Camden Town district of London, working as a freelance journalist and reviewing books. Settle remembers that she had been introduced to her as the author of *Novel on Yellow Paper*, for in those years Stevie was not known for her poetry. "She looked like a little girl left out in the rain", and "a rattled version of Alice in Wonderland".

"Stevie was a total natural comic", Settle recalls, "with a quality like her poetry, funny on the surface, but the source material underneath was dreadful. Dorothy Parker looked sentimental beside her." Stevie told about George Orwell, for example, who had trouble after the war getting protein for the boy he had adopted, because rationing was still on. He bought a hen, but it laid eggs without shells. In an effort to help, Stevie suggested he set the hen on a frying pan. "It was a terrible story", Settle notes, "with that twist at the end which was typical of her humour." Once, Stevie recited in Settle's flat a poem which had appeared in a contemporary revue: her eyes popped as she exclaimed, "Hush, hush, it couldn't be worse / Christopher Robin is having his nurse."

"Right after the war", recalls Mary Lee Settle, "there was a wonderful upsurge in London; people were writing a lot and talking a lot. It was a kind of false energy, and the fatigue didn't hit for several years." That period, with its sense of community and nervous energy, was a busy time for Stevie. As we have seen, from 1945 to 1950 she had four short stories published (with the threats of libel and broken friendships two of the stories entailed), and a fifth finally enabled her to break into broadcasting on the BBC. During those years, too, in addition to her secretarial job, Stevie continued to review books: eleven for the *Tribune*, forty-three for *John O'London's Weekly*, and approximately seven hundred for *Modern Woman*. And, of course, she finally obtained a publisher for her third novel.

Stevie also kept busy socially. In the early 1950s, for example, she was in touch with Rose Macaulay, who once asked her natatorial friend to come to the Lansdowne Club, holding out the lure of its beautiful swimming bath. On occasion, because Stevie's house in Avondale Road was so small and out of the way, and because her energy reserves were so low, she gave joint parties, such as one held in January 1950 at the London home of Sally Chilver. Armide Oppé and Helen Fowler were there, as were John Hayward and the publisher James MacGibbon, whom Stevie had met several months earlier, and whom she eventually made her literary executor. Polly Hill, who was to become distinguished for her work in African anthropology, also gave a joint party with Stevie, either in 1951 or 1952. It was held in the huge room where Dr Hill was living in Hampstead: "So enormous was this party", she recalls, "that we found ourselves obliged to hold it on two consecutive nights; much of the literary world was there, particularly I remember Louis MacNeice."

During the war and until 1951, Polly Hill worked as a temporary civil servant, and often met Stevie for lunch at the Strand Palace Hotel, which Stevie loved and she hated. Her memory of Stevie at that time is vivid:

> I often found dear Stevie's adolescent giggle embarrassing—an unacceptable, and isolating, cry for help. Indeed, friend though she was (and she helped me to get poetry published), I sometimes found her company depressing. Might not suicide after all have made sense? The unfashionable literary figure that she had become (though not of course with me) could hardly have foreseen any sudden burst of fame except posthumously. As for her class-ridden childish clothing, this matched her preference for bad architecture: Avondale road was not so much warmly celebrated for its familiarity as positively loved for its tastelessness.

That Stevie's poetry was out of fashion in the early 'fifties perhaps explains Chapman & Hall's lack of eagerness to publish her fourth volume. In February 1950 she wrote to John Hayward that the firm was "coming up to the point of being about to be prepared in a frame of mind resolute sober and quite without enthusiasm to bring out another book of poems and drawings." *Harold's Leap* was published that autumn, and several weeks after its appearance Stevie remarked it had "probably already ceased selling if it ever did!" In February 1951 she wrote to Daniel George, thanking him for his "kindness to Harold in his leaping". George had given the book a favourable review, and Stevie told him it cheered her up, "for all that the rain falls and falls". She hoped it would also cheer up her publisher, because she feared he had "fallen a prey to morbid fancies, one of which runs to the tune of, Never again".

Stevie thought her poetry had grown "*perhaps* simplified & more direct", and she preferred *Harold's Leap* to her earlier volumes; or so she said later in 1951, specifically citing the title poem, "The Weak Monk" and "I rode with my darling . . ." It is difficult to characterize briefly a book of sixty-five poems, but the three named by Stevie do reflect two important themes in its mix of characters and moods: the admonition to carry on, and the desire to escape. "In my poems I have tried to shift [my] feelings outwards", Stevie once explained, "by fixing them upon imaginary people." Elsewhere she amplified this remark by saying her whole life was in her poems: "everything I have lived through, and done, and seen, and read and imagined and thought and argued." Then why did she turn them all upon imaginary people? "Because", Stevie said, "it gives proportion and eases the pressure, puts the feelings at one remove, cools the fever."

Two of the imaginary people Stevie had in mind are the subjects of "The Weak Monk" and "Harold's Leap", which may be read as admonitory parables. They are good poems, if not among Stevie's very best, but after the continuing strain she underwent in getting her work published, one can easily see why they meant so much to her. The monk writes a book called "Of God and Men" (and the reader may note that *Harold's Leap* contains Stevie's poem "God and Man"), but fearful because it is not in accord with Catholic doctrine he buries it, expecting God to "rescue his book alive from the sod".

> Of course it rotted in the snow and rain;
> No one will ever know now what he wrote of God and men.
> For this the monk is to blame.

In this poem, Stevie explained, she was criticizing whatever vanity, fear of criticism, or tiredness which might keep her from trying to get her poems published.

To keep trying requires courage, and so for inspiration Stevie wrote the tale of brave Harold who, leaping from one promontory to another, falls into the sea and drowns:

Harold was always afraid to climb high,
But something urged him on,
He felt he should try.
I would not say that he was wrong,
Although he succeeded in doing
    nothing but die.

According to Stevie, she had a very isolated office life because her
employers were frequently away in Africa, where they had a gold
mine. So she would tell herself to get on with her writing, and her
conscience raised the phantom of Harold to drive her on, for "he died
doing more than he was able".

Other poems in *Harold's Leap* contribute to the admonitory strain of
its title poem and "The Weak Monk": "The Deserter", for example,
the dramatic monologue of "a wicked poet who has decided to be ill
instead of bothering to work. He is a deserter to ill-health." To
reinforce her point of view, Stevie wrote below the title of this poem,
in a friend's copy of *Harold's Leap*, "(or *Wieder falsch!*)". In a similar
vein she once commented: "The poem is called 'The Deserter' or
'Wrong Again' (because it is no solution)." Also admonitory (and
probably self-admonitory, for Stevie was effectively an orphan and
had a tendency to attach herself to the families of her closest friends) is
"The Orphan Reformed", one of the book's best poems. An orphan
"roams the world over / Looking for parents and cover." She visits
"this pair and that", calling them father and mother, but she does not
find the parental love she seeks. The people are not evil, the poet says,
and the orphan must learn to be alone:

At last the orphan is reformed. Now quite
Alone she goes; now she is right.
Now when she cries, Father, Mother, it is only to please.
Now the people do not mind, now they
    say she is a mild tease.

Another poem of admonition is "Do Not!", which counsels the
reader not to "despair of man", a tendency attributed to "Sadness for
failed ambition set outside, / Made a philosophy of." The poem's last
line slightly undercuts its healthy-mindedness, for it advises suicide to
those who despair of themselves:

Oh know your own heart, that heart's not wholly evil,
And from the particular judge the general,
If judge you must, but with compassion see life,
Or else, of yourself despairing, in death flee strife.

Escape is the other major theme in *Harold's Leap*, and a recurrent yearning of Stevie—especially, during the years she was writing its poems, escape from her secretarial position. She evidently considered becoming an editor, although a piece she wrote in 1949 suggests she concluded that would not constitute a real escape for her. Editorial work offered rich rewards, up to two or three thousand pounds a year, but was fiercely competitive.

> A girl who might be happy as a secretary where the gentler qualities are in demand—not minding about not being important, not wanting to do things on your own, enjoying fixing things for other people and attending to detail—such a girl might be very unhappy in an editor's job which means being on your toes all the time and keeping a step ahead of the rival ladies who edit other magazines.

In any event, editors were not apt to recognize the latent talent of their ambitious secretaries, and help them break into editing: "Editors seldom subscribe to the opinion that the grass grows greenest underfoot." Especially difficult, Stevie observed, is the position of "the girl who has a double job—say a bread-and-butter office job at not too high pressure and a talent for one of the arts that is not sufficiently popular or remunerative to pay her a living wage."

Stevie was in just such an especially difficult position. She once prefaced "I rode with my darling . . .", the third poem in *Harold's Leap* which she specifically cited as among those she preferred, by reminiscing about her life as a secretary. She knew she would not have liked either "the sunny life" of her employers, with its "excitement of the Stock Exchange and the pursuit of gold", or of editors, although she often "cried out" for such a life. "I like circumstances that are proud and eerie", she said, "I like to be hungry". In "I rode with my darling . . .", Stevie said, "a girl rides by, looking wild into a wood. She had meat to fill her, a lover and loving relations, but she left them, she rode away, she chose to be hungry." The poem ends:

> Loved I once my darling? I love him not now.
> Had I a mother beloved? She lies far away.
> A sister, a loving heart? My aunt a noble lady?
> All all is silent in the dark wood at night.

As Stevie noted after reading this poem, "This feeling of wanting to get away to the woods, paths, ponds, lakes and seas of an earlier numinous world is very dominant in the poems in spite of the love of friendship, and in my novels too, and often it may be understood as a soft sighing after shadowy death."

Stevie, of course, was famous among her friends for her own

excursions into woodlands and parks. In fact, the novelist Nigel
Dennis recalls that his tie with her was based exclusively on woods and
forests. He used to carry about *Novel on Yellow Paper* for the sake of its
page or two which described the woodlands round his old home in
Hertfordshire. When he resumed living there in 1950, "the pages
continued into real life":

> Stevie liked to travel out on one line from North London, walk
> through the woods, and catch her train home on another line: my
> cottage lay midway between the two stations, and she and a friend
> would break their walk for tea with me.
>
> I have not the slightest recollection of what was said over tea, or
> of what subjects we discussed. I only carry about in my mind—in
> much the way I carried her novel—the picture of her talking beside
> my log fire and disappearing afterwards into the dark. I know she
> loved the woods dearly and that they were an important part of her
> life, as they were of mine, but how she lived her life generally I never
> could tell—only a hint of an aunt or a mother ever reached me,
> somebody waiting at home.

Clearly Stevie's woodland excursions, unlike those of her imaginary characters, were often social. But this does not contradict her claim
that the characters represent something of her inner life. The theme of
escape in "I rode with my darling . . ." is by far the dominant theme
in *Harold's Leap*. (Stevie may indeed have chosen her title poem in an
effort to restore some balance between the volume's admonitory and
escapist themes.) In "Voices about the Princess Anemone", for example, one encounters another character who runs into a forest. Voices
say "She feared too much", but Anemone sees her fear as "a band of
gold". Again and again one encounters the theme of escape: the
subject of "In Protocreation" is mental escape to a time before humans
walked the earth, while in "Friskers" the speaker does not seem to
mind escaping her humanity itself, when the gods turn her into a cat.
In "Cool as a Cucumber" Mary, who is "unfit for marriage", runs off,
probably never to be seen again, and in "Lightly Bound" a wife and
mother threatens to run off. For the speaker of "The Hat", however,
marriage itself constitutes the escapist fantasy. Stevie prefaced the
poem by saying, "marriage is not always a solution of problems,
though this girl, fancy-fed in a dream of rich hats, seems to think it is."
Elsewhere she noted, "There are a great many hats in my poems. They
represent going away and also running away."

One of the most interesting of these poems, from a biographical
point of view, is "Deeply Morbid", about a girl who types letters in an
office and runs "with her social betters". Her colleagues come "to

Stevie's drawing of a girl being carried off by a hat,
which she placed by her poem "My Hat"

doubt her" because of a "look within her eye" which always seems "to say goodbye". One day, as she gazed upon a Turner picture in the National Gallery, the hold of her friends and colleagues and work grows weaker, and she magically enters the picture's happy light. The poem ends:

> They say she was a morbid girl, no doubt of it
> And what befell her clearly grew out of it
> But I say she's a lucky one
> To walk for ever in that sun
> And as I bless sweet Turner's name
> I wish that I could do the same.

Other poems in this vein, but where the escape explicitly is death, are "Do Take Muriel Out", in which we are told that Muriel will not complain when Death dances her "over the blasted heath"; "The Crown of Bays", in which a writer gladly receives "annihilation"; "I Am" and "A Shooting Incident", in which the subjects deliberately shoot themselves; and "Mr Over", in which a character envies Mr Over's noble death and, as Stevie said, "toys with the thought of suicide". So attractive is death in *Harold's Leap* that the promise of it in "From the Coptic" is the only thing which persuades red clay to become Man. The speaker of "The Wanderer" says the ghost who wishes to return to life is happier where she is, and the message of the ghost in "Le Revenant" is that "It is much better to be dead". Although the people who hear "Le Revenant" stone him from their doors, obviously others in *Harold's Leap* find his message compelling. Indeed, in "God and Man" God must plead:

Oh Man, Man, of all my animals dearest,
Do not come till I call, though thou weariest first.

Of course, not all the poems in *Harold's Leap* are either admonitory or escapist. There are, for example, several satirical poems, and poems which capture what Stevie called the "ticklish comic element in human suffering"—such as "Pad, Pad", "Drugs Made Pauline Vague", and "The Broken Friendship". The last-named may have come, as did "A Mother's Hearse", out of the strains Stevie experienced from putting friends into her writing: in the poem "Full Well I Know", two voices discuss the issue of an author using in her writing the people she knows.

When asked in June 1951 if she could recommend any book or article about her work, Stevie cited the commentary she herself had written to link her poems in a BBC Third Programme in February. In October she broadcast another such reading, and then again in July of the following year. Besides expanding the audience for her work, these readings were lucrative, because Stevie had shrewdly arranged for the reversion of copyright on her books published by Jonathan Cape.★

As early as January 1948 she had written to Cape, saying that a clause in her contract with them seemed to indicate that the copyright on her first two novels and books of poems had reverted to her. "I am sure something must have happened about this clause", she said, "or otherwise we should not have had so much correspondence about reproduction fees, and your share of them. What is the position?" The firm tried to discourage her from claiming copyright, but the following month Stevie wrote a formal letter asking Cape to reprint her books within three months. According to her contract, if the books were not reprinted by three months after her request, the rights would revert. Jonathan Cape replied that the demand for her books was insufficient to justify reprinting, and he explained: "so long as the books are with us we shall not lose sight of them, and we shall be on the look out to do any business in connection with them."

Stevie was temporarily mollified, and thanked Cape for the trouble he had taken, and evidently still was taking, on behalf of her work. She

---

★For example, Stevie's fee for preparing and reading her material on the October 1951 broadcast was only ten guineas, but her copyright fee came to over £60. And when her readings were rebroadcast, she received additional payments. As Stevie said to a friend when she heard that her October broadcast would be repeated: "Financially this is charming as they are extremely generous payers, I will say that for the Third, but of course it was all copyright material, and the copyright mine, so that largely accounts for the size of the cheque."

would not mention reversion of copyright again, she said, unless she ran into a publisher who was "recklessly determined" to have her books. But two years later she again inquired about her copyright, and by the summer of 1950 Jonathan Cape had released the rights to her. He thanked her for her pleasant farewell letter. "I would like to call it a happy letter", he said, "but that would not be so. It is happily phrased, however."

Soon after, Stevie was in touch with Anna Kallin who had coached her in 1949 for the reading of "Sunday at Home" on the BBC. At that time Kallin invited Stevie to contact her if she ever wished to do another broadcast. By November 1950 they had discussed a poetry script written by Stevie, and several weeks later Stevie submitted a revision in which, she said, she "changed the order round a bit and put in a few short happier [poems] as I thought when I was reading the other one over it was almost unrelieved gloom." Perhaps Anna Kallin was not enthusiastic about the discussion of the drawings that went with the poems, for Stevie noted that in the revision she had omitted references to the drawings with the exception of the last poem, "and by that time any listeners who are left will probably be past caring."

Stevie worked hard on the three "Poems and Drawings" scripts produced by Anna Kallin for the Third Programme. She told Kallin that her second script had required a lot of thought and time, and, submitting the third script, she apologized for having taken so long, adding "it is not easy my word it is not". Essentially Stevie saw these scripts as "readings *à thèse*," and the thesis in each case was a view of humans as highly nervous creatures, recently emerged from the primeval cave or Garden of Eden, and not at home as they travel through the world. "Hazards and temptations lend excitement to the journey," she wrote at the beginning of her third script, "arguments, witchcraft, pride, nobility, despair and religion are their solace and affliction." Her poems and linking commentary illustrated the thesis, and because each script employed different poems, each required a new commentary. Attention had to be paid, also, to timing and emotional proportion. As Stevie said when submitting the second script:

it's like mixing a cake, very tricky, trying to get the right balance, not too tearful, not too bright, *you* know. I have stuck in a longish introduction. . . . I think it is right, I mean tells the truth and sets the compass (cakes and compasses, crumbs!) but if you think the lone listener to Third Poetry broadcasts can do without Guidance, by all means cut it. It is on the whole more carefully sectionised than the first one (I think) so cutting for timing, if necessary, should not be difficult.

After Stevie's "Poems and Drawings" broadcast in October 1951, she received a letter from her old Cambridge friend, Rachel Marshall. Mrs Marshall, a specialist in voice production, discussed the way Stevie's poetry and commentary had slid into each other. Stevie wrote back that she and Anna Kallin had

> batted away at it and it is *the snag* with this sort of broadcast, especially for a non-professional, in fact I doubt if even by taking thought I could have set myself a more technically difficult task; I listened to the broadcast fairly objectively—this was made easier by the fact that my voice was to me quite unrecognisable—and on the whole I thought there was a sufficient difference. As you say the comment was conversational but my word the poems were practically incanted, though I agree a bit on the fast side perhaps. . . . I had one card from a lady who said, was good enough to say ha ha, she wished I would come and read to her all night (Woking district) otherwise the comments went through to my producer, a powerful lady called Anna Kallin, Third producer and Planner (crumbs!).

Some insight into why Stevie exclaimed "crumbs!" after mentioning Anna Kallin may be gleaned from a letter she wrote to Kallin after the second broadcast, in which she called her "an angel though I think a severe one", and mentioned her own nerves and embarrassment. (Thanking a friend for a fan-card after her third broadcast, Stevie noted that she always found them "nerve-racking".) It seems some tension had developed between Stevie and Anna Kallin. In September 1952, a few months after Stevie's third broadcast, she wrote to Kallin about a possible fourth. In the margin of the letter, next to Stevie's comment "I am so sorry you are having such a sad time being so ill", Anna Kallin jotted a note that this was a typical lie, and next to Stevie's inquiry about getting together, "you cannot manage the evenings I understand?", Kallin wrote that she could, but not with Stevie. And she jotted her agreement with Stevie's remark, "all this is a long rigmarole and you may already have had enough of the poems."

In the past, the composers Elisabeth Lutyens and Stanley Bate had done musical settings for Stevie's poems, and Anna Kallin may have felt pressed by Stevie's saying she spoke to them about settings for her proposed fourth poetry broadcast. In December Stevie wrote again to Kallin about the proposed broadcast: "You say I have had a lot of priority but isn't that thinking too much of the person instead of the thing?" Anna Kallin sent the letter to her superior at Broadcasting

House, appending a note that she had put Stevie off. She questioned
whether the fourth poetry reading should be done, adding that Stevie
was good but uncompromising, and took quite a long time to do the
programmes.

In May 1953, Stevie wrote to thank Kay Dick for sending a card:
"You really are a comfort, a proper propper-up-er of wilting egos",
she said, "I am *most appreciative*." Getting on to the subject of the BBC,
she explained:

> I did ring Anna Kallin up, and was told, Yes, I might have a stab at
> another poem programme, provided I didn't "commit" them to
> anything. This wild enthusiasm has resulted in my doing nothing, ab-
> solutely nothing, so far—but I will, oh yes, one doesn't put that baby
> on a cold slab, but all the same I should say at least three-quarters
> of me doesn't want to, yes, money or no money, and be hanged.

"It is all very depressing and frustrating", she concluded.

In the same letter Stevie mentioned other disappointing news, that
American editors at the *New Yorker* and elsewhere were rejecting her
poems: "they always say the same thing, they are not quite the right
mixture and the slant etc. and especially the allusions are far too
English, so there it is, and I suppose too they are right and they are!"
But even in England it was a bad time for her poetry. In February she
had learned from Chapman & Hall the "horrifying figures", as her
editor referred to them, for *Harold's Leap*: 231 copies sold in 1950, 112
copies in 1951, and 16 copies in 1952. Much as he would like to publish
another volume, the editor said, he was unable to do so.

Nor was Stevie's disappointment confined to English book publication. She was used to rejections as well as acceptances when
attempting to place her poems in newspapers and magazines, but in
the early 'fifties she had begun to feel locked out. To a friend she
admitted feeling "languid & triste" because *The Listener* returned
some poems: "There's no two ways about it," Stevie said, "the
Mandarins do not like my poems & I am beginning to think they must
be right." Her exasperation is also evident in a letter she wrote to
Naomi Mitchison:

> I now address myself to all Old Pards, including you, love, as the
> Great Unaccepted. Truly I need a shover, a nice honey-tongued
> worm, to belly around for me, some pretty young man, eh? with a
> "theory"! Well your old battle-axe on the *New Statesman* won't have
> me, nor John Lehmann, nor Spender, nor Iain Somebody on the
> *Spectator*, nor Ackerley on the *Listener*. Only *Punch* will sometimes
> if they're funny. Love, love!

"It is shockingly sad really", Stevie wrote to the poet D. J. Enright in the spring of 1953, "how good poetry is when as far as the public goes it might be one of those branch lines scheduled for closing."

There is something funny about this metaphor of poetry as a "branch line scheduled for closing", just as there is something funny about Stevie's metaphor, in reference to her BBC readings, of not putting "that baby on a cold slab". Typical of her humour, such remarks arise out of sadness but transform it by wit. As Stevie once said, albeit slightly exaggerating, she only listened to her Muse when she was unhappy. Appropriately, therefore, she wrote her most famous poem in April 1953, at a time when she felt "too low for words". Writing the poem was a way of working off despondency. "*Punch* like it, think it's funny I suppose", she said, "it was touching, I thought, called 'Not Waving but Drowning'."

> Nobody heard him, the dead man,
> But still he lay moaning:
> I was much further out than you thought
> And not waving but drowning.
>
> Poor chap, he always loved larking
> And now he's dead
> It must have been too cold for him his heart gave way,
> They said.
>
> Oh, no no no, it was too cold always
> (Still the dead one lay moaning)
> I was much too far out all my life
> And not waving but drowning.

Of course, the humour of this poem is black humour, which *Punch* probably sensed. It is funny, in a macabre way, that the one occasion on which "They" might have recognized the speaker's desperation only confirms their view that he was a merry fellow, larking to the end. The dead speaker's incongruous liveliness is also funny, what with his explanations, his moaning, and his cries of "no no no".

Over a decade after its first publication (in *The Observer*, by the way, not *Punch*), Stevie said she was inspired to write "Not Waving but Drowning" by a newspaper story about a drowning man whose friends thought he was waving to them.* "I thought that in a way it is

---

* Ironically, Stevie may have been misremembering a newspaper story headlined "Needless Dash To Bather On Rubber Float", about a man who was thought to be in trouble when he was only waving to a friend on shore. The undated cutting was found among her papers after her death.

Stevie at home in Palmers Green
Photograph by John Goldblatt, © Sunday Telegraph Magazine

The Queen's Gold Medal for Poetry
Photograph by Charles Noble, reproduced by gracious permission of Her Majesty the Queen.

Ethel Spear

Stevie's maternal grandfather, John Spear
Both McFarlin Library, Tulsa

Stevie is the third sitting
figure from the right
*Photograph courtesy of
Mrs Olive Pain*

Inscribed photograph of Sydney Basil
Sheckell
McFarlin Library, Tulsa

L. to r: Kathleen Dykes, Stevie Smith,
Flora Robson, and Margery Haywood
in *Ali Baba*, 1917 or 1918
Photograph courtesy of Mrs Olive Pain

Stevie's mother, Ethel Smith, summer 1918

Stevie's father, Charles Ward Smith, 1919
Both McFarlin Library, Tulsa

Stevie's "Auntie Granma", Martha Hearn
Clode, 1918
McFarlin Library, Tulsa

Stevie at the beach, August 1919
McFarlin Library, Tulsa

Molly Ward-Smith

Stevie in the 1960s by the lake in Grovelands Park, Palmers Green
Photograph by Jane Bown, Observer

Stevie and her Lion Aunt

arl Eckinger (centre, wearing glasses) in
wiss Army officer's uniform.
Photograph probably taken during
World War II)
*Häusermann Collection, Séminaire d'Anglais,
Université de Neuchâtel, Switzerland*

Stevie in the Marshalls' garden (4 Clare Road, Cambridge), summer 1938
Photograph courtesy of Geoffrey and Kitty Hermges

Top left: Undated photograph of the Reverend Roland d'Arcy Preston, vicar of St John's church, Palmers Green

Top right: Professor Hans Walter Häusermann
Photograph courtesy of Mrs Häusermann-Häusermann

Undated photograph of Stevie (*l.*) and Anna Browne walking in London

Stevie reading at the Theatre Royal, Stratford East, 1965
Photograph by Ken Coton

Bottom left: André Deutsch and Stevie Smith at a party to celebrate the publication
of Maria Browne's *Whom the Gods Love* (copy by the candelabra)
Photograph courtesy of Maria Browne
Bottom right: Stevie photographed by Robin Adler (early 1950s)

Stroud Festival, 19 October 1970. *L. to r.:* Mrs Marjorie Evans (chairman of the Stroud Poetry Committee), Lady Moyne, Stevie Smith, Morris Broadbent, and Lord Moyne (Bryan Guinness)
Photograph courtesy of Lord Moyne

Stevie Smith in the last year of her life
Photograph by N. Vogel (National Portrait Gallery, London)

true of life too", she noted. Many people "do not feel at all at home in the world" so they "joke a lot and laugh and people think they are quite alright [sic] and jolly nice too, but sometimes the brave pretence breaks down and then, like the poor man in this poem, they are lost."

Besides discouragements about her poetry, other matters were testing Stevie's bravery in 1953. She was suffering from a painful knee condition which required regular sun lamp treatments in hospital, and she was engaged in a dispute with the Inland Revenue. Fortunately she could report to a friend in May that she had "just won unaided a battle against Mr Bloodsucker Income Tax Inspector who has at last been brought to admit that he has overcharged me something shocking on last year's so called literary earnings." But, in all, the strain of this dispute, and of getting her poems published and broadcast, as well as the pressures of her office work and book reviewing, were taking their toll.

The book reviews, especially, remained a heavy commitment. Although Stevie's final review for the *Tribune* appeared in late November of 1949, she began in 1950 (and continued until the spring of 1953) to review about four novels a month for *World Review*. And she continued reviewing for *Modern Woman* until 1951, and for *John O'London's* until early 1954. Of course, the books she had to read were not always congenial. It is no wonder that, in a capsule review of twenty-three nursery books in the December 1952 *Britain Today*, Stevie exclaimed of a few, "what a treat it is to get these books that grown-ups can read without yawning".

If Stevie's duties at Newnes, Pearson had been arduous, doubtless she would not have been able to do so much reviewing. Yet, however unimposing her secretarial chores, the office routine was tedious and confining. Like most wage-earners, Stevie was not free to take trips whenever she wished. Invited by her friends, the Fowlers, to visit them in Austria in September 1952, for example, she replied she would like to do so "if my dear baronets can loose me round about then, the difficulty being that our Annual Meeting ha ha when we display how rich we are, takes place probably right at the beginning of August or the beginning of September and after that fuss naturally the baronets rather like a few weeks off. Never am I more wanted than when they are 'off' I might almost say only then am I wanted." The following year Stevie had to cancel plans to go to a PEN Conference in Dublin because the baronets decided they wanted her in London during the week in June for which the conference was scheduled.

"Private secretaries are servants of a rather low order", Henry Adams wrote, and after Stevie had been one for several years she copied his view into one of the reading notebooks she kept in the 1920s. Her job had been "an awful strain", she said in 1970:

> I always said to my unfortunate employer, "Well, I'm not really here." Now this is a very profound remark to make because the accent's on the "really". You see, I must have felt all the time, "My real me is not really here." It's a very neurotic thing because actually of course my real me was. I got frightfully tired. I didn't want to be there. As I couldn't actually escape I pretended I had escaped.

In her office on 1 July 1953, Stevie attempted the ultimate escape by slashing her wrists. Ironically, she who had celebrated (and was to continue celebrating) Death's having to come when called, was contrite about her one attempt to summon Him. Stevie wrote to Anna Browne explaining she just couldn't stand the strain of work any more, but expressing deep contrition for what she had done and for the terrible distress it had caused Aunt. She asked Mrs Browne to destroy the letter.

Several years later Anna Browne and Stevie saw the film of Mary McCarthy's *The Group*, and Stevie had to leave, not because it was a bad movie, but because she found it depressing that "all those women had to go to work every day". To another friend Stevie wrote of her enduring distaste for office work: "It's that old long long office job of mine (at Newnes's) that makes me afraid of *jobs*. I sometimes dream I'm back there & last time I did poor ghastly old 'Sir Phoebus' (not entirely the loving chap one made out!!) said to me 'You're back where you were.'" In another letter she told this friend she had been Rolandine, an allusion to her poem "Childe Rolandine", with its complaint about work that is tedious and its protest against the privileged rich:

> Dark was the day for Childe Rolandine the artist
> When she went to work as a secretary–typist
> And as she worked she sang this song
> Against oppression and the rule of wrong:
>
> It is the privilege of the rich
> To waste the time of the poor.

Rolandine offers her tears to the "spirit from heaven" who, she defiantly claims, grows fat on "Mighty human feelings".

Eleven days after her attempted suicide, Stevie wrote to Kay Dick from bed where she had been, more or less, since her last day at work. She reported that her doctor advised her to stay at home for at least a month: "I am a Nervous Wreck, it appears, also anaemic. Hurra! . . . It is heaven having such a long time before I need think of that awful office life again, sometimes doctors are *the thing*." She went on to say she had just begun reviewing for the *Spectator*, "so here I am, the old hack again". She also had written many new poems, but was discouraged because *Punch* seemed to have frozen up on her. "Do write a nice long letter to your poor unbalanced friend."

Stevie's drawing of a typist wearing a large hat,
which she placed by her poem "Childe Rolandine"

In early August Stevie's physician reported to Sir Neville Pearson that she obviously had been under considerable nervous strain for some time, and the condition in which he found her on 1 July was extremely serious. In the past he frequently had treated her for "general debility and anaemia", and she had been able to return to work. However, he no longer considered Stevie to have the nervous or physical strength to continue regular office work. As a result of this letter, Stevie was pensioned off. "I . . . am so glad", she wrote to a friend, "the doctor said 'No more' heaven bless him." And so Stevie's failed suicide succeeded in freeing her, if not from her mortal coil, at least from the coil of Tower House. Perhaps that is what she most wanted. Her unusual choice of the office as a place to attempt suicide suggests, if further evidence is needed, that unhappiness with her job was a major factor in her decision to end her life. Stevie went on holiday for a fortnight with her sister, and soon heard from Sir Neville about her pension. She would continue on full salary until 1 December, then she would receive £5 10s. a week until the age of sixty, and thereafter £7 12s. 6d. a week. This would enable her, he said, to be "reasonably comfortable". But perhaps he suspected that £5 10s. a week would not be sufficient, for he added that Stevie would of course be free to earn anything else she could.

For Stevie that meant reviewing. "If I could afford to cut it out", she told Kay Dick in the autumn of 1953, "I think that is what I would do about reviewing." She reviewed thirty-four books for the *Spectator* before stopping in January 1954, and the next month she began bi-weekly reviews for *The Observer*. One month she said the problem with reviewing is the difficulty in singling out any special novel when there are so many good ones, and the next month she spoke of the repetitive thoughts in most novels, joking that they should all have the same title, "New Ways with Left-Overs". But the real problem for Stevie was the constant pressure of having to write "a piece": as she exclaimed after noting that a particular book looked "O.K."—"oh if only that was *all* one need say about them!"

Reviewing had its bright side, though, such as the opportunity it gave Stevie to praise in print books by friends about which she was enthusiastic. When Stevie read Inez Holden's novel *The Owner*, for example, she told Helen Fowler she thought it "absolutely first rate":

> It is what had to be written and I never thought would be after all the Brideshead and Nancy Mitford books about the grand life, where they are always so much hankering after it and their "fun" so deeply servile. Inez shows the corruptive influence on a middleclass climber in a grand house, it is very witty and poetical and profound too.

Her column in *World Review* enabled Stevie to share this enthusiasm with a wide audience. And to one who in 1953 had been called a good neglected poet (by the reviewer of a poetry anthology), it must have been satisfying to be able to slip lines from her poems, as she did on several occasions in the 1950s, into her reviews of books by others. Another advantage, suggested by the record, is that publications for which Stevie reviewed were likely to accept her poems on occasion. Also, as Stevie said, many poems came to her from the books she was sent for review.

The reviews also gave Stevie a chance to air her opinions on topics of the day. A novel about the futility of war, for example, caused her to reflect that "Perhaps England with her insular fighting history is a difficult country from which to launch an argument for pacifism." "I would not call Miss West a feminist", she wrote in another review, "because this suggests—and is meant to—an aggrieved and strident person. I would say, she is on the side of the women." And Stevie's mixed review of *The Second Sex* opens provocatively: "Miss De Beauvoir has written an enormous book about women and it is soon clear that she does not like them and does not like being a woman." Stevie's reviews are not without aphorisms: "Those who eschew feelings have the great writers against them", and "Satirists must never be naïve." And they occasionally provide revealing asides. Less than a year after her attempted suicide, for instance, she said of a novelist that he "gives a perfect impression of the courage and squalor of the period [the 18th century] and of the nerve-racking lives of men and women who yet did not then seem to have nervous breakdowns."

Far from nervous was Stevie's mood on a soft hazy Whitsun, 1954. She had been staying with the Brownes at their home on the Norfolk coast, which she visited at least once a year throughout the 'fifties and 'sixties. Although she usually relished going down to the sea, on this particular day all Stevie wanted was to stay in the garden, which was filled with cow-parsley. The chestnut tree was in flower and had a wood pigeon in it, cooing softly. A neighbour's two setters, called Honey and Red because of their colour, bounded happily in the long wet grass at the side of the house. It was on this occasion that Stevie wrote her poem "The Old Sweet Dove of Wiveton".

According to Anna Browne, when the poem was written Stevie was very happy, and feeling reconciled to life. The poem radiates a sense of peaceful acceptance, and things coming out well in the end (it is not by accident that both the dove and dogs are described as being old). Stevie was fond of dogs, especially when they were large and noble. In "The Old Sweet Dove", they convey a sense of happiness in action. So still is the day, and so open to it is the poet, that she can "hear the splash / Of the water falling from the green grass / As Red and Honey push by." The poem ends with a description of the dove

> Murmuring solitary
> Crying for pain,
> Crying most melancholy
> Again and again.

The poet celebrates both the happy commotion of the dogs and the peaceful silence which is heightened by the dove's melancholy murmur. The pain which gives rise to the dove's sweet cry of "Love, love" is part of the whole, part of what is accepted.

Poems and Drawings

When Chapman & Hall decided early in 1953 that they could not publish another volume of Stevie Smith's poetry, she embarked on a prolonged and exasperating search for a new publisher. She first approached Chatto & Windus, who were willing to publish the poems without Stevie's drawings, which they thought too comical. "But then so often are the poems", Stevie told a friend, "& the drawings *I* think are not *only* comical." She next tried Duckworth & Company, and was told that, although her drawings helped to realize the elusive madness of her poems, "what is really needed is your presence and voice to recite them." Such news was discouraging, and Stevie was grateful when her friend L. P. Hartley asked to see some of the poems. She had reviewed Hartley's novel *Eustace and Hilda* several years earlier, and what she wrote gave him "the keenest pleasure". They later met and kept up a correspondence. Early in January 1955, Stevie told him:

> I would love you to see some of my poems, & it is more than nice it is positively reckless of you to suggest it. I should think I now have about seventy—all with several drawings *each*. I cannot get any publisher to do this new book of them. Ian Parsons (Chatto) says, Yes, he will do poems without drawings, & Mervyn Horder (Duckworth) says he likes the drawings best. . . I am now going to try André Deutsch, so please keep your fingers crossed for me.

Stevie was all too aware that current fashions in poetry did not favour her kind of writing. She had just been reading a volume of Edgar Allan Poe's poetry, she told L. P. Hartley, and thought that were Poe "writing today . . . he would not get published". Editors at the *New Statesman* and the *Listener* would not be enthusiastic about his "brittle laughter that covers so much *fear* of not being on the right bus". It is necessary to be "*à la* nowadays in poetry", she concluded, "*à la* Eliot, Spender etc. & soon I fear *à la* Dylan!" (There seem to be no references to Stevie in the writings of these poets and, indeed, Dylan Thomas once complained to Elisabeth Lutyens that it was "tasteless" for her to have scheduled Stevie to read her poetry at a fund-raising affair.) Poe may have been on her mind when she wrote, at about this time, "Anger's Freeing Power", a poem featuring a raven, although the theme is not that of Poe's "The Raven". In Stevie's poem, the bird beats itself against the walls of an open cell, and sobs and sighs, making "a prison of a place that is not one at all". It is not the speaker's loving words which rouse the bird to fly free, but the power of its anger when two fellow birds jeer. (Perhaps this "dream" touched upon some truth about Stevie's past situation at Newnes, Pearson.)

"Anger's Freeing Power" was among the batch of about fifty poems Stevie submitted to David Wright, saying no one would publish them. Wright was, in the mid-fifties, an associate editor of the literary magazine *Nimbus*, and in that capacity he had written asking Stevie for a few poems. He had long admired her work and was amazed and infuriated to learn she was "actually finding it impossible to have [her poetry] published in the magazines that were current at the time". His acceptance of about a dozen of Stevie's poems, and his request for drawings to go with them, must have cheered her up. And she must have been encouraged, too, by L. P. Hartley's response to the poems she had sent to him: "They all move me, and many make me laugh. I can't think of any other poet whose tears are so near to laughter—& yet the two are perfectly distinct . . . I do hope that they will be published. It seems so cruel they shouldn't be, (*and* with your drawings)."

Unfortunately for Stevie, the enthusiasm of Hartley and Wright for her drawings was not shared by the firm of André Deutsch, which agreed to publish her fifth volume of poetry, but which, like Chatto & Windus, did not want to include the drawings. She was perhaps a bit desperate when, in May of 1955, she wrote to "Dear Miss Athill" at Deutsch, saying she didn't mind if the poems were published without the drawings. Her change of heart, she said, was "brought about by having another look at them in course of sorting them out for *Nimbus*—I don't think they're as good as all that no I don't."

L. P. Hartley wrote to tell Stevie he was "overjoyed" that Deutsch had accepted her poems, although he was sorry about the drawings: "They have a Lear-like quality which some people might automatically think 'funny'—but then the poems are 'funny' in a funny way, at least a good many of them are, aren't they?—& in just the way your drawings are. It seems a great pity to leave them out, for the essence of the poems is so often in them." David Wright was also disappointed by Deutsch's reluctance to print the drawings, and he sent Stevie pulls of the ones he was publishing in *Nimbus*, suggesting she show them to Deutsch. Stevie did so in November, telling Diana Athill: "oh dear oh dear, how I do wish some drawings could be put in the book. I suppose there is no hope of this—I would gladly give you back the £50 advance & have *no* advance but only royalties if you would only do the drawings."

Not getting a reply to this letter, Stevie wrote again to Diana Athill in mid-December and, in a daring move, sent back her advance asking to be released from her contract: "I absolutely am riveted to them", she said of the drawings. The story has a happy ending. Stevie must have approached next the firm of Victor Gollancz, for on 21 December Hilary Rubinstein, a Gollancz director, asked her for the two volumes of her poems which she "brought into VG the other day": the volumes would help the firm, Rubinstein said, "in working out the rather complicated production problems in the case of your new book." But before matters could go very far with Gollancz, Diana Athill returned Stevie's advance and told her Deutsch would publish her drawings with the poems.

The question arises as to why Stevie, having difficulty getting a publisher and aware that her poetry was neither profitable nor fashionable, was so "riveted" to her drawings that she was willing to give up her contract with Deutsch. Certainly she had no delusions about being a great visual artist. Although she always enjoyed drawing, she was wry about her ability, once telling an interviewer: "I am not a trained drawer, you know. It's rather more like the higher doodling, or perhaps just doodling without the higher." To a friend she wrote of her weakness in depicting hands and feet (see, for example, the "Childe Rolandine" drawing on page 189), and joked that her drawings always reminded her of Milton's line: "With dreadful faces thronged and fiery arms."

Stevie's strong desire to have her drawings published in her fifth volume of poetry, as they had been in all of her earlier volumes, cannot be explained by the fact that the drawings inspired her poems or vice-versa. Although on occasion she would do a drawing for a specific poem or write a poem inspired by a drawing, that was not her usual practice. She doodled on the backs of envelopes, memo pads and other scraps of paper, and if she liked her doodles she saved them in a box. When she collected enough poems to make a book, she would search through the box and try to find drawings to go with her poems.

"I take a drawing", she said, "which I think 'illustrates' the spirit or the idea in the poem rather than any incidents in it."

Stevie's reason for wanting her drawings published with her poems was that she considered the drawings to be very much "part of" the poems: at least that is what she told Diana Athill, and it is what she had said when Chatto & Windus wanted to publish her poems without the drawings. Indeed, she had come to think of some of her drawings and poems as units to such an extent that, during poetry readings, she would describe the drawing as an introduction to the poem. But although Stevie's claim that her drawings were part of her poems seems simple, it requires examination, because the way in which it is true is complicated.

Perhaps the rarest combinations are those in which Stevie's poems seem to be completed by the drawings she placed with them. Consider "Torquemada":

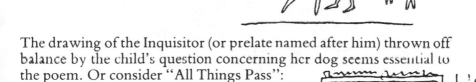

Uncle Torquemada,
does Beppo know about Jesus?

The drawing of the Inquisitor (or prelate named after him) thrown off balance by the child's question concerning her dog seems essential to the poem. Or consider "All Things Pass":

All things pass
Love and mankind is grass.

As one critic has noted, the drawing, with its apparently middle-aged couple and frilly curtains, provides a mundane context for the poem's biblical (Isa. 40:6) portentousness, "a reminder of the banality of truth, of the truthfulness of the banal".

Easier to find than cases in which Stevie's poems seem dependent on her drawings are those in which a poem and drawing mirror each other in some way. Consider the macaronic poem "Ceux qui luttent . . .":

> Ceux qui luttent ce sont ceux qui vivent.
> And down here they luttent a very great deal indeed.
> But if life be the desideratum, why grieve, ils vivent.

Wordplay is primary: the first line's weighty proverb "Those who struggle are those who live", which sounds all the more grand for being in French, is deflated by the second line's matter-of-fact English. The Latinate "desideratum" is also punctured by the flip "why grieve, ils vivent." There is the fun, too, of rhyming across languages, and of a bilingual pun.

Stevie described the drawing she placed with "Ceux qui luttent . . ." as depicting "a Saturday afternoon in a poor quarter", with people gazing out of lodging-house windows above busy shops, and a child below observing "a butcher with a large knife standing in front of his frozen carcases". Upon closer examination one notices that another child, a passerby, and two of the lodgers seem to be looking in the same direction as the butcher, who has a wild look in his eye, suggesting that one of those who "loot" has just run from his shop. But this sounds more grim than it looks: life's struggle consists of shopping and gazing out of windows, as well as of robbery; curls of smoke rise gaily from chimneys; the butcher's appearance is funny; his carcases are decorative; and those glancing in the direction of the implied thief, with the exception of the butcher, seem amused. The poem's proverb and the incident in the drawing are grim, but the wordplay and the visual depiction are jokey. Thus the poem's juxtaposition of content and style has a parallel in the drawing.

The poem "Death Bereaves our Common Mother / Nature Grieves for my Dead Brother" presents another case of Stevie's drawing imitating her poem's incongruities:

Lamb dead, dead lamb,
He was, I am,
Separation by a tense
Baulks my eyes' indifference.
Can I see the lately dead
And not bend a sympathetic head?
Can I see lamb dead as mutton
And not care a solitary button?

The contrast between the poem's mournful content and its comic style (the overblown title, the rhyme of "mutton" with "button" reminiscent of Ogden Nash, etc.) has its analogue in the contrast between the drawing's mournful scene and its comic rendering (the juxtaposition of the woman's soft curve and the lamb's rigor mortis).

The amusing drawing attached to Stevie's poem "Conviction (iv)" presents another kind of mirroring. The drawing is mysterious. Are the lovers in a park with a watchdog on guard, or are they in a jungle unaware of a hungry lion? Stevie once published this drawing without the poem, placing beneath it the caption: "This does not break down any barriers." Presumably the comment refers to a dichotomy between human spiritual and animal natures, in which case the beast could be an emblem of lust.

The poem (see page 139) raises its own questions. Its first line, "I like to get off with people", makes one wonder whether or not the speaker is innocent of "get off" as possible slang for orgasm. (In one of her novels Stevie has a character use the expression in precisely this way.) But does not the poem anyway have an innocence about it,

naïvely celebrating the joy of sex? Perhaps, rather, it is mock-innocent, for its last line raises the possibility that the speaker's desire is too good to be true: "There is no bliss like this" can mean "such bliss is beyond compare" or "such bliss does not exist (except in imagination)". "Conviction (iv)" lends support to Christopher Ricks' claim that Stevie Smith's writing "depends upon its being always in question whether her innocence *is* mock-innocence or not." And the support is strengthened when the poem and drawing are placed together, because a *possible* irony emerges: if the beast in the drawing is a watchdog, the couple may well be "Safe from all alarms", but if the beast is a hungry lion, the speaker's blissful declarations are about to be rudely interrupted. Of course, if the beast is an emblem of lust then the speaker's declarations are twitted. But what is it, and did Stevie even intend the mystery? There remain the poem and drawing to tease, the ambiguity of the one a witty parallel to the mysteriousness of the other.

The drawings described here as mirroring Stevie's poems do not alter one's reading experience, but reinforce it in another way, adding a "thickness" to the text. But although there are other drawings which could be cited as contributing in one way or another to the experience of reading Stevie's poems, it is clear that in the overwhelming majority of cases the drawings are mainly decorative. This is not to say, however, that the drawings are irrelevant, for decoration has implications. Often, next to poems about rejection, loss and despair, Stevie's childlike drawings playfully wink at the reader as if to say, "it is legitimate to express anger or sorrow, but then one must laugh", or, "this is a volume of poetry, but one need not be 'serious' about it".

Stevie's decorating her text was subversive, not that other poets hadn't done so, other "eccentrics" such as Blake and Lear. But doodles are personal, and T. S. Eliot had made popular the notion that good poetry is impersonal. It would have been as unlikely for him to decorate his serious poems with doodles as it would have been for him to sing them. During the heyday of New Criticism, poems were supposed to be well-wrought objects. Such a metaphor minimizes the relevance of things external to the text. If Stevie's poetry is more fashionable today than it was in the 'fifties, so too is the metaphor of a poem as a voice. In terms of such a metaphor an author's drawing need not be dismissed as a frill or crutch, for it can be taken as another voice or statement. Whether a poem and drawing are in harmony, or counterpoint, and how they colour each other become matters of interest.

During the year of Stevie's negotiations with Deutsch about the publication of her fifth volume of poems and drawings, which was to be called *Not Waving but Drowning*, she wrote two stories. The features editor of the *Evening Standard* had asked her in March 1955 to contribute to a series called "Did It Happen?" (The paper's readers were asked to guess whether the stories in the series were fact or fiction.) Stevie submitted "To School in Germany", and it appeared in the *Standard* in May. Perhaps Stevie had Karl Eckinger in mind when she created the character Maxi, a university student the English schoolgirl narrator meets in Germany before the war, who declares his love for her. Though the narrator finds Maxi handsome, his tendency to go on about the inferiority of the Jews, the English, and others, puts her off. After the war she encounters Maxi in a restaurant near Berlin and introduces him to her cousin. The cousin, who is in the army, immediately recognizes Maxi as a war criminal and arrests him. "I hate this 'holier than thou' situation we are all in now", the narrator declares, to which her cousin replies, "We ARE holier than him."

Stevie wrote another fictional story for the "Did It Happen?" series in August and, again, it was published two months later. In "Getting Rid of Sadie" two children, the narrator and her brother, plot to get rid of Sadie, a sadistic governess who is making life miserable for their cousin. They plan to tie Sadie up and threaten her life unless she writes them a cheque for £400. They will return the money when Sadie gets another job. But their rehearsal does not go well, and they drop the plan. The cousin is sent to boarding-school, however, so Sadie leaves anyway. Years later the narrator and her brother read that Sadie has been convicted of "extorting money by menaces". "Perhaps we were rather alike, us and her", the narrator tells her brother, reflecting Stevie's belief that the hearts of children are not so different from adult hearts.

One finds this conviction elsewhere in Stevie's writings, such as the review she wrote a few years later in which she pointed out that

Even modern writers so good as L. P. Hartley, Miss Bowen and Miss Rosamund [sic] Lehmann . . . are not free from romanticism, in the sense of using the child . . . as in some way innocent . . . how rarely in serious literature, how very rarely, is the child presented as a nasty little thing. Yet we are often nasty, and we are as much the child's old age as he is our youth.

And again, in Stevie's poem "But Murderous", there are the lines "Oh the child is the young of its species / Alike with that noble, vile, curious and fierce." "But Murderous" seems to have been written around the time of "Getting Rid of Sadie", for on 28 August 1955, L. P. Hartley asked Stevie if she'd really sent him "the poem about the lady who wanted the abortion". Indeed, the poem was probably inspired by a pacifist novel Stevie reviewed a few months earlier. She did not find appealing the hero's wife, noting that "she has a criminal abortion because she does not want their child to be born into a 'murderous' world (should human infancy be so superior?)." Stevie's parenthetical remark, which became part of her poem, turns upside down the usual argument against abortion, that human life is sacred:

> And why should human infancy be so superior
> As to be too good to be born in this world?
> Did she think it was an angel or a baa-lamb
> That lay in her belly furled?

Stevie's review of the novel which seems to have inspired "But Murderous" appeared in *The Observer*. Terence Kilmartin, that paper's literary editor, selected the books she would review, and sometimes accepted her poems for publication. Stevie could be "bitchy in a funny way", he recalls. Sometimes the reviews she wrote were too whimsical, or even rude, and he would telephone her to suggest changes. In Kilmartin's view, Stevie was a "highbrow, though not socially a snob". A lot of her reviews he thought too highbrow, however, "caviar for the general reader". In May of 1955 he wrote to explain that, although he was "an utterly devoted admirer" of her reviews, "many of 'our readers' find you a little difficult to follow". Later that year he had to tell her that some poems she submitted left him "completely flummoxed".

Kilmartin remembers once being invited to Stevie's house for a lunch of junket, and on another occasion he accompanied her to a lunch with Kay Dick, Kathleen Farrell, and Ivy Compton-Burnett. Dame Ivy was one of the novelists whose books Stevie reviewed, and their names have become linked by the catchy title of Kay Dick's *Ivy & Stevie*. Stevie admitted Ivy's writing ("*Faites attention*", was her advice to readers of Ivy who, if they did, would have as a reward "much pleasure"), and their theological views were often compatible. Perhaps mutual friends imagined that these two writers, unlike in so many ways, might find enough in common to become friends. Dame Ivy's friend and executrix, Hester Marsden-Smedley, who considered Stevie to have had "a kind of shabby chic", remembered that she drove Stevie home three times after teas at Miss Compton-Burnett's. Kay Dick thought the lunch party she gave succeeded because Ivy recognized Stevie to be someone special—"another genius". But other efforts were less successful. Anna Browne once took Stevie to tea at Ivy's, although the latter had not been keen on the idea. Stevie felt that Ivy, whom she had met only once before, disapproved of her. As a result Stevie behaved in an untypically constrained way at the tea. "Ivy liked people to be socially top drawer", according to Mrs Browne, "and Stevie didn't care a fig for social background". (Anna Browne thought of Stevie as descended from a "middle-class aristocracy . . . *noblesse de robe* if you will. People who had had certain standards for one or two hundred years, were stable, educated, and responsible.") Ivy, who was unusually pro-establishment for an artist, felt too (although wrongly) that Stevie was anti-, and so disapproved of her.

Elizabeth Sprigge was another friend who invited the two writers to tea. Arthur Waley was there, as well as Ronald Orr-Ewing, who thought of Ivy as a battleship and Stevie, light craft. As the guests talked, Orr-Ewing said how little he liked George Eliot's *Scenes from Clerical Life*, and Ivy objected. Stevie chimed in: "I quite agree with you. Pass the honey." He felt that beneath the chat war was being waged.

Stevie must have visited Ivy on at least one other occasion because near the end of her life she told Naomi Replansky how amazed she was, on entering Dame Ivy's elegant flat, to find the floor covered with linoleum. As she advanced to greet her hostess, Stevie skidded on the slippery surface and fell. Laughing hard, she said, "I went skating across the floor and landed right at Ivy's feet." Perhaps it was the recollection of this that led Ivy to seem "frosty" when Rosemary Cooper once mentioned Stevie's name to her.

Stevie's last bi-weekly column of novel reviews for *The Observer* appeared on 1 January 1956, and three days later she wrote to "Dear Terry" to say she felt "desperate . . . at being told so often I couldn't be understood (both poems & reviews). One feels defeated. *Is* defeated I daresay. Lord God, you may be right." But Stevie's defeat was not total, as she began to review non-fiction for *The Observer* and continued to do so, on a less regular basis, until 1969. And she must have been cheered up, when she retired from reviewing fiction, by a letter she received from the novelist John Rosenberg, who wrote to thank her not only for her recent review of one of his books, but also

> for all the sensibility and all the enlightenment you have put across in your novel-reviewing in the last two years. What I have so much admired in your reviewing is the lack of a conventional cataloguing . . . and that you do novelists the unusual justice of setting forth the impressions that make up your value-judgements, not merely giving the sum pronouncements, so that the reader of your reviews has recreated in miniature for him, for his own powers of criticism and understanding, the elements and essentials of the book in question. It is creative reviewing.

The discontinuation of Stevie's bi-weekly column did result in a trip abroad. "I see with much regret that you are leaving the columns of *The Observer*," the Secretary of the Danish–British Society wrote to her on 2 February, "but it gives me the opportunity of inviting you to come to Copenhagen to read a lecture on modern English Literature to our members." Stevie replied that she would prefer to speak about and read her poetry, but when she learned that would not be possible she accepted the invitation anyway, and spoke about the work of some young novelists writing in England. "Time, mercifully, seems to be a proper Highbrow", she told her audience, and although she discussed the themes of Kingsley Amis, Olivia Manning, Angus Wilson and others, she did not claim to know how they would stand up against such "established great ones" as Ivy Compton-Burnett, Evelyn Waugh, and Henry Green.

Stevie's address to the Danish–British Society was in March, a particularly busy month for her. One of the ways she supplemented her income after leaving Newnes, Pearson was by writing reader's reports for The Bodley Head, and on 21 March she jotted a note at the bottom of one report, apologizing for being late and explaining that she'd been in Copenhagen " 'lecturing' on Modern English Writers —my word!" ("I do so hate lecturing, & have only done it once", Stevie later confessed to a friend, referring to her experience in Denmark.) Stevie also mentioned, in her note to The Bodley Head, that she'd "said quite a lot about Inez Holden", so they should not be surprised to hear from *Statsradiofonien* about broadcasting some of her stories.

The weekend before Stevie left for Copenhagen she took part in a panel on literature with sixth-formers in Salisbury. The discussion was recorded by the BBC for broadcast on its Light Programme the following weekend. Stevie spoke about her own work, reciting her

her ideas on writing

poem "The Actress" and revealing that it was originally a 5,000-word story which she'd scrapped. And she delighted in the programme's give and take, playfully drawing inspiration from a panellist's claim that a composer can more easily avoid litigation than a poet. "I was just thinking of a title you might give if you wanted to bring a libel action to a symphony," she said. "You could say, 'A symphony written to show what I think of Mrs Snooks,' and it would be full of the most fearful discords and things." She repeated some of her long held beliefs, such as the value of learning poetry by heart at school, and the notion that authors get most of their writing out of bad temper, irritation and mental suffering. And she speculated that the reason women did not in the past make as great a contribution to the arts as men was that women's traditional work involved constant inter- ruption, and artistic achievement often requires long periods of concentration.

In late March, after her return from Copenhagen, Stevie again recorded for the BBC, but this time the recording was of another Third Programme poetry reading, her first since 1952. The pro- gramme, produced not by Anna Kallin but by D. S. Carne-Ross, was broadcast on 12 April. Kay Fuller of the BBC did not recommend Stevie's programme for overseas, finding it melancholy and evidence that poets are seldom the best readers of their work. She was particularly irritated by Stevie's impure vowels. Stevie may have been unaware of this evaluation, but she did say, in a letter of thanks to a friend for her encouraging words about the broadcast, that she'd had trouble with her s's: "'S' is difficult, isn't it, & there are such a lot of them in our language! I've often noticed it in theatres when the interval breaks out—it's like the sea rushing in."

The following month Stevie had bad news from Diana Athill. Because of a formidable queue at the printers, Deutsch was putting part of their autumn–winter list forward into early 1957, and *Not Waving but Drowning* was among the rescheduled volumes. Stevie had already been consulted about biographical information to accompany the book's publication, and so she must have assumed all was going well. She mentioned that she did not want the advertising for *Not Waving but Drowning* to stir up Palmers Green, "Not only for reasons of a becoming modesty—but because I want a quiet life & a lot of the people here are in the novels & poems. Very few in this suburb know me as Stevie Smith, & I should like it to stay that way, if you don't mind."

Reading was a large part of that quiet life, of course, and not just the books about which Stevie had to write reader's reports or reviews. In the summer of 1956 she enjoyed reading Rose Macaulay's *The Towers of Trebizond*, for instance, and wrote to tell her so (although in later years she was to call it "noveletish"). A fascinating fragment has survived of Rose Macaulay's response:

> You are very sagacious about my hero-heroine. The fact is that, in order to put a rather (in some ways) autobiographical book at one remove from myself, I did two things, one was (or I meant to at first) to make my narrator masculine. Then I decided not to do this, but to leave the sex uncommented on. . . . The other step I took . . . was to think up a style not normally my own—a kind of rambling drool—as you say, *not* Gibbonesque. I suppose it *is* a kind of "lost" voice: and, indeed, Laurie felt lost, [when?] not occupied in speculations about the world, etc. I have left her in a sad plight. But in the end I think she rejoined her Church.

Another book Stevie read that summer or perhaps autumn was Littleton Powys' autobiography, *Still the Joy of It*. Stevie had long admired the writings of Llewelyn and John Cowper Powys, and had written enthusiastic reviews of their books. John Cowper Powys, in turn, found Stevie's reviews of his work penetrating and wrote to tell her so in the early 1950s, and to praise *The Holiday* (in which he is praised), as well as the poems and drawings in *Harold's Leap*, which contains Stevie's four-line "Homage to John Cowper Powys." But, however heartfelt that poem, it is not one of those in which Stevie's strengths as a writer are manifest. Ironically, the writing of Littleton Powys, which Stevie did not admire, inspired such a poem.

In *Still the Joy of It* Littleton Powys told, in part, of meeting Elizabeth Myers in the early 'forties, when he was a "lonely widower of sixty-nine" and she "a little-known writer aged thirty". They soon married in Hampstead and, according to Littleton, their marriage was "like one prolonged honeymoon" until Elizabeth's death from tuberculosis three and a half years later. Contributing to her happiness during her last years was the success of her first novel. *A Well Full of Leaves*, published about a month before her marriage and "received everywhere with acclamation". One of the approving reviews, in fact, was written by Stevie who called the book "bravely unusual". But writing to John Hayward, Stevie questioned the latter part of the novel, and sent up a passage in which Elizabeth Myers describes blood "quietly" streaming into the handkerchief of her tubercular protagonist.

Given her admiration for the Powys family, and her familiarity with Elizabeth Myers' first novel, Stevie must have read Littleton Powys' autobiography with some interest. Certainly she was fascinated by a sentence in which he described the night of his wedding. "Our wedding night coincided with the most spirited German air raid that had been experienced in London for a long time; and the confusion was increased by a very large fleet of our own bombers passing over London on their way to Germany at the same time." This account became part of Stevie's poem "I Remember". Perhaps only she would have introduced it during poetry readings as "a happy love poem". In this spirit, in a letter written to Sally Chilver on 20 November 1956, she enclosed the poem along with two other newly written ones to show, she said, that if she could not manage *à deux* love, she could at least "have a boss shot at a general feeling of warmth & affection".

"I Remember" does express affection, or at least its speaker does, but it is dotty, having a funny sad tone characteristic of many of Stevie's best poems:

It was my bridal night I remember,
An old man of seventy-three
I lay with my young bride in my arms,
A girl with t.b.
It was wartime, and overhead
The Germans were making a particularly heavy raid on
     Hampstead.
What rendered the confusion worse, perversely
Our bombers had chosen that moment to set out for Germany.
Harry, do they ever collide?
I do not think it has ever happened,
Oh my bride, my bride.

The poem's humour is the sort one critic has called Stevie's "gleeful macabre". It arises from her rhymes, and from her artful presentation of incongruities: old groom and young bride; wedding couple and bombers; crossed flight paths; the tender coming together below and the imagined explosive collision above.

When Derek Parker presented a radio programme about Stevie Smith several years later, he included a recording of her reading "I Remember". Writing to thank him for the programme, Stevie confessed that the poem is almost a transcript from Littleton Powys' description of his bridal night: "I altered the setting of the words slightly and brought in some rhyme, but what fascinated me about L. P.'s passage was the way the *utter* banality and really dotty pomposity of the language lent such horror to the events he was describing, *really* a last turn of the screws." Of course, Stevie had made up the conversation at the end of her poem in order to point up the dottiness, and she slightly exaggerated the difference in age between the bride and groom for effect. "I hope the shade of poor L. P. did not turn a pale sort of green colour at the sound of that very nervy applause", she told Parker.

There was plenty of time for "I Remember", and the other new poems Stevie sent to Sally Chilver, to be included in *Not Waving but Drowning*. Indeed, before it appeared there was time for Stevie to take two more trips abroad and to give another poetry reading on the BBC, because the book was further delayed until the autumn of 1957. Stevie went to Amsterdam to view a Rembrandt exhibition with her friend Armide Oppé on the first of those two trips. During that weekend Stevie declared, "I must look west across the North Sea!" So off they went, on a cold evening (it was September 1956) to a completely deserted seaside place. "It was just as desolate as it was to look east", Armide Oppé recalls. But Stevie wrote triumphantly, in a postcard to her sister, "Wonderful to see a sun*set* over the North Sea."

When she returned from Holland Stevie got back to work on a programme she had proposed to the BBC in June, to be called "Too Tired For Words". That was also the title of an essay she published in *Medical World* in December of 1956, a *"cri de coeur* from the doctor's waiting room" she called it. Both the essay and her BBC programme, which was recorded in February 1957, have commentary linking several of Stevie's poems. But Douglas Cleverdon, who produced her "Too Tired . . ." broadcast, suggested that some of the poems be sung (and they were, by professionals), so Stevie changed her script to include poems which she had written with tunes in mind. The programme "was difficult to do & took months", Stevie told Rachel Marshall, "& even now I'm not sure it wasn't a shade on the melancholy side". But at the party a friend gave for the rebroadcast everybody laughed a lot, and so Stevie supposed it was not as bad as she feared. To another friend she mentioned her "angel of a producer" and the very nice actors and actress who offset her "gloomy drawl".

The programme was lucrative, as always, because of the copyright fee, but questions had been raised at the BBC about the amount of Stevie's payment, and she had to write in her defence:

> I do not myself think that this broadcast differs from the other poem-broadcasts I have done . . . [with the exception of] this difference only—that the commentary in the earlier ones stresses the fixing of thoughts and emotions upon imaginary people to make the thing less personal, whereas in this one the stress is upon tiredness acting in various moods of despair, hilarity, arguments and religion, upon these imaginary people. But in no instance ever is

there to be found one word in any of the narratives that could be considered as a criticism or review of the poems, in the way that a reviewer or critic might—and should—deal with them. . . . But if, where a broadcast is long and the copyright fee mounts up to something that is considered excessive, the B.B.C. usually does ask poets to accept a cut, then that is another matter. Is this what usually does happen?

One supposes Stevie knew the answer to her question was "no".

Stevie's second trip abroad, before the publication of *Not Waving but Drowning*, was in May. "For three long weeks, and long weeks they were, I stayed in Milan." Her hostess during that rainy and cold period was the Brazilian Consul in Milan, Margarida Guedes Nogueria, who was married to Stevie's friend, the composer Stanley Bate. The marriage was not generally known, as evidently it would have jeopardized the Consuelessa's career. In order to prevent or at least discourage her husband's heavy drinking, Margarida Nogueria kept money from him and also removed the money from Stevie's bag, including perhaps the £10 the British Council in Milan had paid her for reading some poems there. So Stevie and Stanley were left to wander about with little money, once accepting a ride from the servant, Cesare. Stevie wrote a funny account of her adventures in Milan in a letter to Barbara Clutton-Brock, and also shared her anecdotes with the Brownes and other friends. One of her stories was about the time she and Stanley decided to swim in a lake, and Stanley, knowing little Italian, made signs to show he wanted to hire a pair of bathing trunks. The vendor thought he was making a lewd proposal.

Out of this holiday Stevie wrote her poem "On the Dressing gown lent me by my Hostess the Brazilian Consul in Milan, 1958". (The year, in fact, was 1957.) The speaker of the poem admits she ran in the streets with the Consuelessa's husband, growing "more English with each drink":

> "Give us money" we said, "you have not given us much."
> We were your kiddies, Consuelessa, out for a touch.
>
> Yet I admit your dressing gown
> Wrapped me from the offences of the town
> But never from my own
> Ah Consuelessa, this I own.

The poem is perhaps more effective for not stating the circumstances in which it was necessary to wrap the speaker in a dressing gown. "On the Dressing gown . . ." was not published during Stevie's lifetime, and perhaps she never decided to publish it. In any event, it must have been composed too late for inclusion in *Not Waving but Drowning* if, indeed, it was written before the autumn of 1957, when the volume appeared. (The poem was certainly written by the summer of 1959, when Stevie referred to it in a letter to a friend.)

More attention seems to have been paid to *Not Waving but Drowning* than had been paid to *Harold's Leap* and, in general, the reviews were enthusiastic. Roy Fuller's review, in the *London Magazine*, was an exception. He cited Stevie's "I Remember" as "just about the best she can do", but complained that usually in her poems something "seems to go soft at the centre or round the edges". Diana Athill wrote to warn Stevie about this review, and she replied:

> It sounds awful and I promise not to read it when it does come, always best I think, as these things make one cross and one goes and does something silly, even *I* might, pickled as I am in age and accustomedness. I dont know Mr Fuller, except I think by sight but it may be somebody else, so if he is vicious it cant be personal. All the same, *there* is [sic] Those enemies we have heard of . . . the Chatto–Lehmann–John Hayward and I fear also Spender link up.

It is interesting to see Stevie by this time including John Hayward among her enemies.

Not her enemy, at least on the basis of her review of *Not Waving but Drowning*, was Muriel Spark, who noted that Stevie's art "is most of all dependent on the curious chit-chat rhythms, elongated lines, comic metrical arrangements and mordant phrases which are used in the volume with particular skill and point." She singled out "In the Park" as "one of the most fascinating poems" in its portrayal of "two old gentlemen walking by the lake, one of whom mournfully bids his companion to 'Pray for the Mute,' whereupon the other, who is deaf, proceeds to praise the newt. 'So two, better than one, finally strike truth in this happy song.'"

"In the Park" was also admired by the *TLS* reviewer: "a poet capable of writing 'Praise is the best prayer, the least self's there' must be allowed her waywardness." The waywardness referred to was Stevie's tendency to risk "triviality and whimsy, from which it can hardly be said that her little drawings, half wistful and half skittish, help to divert attention." But *Punch*'s reviewer found himself "quite liking some of the pictures", and *The Listener*'s review defended Stevie against the charge of whimsy, in addition to making some insightful observations about her technique. Noting that "the obscure forces governing poetic reputation have shown little interest in Miss Stevie Smith's poems", the reviewer went on to say that Stevie has been driven to "a perilous undertaking":

to de-whimsy the funny-peculiar or, as the serious French call it, the Absurd. The horrible thing about the funny-peculiar is that if looked at too closely it becomes very sinister indeed and everyone rushes for shelter—the French to a metaphysic, the English to the creation of amiable grotesques. But Miss Smith does not take shelter. . . There is a poem in her new volume called: "Will Man Ever Face Fact and not Feel Flat?", and this question throws a good deal of light on the poetic motives which dictate her *art poétique*, the control of the off beat and the half-rhyme she has brought to such refinement:

> "Oh to be Nothing", said Eve, "oh for a
> Cessation of consciousness
> With no more impressions beating in
> Of various experiences".

The catastrophic let-down of the last line is an example of the precision-instrument which Miss Smith uses to reduce the funny-peculiar to flat fact. If this were all there was to it she would be no more than a virtuoso executant of ruthless rhymes; she is a poet because she discovers how to give the flat words a texture of innocence and tenderness.

In contrast to the reception of *Not Waving but Drowning*, Stevie's next book, published a year later, went almost unnoticed. It was a volume close to her heart. As far back as the publication of *Mother, What Is Man?* Stevie had wondered if she could have a page or so of drawings with no poems, but only captions. By the spring of 1958 she had found a "dashing publisher" who wanted to do a whole book of her captioned drawings. The publisher was Gaberbocchus ("Jabberwocky" in Latin) and the book, *Some Are More Human Than Others*. It was noticed in *The Daily Telegraph* by John Betjeman, but "owes much to Thurber" was all he said. Oswell Blakeston, in *Art News and Review*, was more enthusiastic: "these are not random jottings from a notebook", he said, "but a profound philosophy in doodle drawings: death is not easy and life generally wins by imitating death." Blakeston described a few pages—"I'm devoted to the scribble pig and his hand-written caption: 'Who knows what he is thinking? (He is not a child)'; and the puss with upraised eyes saying 'I sometimes thought my mind would give way'; and the girl who finds her gas stove such a friend with its funny little humming noise: 'Now Agnes, pull yourself together. You and your friends!'"—and he concluded that the book was "A treasury to be treasured." More restrained was the brief *Spectator* review, which found *Some Are More Human Than*

Others to be "awfully like something done for an artistic analyst". One extensive review, if it can be called that, was published in the *Hampstead and Highgate Express*, an amusing run-on gossipy piece. In the form of a letter, "letitia" tells "athalie" about the book, mentioning, at one point, the observation of a friend called "stepan":

> so I asked him what all that boiled down to and he said: "we all have these thoughts sometimes," I replied: "I bet there's a drawing for that"—and there was: a sort of kafkaesque beetle, he said it was achingly funny but I said how could it be when one couldn't say whether it's the picture or the words that make you laugh? he said did it matter and anyway people would buy it just to make up their minds, so I said: what! at eighteen bob?
>
> so there we are. some people said she'd never have another success like her "novel on yellow paper." she's coming to our little gathering next week. I shall feel positively *naked*

Abroad *Some Are More Human Than Others* was noticed, oddly enough, in the *Natal Witness*, published in Pietermaritzburg, South Africa. But the review is only a few sentences. It was also reviewed in another country, however, and by a new friend of Stevie, who was also her champion, and about whom we shall now learn more.

15

"Oh Christianity, Christianity"

In September 1957, a few days before her fifty-fifth birthday, Stevie dined with her friends the Russell-Cobbs in their Chelsea home. Among the guests was a Swiss friend of her hosts, Hans Walter Häusermann, Professor of English Literature at the University of Geneva. He and Stevie liked each other, and he returned with her to Palmers Green and was introduced to Aunt. He jotted down impressions of Stevie, noting her "beautiful, large eyes . . . fine intellectual nose, [and] artistically shaped mouth. But her skin is flabby, excessive for the delicate skull and jaws." He noted as well, her "childlike posture" and characterized her "demeanour" as "not that of a mature, responsible grown-up. Her demeanour can neither be measured by the academic criteria à la [C. S.] Lewis, according to which maturity is an ethical condition, nor by T. S. Eliot's idea of wisdom which . . . is an attribute of great poets. One could perhaps define Stevie Smith's posture as existentialist." In describing Stevie as a child trying to forget her pain, he noticed what others had: her curious and often critical awareness of children. She had spoken to him, after their dinner at the Russell-Cobbs', about a friend's young son who had " a cold, domineering egotistical gaze; he likes to drink and to smoke Woodbines", she noted.

The following month Stevie wrote to Häusermann in Switzerland, beginning a correspondence that continued for a decade and is, on her part, richly autobiographical. She approved his estimation of her writings and attitudes, cheerfully sent him copies of her books and photographs, sometimes glossed her texts to correct printer's errors, and generally took advantage of a sympathetic and perceptive ear to complain about the failure of certain editors and critics to appreciate her work. More than that, she gloried in him as a champion, as when he published a highly favourable review of *Some Are More Human Than Others* in the Swiss paper, *Neue Zürcher Zeitung*.

Complaining to Häusermann that many literary journals were refusing to print her poems, Stevie described herself as finishing a lecture that the Cambridge Humanist Society had asked her to deliver. Called "The Necessity of Not Believing", it was about her life, she said, and how first she was an Anglo-Catholic and then was not. She told Häusermann it explained her religious position, and described the talk as illustrated "with some poems that are anti-religious". She presented the talk on 20 November, the very day on which an editor at *The Observer* wrote to tell Stevie that the lecture was vivid but far too long and too talky for publication in a newspaper. His deeper concern, though, seemed to be the "need to go a bit more cautiously and take more account of opposing views". He suggested she send it elsewhere, although he doubted *Encounter* "would stomach it".

In December Stevie told Häusermann she could not find a publisher for "The Necessity of Not Believing" and was still "hawking it round". But a few weeks later she sold it for twenty guineas to "the kiddy-press", as Stevie called the undergraduate journal *Gemini*, edited by the Indian poet Dom Moraes who was then a student at Oxford. Probably because of its appearance in a relatively obscure publication, the essay made little stir. Stevie subsequently delivered the talk to a student society of "rationalists" at University College, London, and for a time Gaberbocchus Press was planning to publish it, "illustrated with harrowing Goya etchings". In a later version of her essay, Stevie mentions the attractiveness of the notion that a loving God is in charge of the universe. "I do not think there is any harm in trying to behave as if this were the case", she says. But she opposes any effort to assert that it is certainly so. Such assertions tempt believers to define "the nature of God and Goodness, and [to] be angry with those who do not agree". Besides, she does not find the world of doubt loveless or deprived of joy.

"The Necessity of Not Believing" gives a memorable picture of Stevie's churchy childhood and all that religious ritual meant to her then and subsequently. Later she came to see cruelty in some of this, admitting that "the nerve of cruelty . . . lies in all of us". Hell was only rarely mentioned in her childhood but, Stevie said, "I could not forget Hell. And I thought: How could a God of Love condemn anybody at all?" In "Thoughts about the Christian Doctrine of Eternal Hell", a poem she included in her later version of "The Necessity of Not Believing", she concludes:

The religion of Christianity
Is mixed of sweetness and cruelty
Reject this Sweetness, for she wears
A smoky dress out of hell fires.

. . .

This god the Christians show
Out with him, out with him, let him go.

Stevie confessed herself torn, as time passed, because the beauties of
Christianity were so apparent and so dear to her and so woven into her
identity as "a child of Europe": "Some people may not find the tug of
this sort of thing very strong, but I did." The history of the Church,
particularly in the Inquisition, appalled her, and her conscience found
itself at odds with some of the positions taken by those writers she had
admired for decades: Ronald Knox, Chesterton, and Belloc. Against
them she balanced Lecky, Acton, Shelley, and Gibbon. Vera Brittain
was only one of her contemporaries who came in for criticism in this
regard. "The Gospels, from which as a pacifist Miss Brittain draws her
chief strength . . . are not noticeably tolerant of persons out of line
with their teaching", Stevie wrote. The pamphlets of The Catholic
Truth Society afforded Stevie her best evidence of the cruelty in
Christianity, especially its "hideous" teaching about hell. Eventually
she came to the conclusion that Christianity is immoral, although she
admitted that her certainty of this was not sustained and that often, in
later life and in poems, she hovered between belief and unbelief.

Her thoughts about Christianity were confused by her feelings, she
later maintained: "My feelings fly up, my thoughts draw them down
again, crying: Fairy stories." But as her poetry demonstrates, she was
attracted to fairy stories. Yet another aspect of her mind, not unrelated
to her art, filled her with "a determination to make words mean what
they say". In "Why d'You Believe?", a poem published in "The
Necessity of Not Believing", the speaker asks, "Why call Belief what is
not more than Hope?", and later in the poem insists, "Write that word
right, say 'hope', don't say 'belief'."

The path of the unbeliever with a religious temperament, Stevie
acknowledged, "is fraught with the perils of flatness and ennui, and
religion by contrast, particularly the Christian religion, is so dramatic
and exciting . . . above all [with] its sweet promise of a heavenly
father." So she felt it was "deeply human" to love God and, at the
same time, useless. Putting up with flatness, Stevie felt, was man's
fate—a notion she had already expressed in her poem "Will Man Ever
Face Fact and not Feel Flat?" Stevie described herself as cheered up by
the degree to which the doctrine of eternal hell and related notions

were denied in her day, except perhaps by Roman Catholics. Part of her stern disapproval of Catholicism was rooted in its insistence on the doctrine of eternal punishment: "One may like the Catholics for their logical adherence to Christ's words . . . when they teach that 'out of hell there is no redemption'. But I prefer the Anglicans who at least try to find a way out."

The paradoxes of Christ also fascinated Stevie, and finally she could conclude only that we should cling to his loving words and let the others go. In her poem "Was He Married?", a dialogue between an ingenuous youth and a less kind older person, Stevie raised what for her were contradictions: if Christ is divine then he cannot have known guilt for sin and other earthly conditions, and so could not be, as Christianity teaches, human as well. A god, the poem's older speaker concludes, "is Man's doll". It represents some advance that man has fashioned a god of love, but the advance will be even more valuable when men can love love but not deify it.

Another poem on this theme is "Oh Christianity, Christianity", which Stevie incorporated in the essay derived from "The Necessity of Not Believing". In the poem, a speaker rather crossly asks why this religion does not respond to our difficulties in understanding doctrines such as the dual nature of Christ. In another mood, however, Stevie perhaps felt the need to salvage from her passionate rejections "The Airy Christ", as she calls him, in her beautiful poem of that name:

> Who is this that comes in splendour, coming from the blazing
>   East?
> This is he we had not thought of, this is he the airy Christ.
>
> Airy, in an airy manner in an airy parkland walking,
> Others take him by the hand, lead him, do the talking.
>
> But the Form, the airy One, frowns an airy frown,
> What they say he knows must be, but he looks aloofly down,
>
> Looks aloofly at his feet, looks aloofly at his hands,
> Knows they must, as prophets say, nailèd be to wooden bands.
>
> As he knows the words he sings, that he sings so happily
> Must be changed to working laws, yet sings he ceaselessly.
>
> Those who truly hear the voice, the words, the happy song,
> Never shall need working laws to keep from doing wrong.
>
> Deaf men will pretend sometimes they hear the song, the words,
> And make excuse to sin extremely; this will be absurd.

Heed it not. Whatever foolish men may do the song is cried
For those who hear, and the sweet singer does not care that he was   X
    crucified.

For he does not wish that men should love him more than
    anything
Because he died; he only wishes they would hear him sing.

Stevie insisted that she felt liberated by "the icy indifferent wind that blows across the flat fields of geological time . . . and the years that go on for a million million years." She wrote scornfully of the "hurrying back to religion" in the post World War II world, and attributed religious conversion to degeneracy of the nerves. Yet her attention to religion never flagged. Sally Chilver remembers how Stevie pressed on her Miegge's book about the cult of the Virgin Mary (a work she mentions in "The Necessity of Not Believing") and insisted that Sally read it. She also recalls Stevie's crack about the Assumption of the Virgin: "If the universe is expanding is she still Going Up?"

"The Necessity of Not Believing" did draw responses from some readers, and among them *Gemini* published lengthy comments by Mervyn Stockwood, then Vicar of the University Church, Cambridge, and Simon Barrington-Ward, Chaplain of Magdalene College, Cambridge. Both congratulated Stevie on having surrendered a religion which was distorted and false. The essence of Christ is not contained in the evils men have committed in his name, they argued, and the objections Stevie had to certain doctrines result from the limitations of the interpreters she cited rather than the doctrines themselves. But they could not quite control their crossness. *Gemini* published, as well, the letter of an Oxford undergraduate who subsequently engaged Stevie in a correspondence concerning religious belief.

Late in 1957 and in the next several years Stevie's attention was very much directed towards religion, especially by the books which the *Spectator*, *The Daily Telegraph*, *The Observer* and, throughout the 1960s, *The Listener*, sent her to review. For instance in December 1957 Stevie wrote a piece for the *Spectator* entitled "Unbelieving" in which she reviewed *God and Us* by Jean Daniélou, *Christianity and History* by Herbert Butterfield, *The Gospels in Modern English*, and five other books about religion. This was typical. If we take her at her word she was not unhappy to be assigned these books. "It is with great pleasure and curiosity I approach these books of Christian thought—pleasure to be able to read what is often so beautifully and ingeniously expressed and to share, if not the beliefs, at least the humanity that drives beneath; curiosity to learn afresh what these beliefs are." Her refrain is familiar:

> All that the Christians write bears witness, as it must, to the Covenant between God and Man (the Monstrous Bargain, the Writ of Scandal to the unbelieving) by which the Father God accepts the death of His Son, to pay for the sins of the world and save its children from everlasting death . . . monstrous it must seem to unbelievers and a cause of sickness in the minds of Christians. For this doctrine bites secretly and festers in the heart, and out of it comes that dual aspect of Christianity, the sweetness and the cruelty, that marks with so heightened a nervous temper its passage through history.

Stevie ends her review wondering, "Will people always hanker after religion, must they always have it, will they never for conscience's sake, put it away, be good for goodness' sake, not God's?"

Stevie's speculations in many of these reviews continued to afford self-revelations. About George Bernard Shaw's friend, the Abbess of Stanbrook, she expresses surprise that Dame Laurentia "was at home in life and never suffered from doubts and melancholy". "Holy people", Stevie says, "do not usually feel at all at home: one always thinks of them as awkward, with a look coming out of their eyes to say they wish they were elsewhere." This is a description that will do well for herself. She also seems to identify with the unmusical nuns who must have found oppressive the Abbess's stress on plainchant and rigorous practice. Stevie seems to have had herself in mind in another regard when she went on in the same review to write about "that sainted, awkward poet G. Manley Hopkins, always so tormented in his spirit . . . with never a beautiful word he wrote understood, until he was dead, and everyone preferred before him."

In 1957 she reviewed books by and about the French Catholic poet and philosopher, Charles Péguy. A life of the Italian priest Don Orione (whom Stevie wittily mentions will in time be canonized, "if all goes well") offered much to praise, but Stevie spends the concluding paragraph pointing out the superior moral stance of Humanism which "has as much to boast in the field of charity as religion and is strong with the truer strength of something that has not yet been quite worked out and is growing." In her review of Simone Weil's *Intimations of Christianity Among the Ancient Greeks*, Stevie accuses the author of approaching her subject not with the humility of a scholar but with the preconceptions of an enthusiast, and discounts many of her assertions. According to Stevie, Simone Weil failed to confront a knowledge the Greeks carried in their hearts and which Stevie herself embraced: "there is no part of Nature or of the Universe that is not indifferent to man." The desolation Stevie felt in facing life, her longing for religious consolation, and yet her brave insistence on what she thought to be the truth, all surface in the reviews she was publishing in these years. "All religions have martyrs", she wrote in another review, "and science and art have them too. Bravest of all, perhaps . . . are the martyrs to atheism. They cannot thrill people much in this life, as the saints did. . . . I suppose they die in the hope that earth's heaviest sentence—*mundus vult decipi*—may one day die too. Theirs must be a hard death."

Stevie's review of Simone Weil's book elicited another letter from Mary Renault who agreed that though the *Iliad* does not feature church Christianity (as Weil seemed to argue), "it's not atheism either" (as Stevie seemed to argue). Miss Renault hoped that Stevie would remain a true agnostic, one who has an open mind, rather than a partisan atheist.

In March 1958, Stevie told Helen Fowler that the number of poetry readings she was doing just then brought her very little income and she was, consequently, getting poorer and poorer. "I suppose I ought to snatch round for a bit more reviewing," she said, explaining the principal reason for the amount of literary journalism she took on at this time. But the subjects of the books obviously suited a life-long preoccupation which grew more intense with time and fitted in with the publishing projects she was entertaining just then. With religious conversion so much in the air, books about the subject were apparently much sought after, as a list of Stevie's reviews reveals.

Seemingly at her instigation, Deutsch wanted Stevie to do a book about belief to be called *What Men Live By*, and, in March 1958, letters about this unrealized project passed between Diana Athill and Stevie. A book published in 1940 and called *I Believe* was proposed by Diana Athill as a model, but when Stevie looked it over she thought the 1940 publication wandered too much from a religious line. "I suggest that what we want is a book about the Christian religion and what people think about it in the world today and just that", wrote Stevie. "We must have writers who in their books and writings have shown a definitive involvement with the Christian religion either for or against. I should like some . . . [of] those who lapsed and went back again . . . J. D. Bernal, for instance."

Stevie had never forgot the account Bernal gave her of his religious formation with the Jesuits, and indeed had incorporated it in *The Holiday*:

Tengal [her name for Bernal] was a very pious boy and when he was
at Stoneyhurst he set such a pace in piety and religious observation it
was an embarrassment to the priests, he knew all about this and that,
and serving mass, and singing seconds in the choir, and he said it
was very beautiful and very, very rich in emotion. . . . He said that
once every year they had a Retreat, and the boys were so worked
upon by the sermons and the silence . . . they used to kneel there
crying. And then the priest would say: "Eternity, eternity, eternity,
and in hell not one moment is free from pain, for do not suppose that
the senses are dulled, as by the mercy of Our Lady they are dulled
for us who are in pain in the earthly body. But in hell the senses are
eternally revived to receive an eternal pain that is eternally as a fresh
pain freshly received, except that the weakness and the memory
of past pain, that is so actual a part of pain, is there too, so
that the quivering nerves receive it as a fresh pain, but the mind
knows it as familiar, acute, and eternally endurable by reason of the
vigour that is poured into the nerves and bones and arteries of the
damned."

One might suggest that the hellfire sermon Bernal heard as a schoolboy was more a product of the preacher's lively imagination than of
orthodoxy, but Stevie would object. After quoting this passage from
The Holiday, Stevie reminded her readers, in "The Necessity of Not
Believing", that "these things . . . are being taught today and that
they go no whit further than Christ went when he said, 'Depart from
me, ye cursed, into everlasting fire' and when he spoke of hell as a
place where their worm dieth not and the fire is not quenched."

Inspired by her undiminished horror over all this, Stevie wrote
"Thoughts about the Christian Doctrine of Eternal Hell"—an "anti-
religious poem", she called it, "which was the result of a talk I gave to
the Cambridge Humanists, and the long correspondence that came
out of it about Hell." It begins:

> Is it not interesting to see
> How the Christians continually
> Try to separate themselves in vain
> From the doctrine of eternal pain.

Stevie told Professor Häusermann that "interesting" in the opening
line "has as loud a sneer in it as you can get in English".

"Our temperament rules us", she wrote, and evidently she thought
to come at individual truths through the accounts of Bernal and
others. "I daresay my own temperament, preferring this emptiness of

an indifferent universe, is no more a pointer to absolute truth than the narrowest of orthodox religion", Stevie admitted. But one must again emphasize the extraordinary passion inherent in her religious concerns. Her friend Father Irvine later likened Stevie to Jacob "who could not let God go for the whole long night of life". Watching her fierce pursuit of truth about the transcendent, one is immensely impressed by Stevie's integrity. As we can see from some of her writings, she seems to have identified with holy women—especially in their obsessive inquiries into the supernatural. Helen Fowler felt that had Stevie lived five hundred years earlier she would probably have been a Mother Superior in a convent of nuns, and Stevie's intellectual vigour, natural authority, and detachment make Helen's speculation plausible.

Sometimes Stevie was mistaken for a witch, as also happened with holy women earlier in history (she once wrote of Saint Teresa of Avila's "reasonable fear that her visions might bring her under charge as a witch"). But as she told Helen Fowler, "No, I do not—& never have at all—thought of myself as a witch. I know too much about witchcraft & its ghastly history", she added, "to do anything but loathe it". In that same letter Stevie mentions that "The Necessity of Not Believing" is to be published as a pamphlet and tells Helen that when she reads this essay she will know how unsuitable Stevie is to be the godmother of Helen's daughter. She had expressed misgivings several years earlier, when the Fowlers asked her to be Lucinda's godmother. "I am not a really proper practising Christian you know", she warned them. But she was honoured and delighted to have been chosen, and hoped Anna Lucinda would howl like anything at the christening, "to show Mr Old Sooty what she thinks of him and his works".

Hans Häusermann was another of Stevie's readers who noticed "how closely connected belief and unbelief are" in her writing. His warm, personal support and grasp of her artistic intentions were a comfort to her at a time when her statements about religion were drawing fire. Stevie wrote to tell him how much she admired a piece on her novels he had recently published in *Neue Zürcher Zeitung*. He had "seen so much of what I was thinking, but so often saying only slantwise & imperfectly. . . How right you are about the nervousness of the writing," she told him. "There is some dreadful fear that pursues always or that has no form or substance." Stevie seems to describe herself when she comments to Professor Häusermann about Celia in *The Holiday*: "In the depths it is very cold, & it is that which is in herself she fears & flees, but cannot flee, except in death. All the love in the world, to such a character, is useless, she cannot respond & must flee." Stevie depicts herself to Häusermann in the act of writing novels, especially, as penetrating to cold, black depths where the pressure is intense. "Naturally, I do not like it very much," she says. "But you have put your finger on it all right!" Reviewing, and its requirement that she think about other people's books, kept her cheerful, Stevie said. It was writing novels that disturbed her profoundly: "Even thinking, as I read your beautiful sensitive warm-hearted & generous review & critique of my novels, makes me feel suicidal, because of the ghastly human confusion & chill that underlies them." Poems troubled her less, and she told Häusermann of her resolve to "stick to poems. And reviews."

When 1958 began, Stevie had been without a charwoman and "caked in ashes & cross with trying to cook". Also, she told her Swiss friend, "bits keep falling off the old house". And she continued to collect "rejecting letters from *Our* Editors". Professor Häusermann was making every effort to have Stevie's work translated and published in Switzerland, and Stevie had persuaded Michael Hamburger to translate "Not Waving but Drowning" into German. But on the whole Stevie felt "lackadaisical . . . [with] the mulishness of the languid". She thought to imitate one of the Stracheys who, when she was at Oxford, hired someone to come in regularly and say "Barbara, get on with your work." Stevie encouraged Häusermann to write a letter to the *London Magazine*, defending her against Roy Fuller's withering review there of *Not Waving but Drowning*, and was grateful when he did so. She complained to him that she had recently read at the PEN Club, and Fuller had been among her "co-readers": "He looks a most awful ass, if that is any comfort, read his dull poems dully & kept gulping down masses of water. *Water*, help!" Stevie also complained in these weeks about the *New Statesman* and its literary editor, Janet Adam Smith, who didn't dislike her but only her poems. This wasn't a simple matter, Stevie said, "because I don't always like 'em much myself".

That spring she began a stint that lasted through December as a reader of manuscripts for Curtis Brown. "They say it only means going in [to their office] for half a day a fortnight", she told Hans Häusermann, "but I think it means a bit more than that, unless I am going to bring a donkey load away with me to read at home." The need to earn more money appears to have been an important concern to her in 1958. The Curtis Brown job, she said, afforded "a little extra money" although she predicted accurately the job would probably not last long. To Rachel Marshall she complained that "the stuff one has to read [for Curtis Brown] is incredibly awful" but that she could keep her own hours. None the less, Helen Fowler recalls that Stevie was considered "too kind" as a reader and that she often recommended Curtis Brown take on books which were rather second-rate.

Stevie thought the London literary world in 1958 "awfully racketty", a place where people shuffled and pushed for places "and the publishers and editors as O.W. [Oscar Wilde] said, knowing everybody's price but nobody's value (or so, naturally, the neglected ones feel)". She complained about the timidity of literary editors concerned "not to back the wrong horse, in fact . . . they really only like to place their bets after the horse is home." *The Observer* she found to be nervous "about the awful possibility of slipping behind the fashion line, and all the rest of the silliness that goes on, politically as well as literureally." Stevie remained, as she had always been, remarkably impervious to fashion.

Throughout 1958 she continued to write reviews "but on the whole I don't feel v. inspired", she claimed. Her regimen at Curtis Brown sounds reminiscent of the one she followed at Newnes, Pearson: "I am typing this letter in my new office . . . but today there don't seem to be any [manuscripts to read], which is odd as I only come in about once a week, if that." The job was

> pretty dull & if you look closely at it pretty depressing too. They, like most publishers, are such embattled middle-brows, so stupid, so conceited . . . also so cross. And I *won't* be badgered & bothered now I don't have to be (Shades of a previous office life!) as it's all the same to me whether I have the job or not. Generally they are awfully nice & they do have masses of quarto sized paper. . . . All the same these middle-brows, with their regrettable—deplorable—bounce & energy are England's Bane, the walking illustration of that bloodcurdling truth—that the Better is the Enemy of the Good.

In May Stevie had been staying with the Brownes in Norfolk, and she lovingly described to Hans Häusermann the journey back to Palmers Green:

> I adore these trips. . . . Yesterday we had Michael and Anna in the car along with their child Alice, two cats, a sweet pug dog and a hamster. I can't tell you how pretty it looked coming down through Norfolk, a bit of Suffolk and Cambridgeshire, then my own sweet Hertfordshire, v. green after so much rain and the trees v. full and all the may, masses of it, out and smelling like heaven. I do so like the north of Eng. better than the south, it is so empty and sweet.

In her description of the journey from Norfolk, one feels again the comfort Stevie got from being cosily *en famille*.

Late in May she travelled by train to Scotland ("I have a passion for

looking out of trains") to join the Fowlers there, and told Hans
Häusermann, "Papa in the Norfolk family is a barrister, and papa in
the Scots one is a soldier, so you see how un-literary I am." Two
reliable papas (both in fact her juniors) who helped Stevie to stem the
neglect she felt from her own "unrespected papa". She stayed at the
house Laurence Fowler held as headmaster of Queen Victoria School
in Perthshire. "The dear boys do an *awful* lot of piping and bugling
& are simply splendid on Church Parade," she wrote to a friend. "It
is a pretty part of Scotland & quite near some fairly spectacular
scenery."

It may have been on this visit that Helen Fowler took Stevie to see an
historic house. Stevie was determined to have a cup of tea, although
they had ascertained that none was available. Helen went off to post
some letters in the village and returned to find her friend in a grocery
shop sipping tea an obliging shop assistant had brewed for her. To
Helen the incident illustrates the unusual blend of charisma and
childlikeness which often persuaded people from very different back-
grounds to look after Stevie. During her 1958 holiday with the
Fowlers Stevie lazed in the sun and acquired an "odd Scottish sunburn" that, she jokingly told Helen, led people to say, "there's some
Indian connection, if one went far enough back!"

Stevie's holidays were a great contrast to her job which, however
minimal its demands, was becoming more and more tiresome to her:

> Don't really think I *can* go on much longer with this Curtis Brown
> job, the MSS are too *dingy* for words & sort of make one feel sick,
> sick of all written words. £8 a week for roughly one day's work a
> week isn't bad pay but I got on all right without it before I had it &
> already I think ecstatically of walking out & never going back!

Stevie appeared on the television programme "Brains Trust" early
in August of 1958, a "silly programme" she thought, although she
was to appear on it several more times. Just before the programme
began she got "awfully nervous". She wrote a bit complainingly
about "the old hands" who appeared with her, Professor Freddie
Ayer, and her friend, Marghanita Laski: "how [they] *do go on*". Soon
after, she went to Norfolk for another visit with the Brownes. "Far
more social life than we ever seem to get down here—also some nice
bathing—well, perhaps a bit *cold*—& picnics", she wrote to Helen
Fowler. "When we were coming back from the Point one day there
were 8 grownups, 8 children & 7 dogs all in the same small boat.
Happily, the dogs got on all right." Stevie mentions in the same letter
that she had found for Curtis Brown a "new young author—Maria
Browne (Anna & Michael's child) . . . [whose] book is really v.

good." Maria was nineteen when she wrote *Whom the Gods Love*.
When it was published in 1959 by André Deutsch, Stevie attended the
candlelit celebration party to which some guests came in togas because
of the novel's setting—Rome under Nero. (Stevie listed *Whom the
Gods Love* in *The Observer* among her "Books of the Year" for 1959.)
Generous, as always, towards friends, Stevie was trying as well, in
1958, to interest Curtis Brown in a book Laurence Fowler had recently
completed.

Although she was successful in helping some friends publish books,
Stevie was less so at that time with her own work. Alan Pryce-Jones
had kept for more than a year several poems she had submitted to the
TLS, and then returned them. ("Bliss, isnt it?", she wrote to a friend.)
And a few months later an editor at *Encounter* wrote to say that Stephen
Spender was in America but "I do not think the batch you recently sent
in is up to your best standard. Some of the poems are very slight, and
others I feel are rhythmically rather awkward. I think Mr Spender
would agree with me about this." Nor were her efforts to find a
German publisher successful. Perhaps discouraged by rejections,
Stevie said she was writing very little, except for reviews. "As a matter
of fact I have rather turned from Lit. to the Stock Exchange. I have
acquired a new financial advisor—Czech—who is making quite a lot
of money for me (cross fingers!). I must say . . . the Stock Exchange,
compared with the literary world, seems an absolute *model* of recti-
tude, simplicity & affection." Ladislav Horvat was Stevie's broker,
and they wrote detailed letters back and forth in the autumn of 1958
and for some years thereafter about the merits and risks of particular
investments. The sums were comparatively small: for the last quarter
of 1958 the £979 which Stevie invested earned her a profit of £84, but
the security Stevie reaped was not only financial. Besides Professor
Häusermann, who continued to write encouraging letters about her
work, she now had a second protector who was building up, little by
little, the money she managed to save.

Along with stocks and bonds Stevie had also taken up sewing,
having made "a beautiful red petticoat out of 1 yard of silk material
—all hand sewn & so ravishing I should like to wear it as a dress.
Sewing is a great relief from reading," she told Hans Häusermann.
"You ought to try it."

/ sounds like me !

Stevie's reviews in the late 1950s managed to enrage some of her
readers, a few of whom wrote letters of protest. She exchanged
opinions in the pages of *The Observer*, and then, in a private correspondence, with an Anglican canon. A parson wrote to her from
Dorset, provoked by her "little cracks at the conventional weak points
of uncritical Christianity". A former Catholic priest defended Stevie

in her dispute with Canon Scrutton in a letter entitled "Cheers for Miss Stevie Smith", but *The Observer* chose not to print it. Stevie seemed obsessively fixed on the cruelty for which she believed Christianity was responsible in history, and repeatedly raised this theme in her reviews. Some of the protests came from Catholics and concerned her praise of Elizabeth's triumph over Mary Tudor and the Church of Rome. Her review of a selection of Charles Williams's writings also elicited protests. In it she criticized this writer, whom T. S. Eliot so greatly admired, for his "boyish tastes" which could "grow sickly, especially sexually sickly. Love is difficult enough without having High Romance perched on the bed-rail to make everything so complicated and unecstatic." She praised Williams for recognizing the poetry of the Athanasian Creed, "though every clause is soaked in blood". Stevie probably had in mind some contemporary novelists and poets (C. S. Lewis, Tolkien, T. H. White—all writers she rejected as inferior to J. C. Powys), as well as certain Christian apologists, when she concluded her review on Charles Williams by saying, "these boyish dreamers of today, with their grails and their secrets, are weak, pithed at the core and soft; not what one wants."

In a poem from this period, "Why do you rage?", Stevie asked herself:

> Why do you rage so much against Christ, against Him
> Before Whom angel brightness grows dark, heaven dim?
> Is He not wonderful, beautiful? Is He not Love?
> Did He not come to call you from Heaven above?
> Say, Yes; yes, He did; say, Yes; call Him this:
> Truth, Beauty, Love, Wonder, Holiness.
>
> Say, Yes. Do not always say, No.
>
> Oh I would if I thought it were so,
> Oh I know that you think it is so.

T. S. Eliot is another of the writers who did not escape Stevie's censure—especially when it came to his views about "the sickness of states and the lies of statesmen", which she found to be "childish". Early in 1957 Neville Braybrooke had written to her about a book that would be an "Eliot Celebration", a commemoration of the poet's seventieth birthday. Apparently they had discussed her contribution, as Braybrooke said, "I gather that you want to say something about 'Murder in the Cathedral' and perhaps this might be the best place to say it." Stevie's essay, probably the most stunning piece in the collection ("a very un-birthday present" she called it), is "History or Poetic Drama?", and in it she describes *Murder in the Cathedral* as "a remarkable evocation of Christian fears". She admires Eliot's poetic gift, the power of his imagery to evoke fear and disgust, and, indeed, she finds the sense of disgust in the chorus to be "the most living thing out of all the play". But Eliot's purpose is to move us "from human dignity to degradation", she asserts, for "fear is degrading, and we are counselled, for our soul's good, to fear". His effect, however, is to create "the uneasiness of dubiety". Does the play present the truth of history, philosophy, or the Christian religion? Stevie asks.

"Back to the Church", Eliot seems to cry, but whether or not Thomas Becket was as truly good as Eliot portrays him, Stevie cannot forget that the Church at the time of Becket was "somewhat stained with blood and no less greedy for political power than the State". "Mr. Eliot makes Henry and the State his villains, and what is he after? It is something that at first sight looks noble. But is it? Is it not rather something ignoble, a flight from largeness into smallness, a flight in fear to a religion of fear . . . ?" She found the play to be "beautiful and strong in its feelings", but also "abominable". Eliot's "terror-talk of cat-and-mouse damnation" appalled her, and she condemned him and other writers of the day who found "their chief delight in terrifying themselves and their readers with past echoes of cruelty and nonsense".

Sounding a favourite theme, "Truth is far and flat", Stevie concluded, "and fancy is fiery; and truth is cold, and people feel the cold, and they may wrap themselves against it in fancies that are fiery, but they should not call them facts; and, generally, poets do not." In what she says in this essay, and indeed in so many of the reviews she wrote in these years, Stevie is honouring a principle she practised in her critical writing and, in the autumn of 1958, offered to the readers of the PEN newsletter: "we [critics] may turn the sharp edge of our own opinions against the author's argument . . . provided we do not use the book as a mere peg."

One can disagree with an author's opinions and still enjoy that author's writings, as Stevie enjoyed Eliot's play, and as one who disagrees with Stevie's arguments against Christian doctrine can enjoy the poems which contain them. There is a zest, after all, that one can take and that Stevie took in argument. About hell, for example, one might have said to Stevie that the doors are locked from the inside. If one wishes to engage such a doctrine seriously, one should go to better sources than sermons intended for schoolboys and pamphlets written for popular consumption. Father Faber, who wrote the Catholic Truth Society pamphlet describing hellfire, clearly had different notions from those of the priest Flannery O'Connor cited, in a letter to a friend:

My reading of the priest's article on hell was that hell is what God's love becomes to those who reject it. Now no one has to reject it. God made us to love Him. It takes two to love. It takes liberty. It takes the right to reject. If there were no hell, we would be like the animals. No hell, no dignity.

Such an explanation was advanced once by Dorothy Sayers and produced a counter-argument from Stevie. Sayers' argument did not change Stevie's mind, no more probably than her own arguments alter her readers' convictions, for the religious belief or doubt of most people is probably not based on argument. In the end one can perhaps do no more (and a lot less) than the two speakers in Stevie's poems "I. An Agnostic" and "II. A Religious Man", whose friendship is not diminished by their differing views. The purpose of a religious poem, as T. S. Eliot said, is not to convert, but to let us know what it feels like to believe a certain way. And Stevie's poems on Christianity certainly do convey her passion. "You may think it is curious to write poems so much on argumentative subjects," she once noted, "but so did the Arians, you know, when they rushed about Alexandria, singing their popular song: 'There was a time when the Son was not'—what was this but a poem?"

A few years after her essay on *Murder in the Cathedral* appeared, Stevie visited Canterbury Cathedral and "went up the steps from where Becket was despatched". She was surprised to find Cardinal Pole buried there, and amazed the RCs hadn't claimed him. The postcard she sent to Helen Fowler depicted not Thomas Becket but Edward, The Black Prince.

Several years later, perhaps not inappropriately, the *Guardian* commissioned Stevie to write a poem for Whitsun—the great feast of the Holy Spirit. She sent them "How do you see?" in which the speaker, who seems plainly autobiographical, sees the Holy Spirit of God "as the holy spirit of good".

> But I do not think we should talk about spirits, I think
> We should call good, good.

She admits that Christianity is productive of good, and so is loath to speak against it for fear of diminishing the good in the world. Christianity is

> A beautiful cruel lie, a beautiful fairy story,
> A beautiful idea, made up in a loving moment

And Stevie poignantly portrays her own divided heart when she describes the "child of Europe"

Tearing away his heart to be good
Without enchantment. I heard him cry:

Oh Christianity, Christianity
Why do you not answer our difficulties?

The consolations of religion she finds "beautiful, / But not when you look close."

One week later the *Guardian* published a reply in verse written by a Dominican priest, along with eleven other letters, only one in support of Stevie. Father Edmund Hill's poem was fiercely critical of her: he called her arrogant and "not a fair questioner or a fair listener". A pleasanter response came from the Benedictine editor of *The Ampleforth Journal* who asked (and got) Stevie's permission to reprint her poem, along with Father Hill's, in the paper published at that distinguished Roman Catholic school. Stevie wrote warmly to the editor and perhaps this correspondence marked the beginning of her interest in Ampleforth, where she several times visited and read her poems. The Benedictines were the Roman Catholic priests Stevie was to know and like best. In many visits to her sister, who settled in the early 1960s at the edge of the Benedictine monastery in Buckfast, Stevie came to be a friend of several of the monks.

Just as Stevie was writing "How do you see?", an American priest, who was living by the Rule of St Benedict, wrote to a friend that he was reading Stevie's poems. "I love her, I am crazy about her", he said. This was the Trappist writer, Thomas Merton. One wonders if Stevie was surprised—when Merton's friend and publisher, James Laughlin, told her of the letter—to learn that the Trappist found not only "a lot of pathos under the deadpan sad funny stuff", but, as well, "a lot of true religion". Probably not.

Thomas Merton loved her - !!!

16

No Protective Curve at All

The close of 1958 was a desolate time for Stevie. She felt lonely and had some regrets about surrendering her job at Curtis Brown and the £400 a year it brought her. Her sketch book, *Some Are More Human Than Others*, had earned little critical attention and was selling badly. She was "getting rather sick of [books about saints and early church fathers] and I think I must now ask for something else." The poems she sent out to editors were regularly returned, and Stevie admitted they were "hardly Christmas fare". Mostly she wrote in glum moments: "misery alone seems to stir my sleepy old Muse into action", she said. Writing poems did, sometimes, lift "the glumness", as did the surprising alternative of reading either Agatha Christie or Roman Catholic theology. She especially enjoyed a book of theology called *Who Is The Devil?* which she reviewed for *The Daily Telegraph*, a paper that paid well, Stevie thought, but was "dodgy . . . politically".

All that winter Stevie worked at reviewing books and writing poems, among the latter "Was it not curious?" with its lyrical opening:

> Was it not curious of Aúgustin
> Saint Aúgustin, Saint Aúgustin,
> When he saw the beautiful British children
> To say such a curious thing?

Stevie called it a "wicked" poem in which she began by "forgetting it was Gregory & of course *not* Augustine. But then that soft sort of off-rhyme is so sweet I had to keep it & try & get out of it by a last verse recollection! V.v. tricky."

Was it not curious of *Gregory*
Rather more than of Aúgustin?
It was not curious so much
As it was wicked of them.

The famous quotation of Pope Gregory the Great on seeing British youths for sale in the Roman slave market was, of course, *Non Angli sed angeli* (not Angles but angels). In Stevie's poem, however, the curious thing said is that the gospel must be sent "At once to them over the waves." But the censure is the same: "He never said he thought it was wicked / To steal them away for slaves."

"Thoughts about the Person from Porlock" was first printed in November of 1959, but early in the spring of that year Stevie sent Professor Häusermann a copy describing "a new idea—if that's the word for it—I've had about stringing several poems together on a theme. For instance the Person from Porlock begins with (1) The Story (for ignorant types who havent heard of it!) (2) Research Material & how assinine [sic] it can be (3) the glum personal thoughts striking up (4) Wishing they werent so glum."

The speaker of the poem accuses Coleridge of being "already stuck / With Kubla Khan" when the Person from Porlock comes along and is made to take the blame for the poem's being unfinished. The verses in part describe Stevie's own loneliness (she wrote of herself just then as "awfully sad . . . quite *empty*" and "perpetually suicidal"), and in this mood even her poems seemed "pretty silly" to her.

I am hungry to be interrupted
For ever and ever amen
O Person from Porlock come quickly
And bring my thoughts to an end.

In the last stanza she gives herself a pep talk: "Smile, smile, and get some work to do / Then you will be practically unconscious without positively having to go."

This particular depression of Stevie followed on her return from an Easter holiday with the Clutton-Brocks at Chastleton, a "Jacobean Old Treasure", she called it, adding that it had "no electric light & at night, over the 2nd merry bottle of gin and the paraffin lamps, the old place does rather creak." She loved staying there, where she enjoyed "the sort of life & the sort of talk I like". Back home in Palmers Green Stevie felt engulfed by despondency.

In April Ladislav Horvat went to Switzerland to ski, and Stevie hoped he would meet up with Hans Häusermann. Stevie described Horvat to her Swiss friend as "very nice & a wonderful financier, but the bore is he likes one to follow the transactions with him & my arithmetic is not up to it—neither is my interest." She said Horvat was making money for her, but in her suicidal mood she didn't really care whether he did or not. In fact, Horvat found Stevie's "neatly ruled pages from grandmother's small cash book" quite clear and correct and dismissed her "self-deprecatory remarks".

From Switzerland Ladislav Horvat sent Stevie a "cat-card" that she liked very much, and she asked Professor Häusermann to send her another from the "series of Swiss sporting cats". By this time Stevie was assembling *Cats in Colour*, a picture book Batsford published late in 1959. Muriel Spark praised Stevie's introduction and humorous captions, adding, "It is very much a book for Stevie Smith fans as well as cat fans." The book did well in England and sold out in America. It did not seem to cast a shadow, as Philip Larkin later said cat books often do, "over even the most illustrious name". Stevie's introduction begins with mention of "sweet little catsy-watsies" but quickly shifts to stern admonitions about anthropomorphism and the way cats, like children in some of her poems, can be oppressed and made neurotic by the egocentric projections of adults: "what mind have animals? We do not know, and as we do not like not to know, we make up stories about them, give our own feelings and thoughts to our poor pets, and then turn in disgust, if they catch, as they do sometimes, something of our own fevers and unquietness." Stevie quotes D. H. Lawrence saying, "in his ratty way", that so much of modern life is fixed, "and our animals are most 'fixed' of all". Like all tamed animals, cats are nervous creatures, Stevie maintains: "we have given them reason to be, not only by cruelty but by our love too, that presses upon them."

Stevie then tells of a visit she made to the Edinburgh zoo while in Scotland with the Fowlers during the summer of 1958:

> I stopped outside the tiger's glass-bound cage. He was pacing narrowly, turning with a fine swing in a narrow turn. Very close to me he was, this glass-confinement needing no guard-rails. I looked in his cold eyes reading cruelty there and great coldness. Cruelty? . . . is not this also a romanticism? To be cruel one must be self-conscious. Animals cannot be cruel, but he was I think hungry. To try it out, to see whether I—this splendid human 'I'—could impinge in any way upon this creature in his ante-prandial single-mindedness, I made a quick hissing panting sound, and loud, so that he must hear it—hahr, hahr, hahr, that sort of sound, but loud. At once the great creature paused in his pacing and stood for a moment with his cold eyes close to mine through the protecting glass (and glad I was to have it there). Then suddenly, with my 'hahrs'

increasing in violence, this animal grows suddenly mad with anger. Ah then we see what a tiger—a pussycat too?—driven to it, can do with his animal nature and his passion. Up reared my tiger on his hind legs, teeth bared to the high gums, great mouth wide open on the gorge of his terrible throat. There, most beautifully balanced on his hind legs he stood, and danced a little too on these hind paws of his. His forepaws he waved in the air, and from each paw the poor captive claws scratched bare air and would rather have scratched me. This great moment made the afternoon for me, and for the children too and for my old friend, their mama (and for the tiger I daresay) and cosily at tea afterwards in Fullers we could still in mind's eye see our animal, stretched and dancing for anger.

The director of The Royal Zoological Society of Scotland objected to what he took to be Stevie's criticism of conditions in the Edinburgh zoo and argued that the tiger could not have been behind glass. Stevie wrote to Helen Fowler to say "Help! But it *was* behind glass, wasn't it? Dont you—& the children—remember how I teased the poor thing by painting at it?" They did, and wrote saying so to Stevie's publisher, who sent their confirmation to The Royal Zoological Society. In reading the passage, one recalls the sufferings of "Flo", the tiger to whom Pompey compares herself in Stevie's early fiction.

In her introductory essay on cats, Stevie has an eye for "little common cats" as well as for cats "of family", and her "ash-cats going sorrowful about the palings of a poor London street" are reminiscent of T. S. Eliot's grottier feline specimens. In a review of a collection of poems, Stevie once quoted an anthologist as saying: "'a good armful of flowers . . . should include not only tall lilies and perfect roses but some simple daisies and pungent field flowers as well.' Perhaps as a pungent field flower I might observe," said Stevie, separating herself from fancier specimens. In *Cats in Colour* her loyalties seem to gravitate, too, towards the less elegant species.

Among artists who have drawn cats, Stevie mentions in her essay the English painter Louis Wain, "who while he was residing at the Maudsley Hospital for his madness, drew all the nursing staff and the doctors and psychiatrists in cat forms but true likenesses." She does not mention, though, what she once told Anna Browne in a hushed voice, that while an inmate at the mental hospital, Wain painted a version of "The Last Supper" in which Christ and the Apostles are depicted as cats. "Stevie was shocked", Anna recalls, "she kept saying, 'Oh, but it's such blasphemy!'"

The only cat villains Stevie could "think of" in *Cats in Colour* are those that were the embodiment of the devil in witch trials or that the devil gave to witches to be their familiars. She briefly detours to touch again on recurrent themes, especially the cruelty of the Scots who burned witches, whereas the English less cruelly hanged them. "Witch cats . . . haunt my memory", she says, quoting two of her own poems about them: "My Cats" and "Great Unaffected Vampires and the Moon".

Sally Chilver remembers that Stevie enjoyed choosing the photos for *Cats In Colour*, and she probably had fun forming the descriptions that went with them. After her admonition in the introduction about anthropomorphizing cats, Stevie wrote such jokey captions as, "I have fine flanks and a good tail but I don't much care for music", and, "My maid's sent me out without my lip-stick. She'll be sending me out without my dress next!" Stevie says she once had a cat called "Tizdal" ("just such a kitchen fat cat as I love"), but she seems not at all to have fancied cats in the way some people imagine. The title of a review she wrote for *The Observer*, "Keeping cool about cats", describes her sentiments. "I am not a Cat Lover", she wrote unequivocally in 1963, adding, "I like cats. There is more attraction and peculiarity in a plain kitchen tabby than in all the god and devil cats of fancy. Looked at as an animal, it takes the mind off and warms the heart. It is not right to look at it in any other way." In later years Stevie often found herself irritated with her sister's cat, Ming, but on the other side, she was active in trying to help a friend find a home for "three top cats that poor Edith Sitwell left behind . . . We can't have one because of being too much of a tie." After *Cats in Colour* appeared, literary editors connected Stevie with pets and she reviewed a fair number of books about cats—a few of them concerned both with cats and witches.

In May of 1959, Stevie's radio play, *A Turn Outside* was produced for the BBC Third Programme with professionals acting the parts of Stevie and the Interlocutor. The play is much like the poetry readings Stevie gave, but with a macabre twist when the Interlocutor is revealed to be Death. The producer was Douglas Cleverdon who, with his wife Nesta, had been friends of Stevie since the late 1940s. Many of Stevie's friends heard the broadcast and found it to be effective—so much so that Helen Fowler wrote with some alarm to see if Stevie was all right. She reassured Helen that *A Turn Outside* was more in the way of her getting "romantic about old Death . . . it's in all my poems, *passim, ad nauseam*, I fear. No doubt I shall creak on to be 100. But it is sweet of you to worry."

A lovely and hitherto unpublished poem seems to be among those in which Stevie becomes romantic about death. It is undated, but its description of the lover being "as tall and gray as the morning" is reminiscent of Stevie's late poem, "Black March":

I

He comes to my room
He is as tall & gray as the morning
He stands silently
He has a silent look
He stands by my bed.

He takes my shoulders in his hand
His hand is as flat and gray as the morning
Speak to me!
He does not say a word
I have never heard his voice.

2

In the avenue of my faubourg
The white flowers shine under the green leaves
At night time.

The summer sun falls quickly
But not so quickly as your kisses fall
And the flowers
Are not so light as your eyes.

3

I love you darling
The moment before it is absolutely certain that you love me
Is flying.

This moment
Is like a dandelion puff before a wind that is rising.

The same month that *A Turn Outside* was produced Stevie went to Oxford with Elisabeth Lutyens for a performance of *The Bacchae*, a play which, since her days at The North London Collegiate, had absorbed her, partly because she thought that in it Euripides was "sailing as near the wind of atheism as he dared". Miss Lutyens, who did a lot of work for the Oxford Playhouse, drove Stevie to the performance. She remembered "an awful row about sausages" Stevie had left in the car. According to Liz, Stevie decided to return home with "a dishy young man with a smarter car". The sausages became a major matter, with notes and phone calls from Stevie, who tried to recover them. "It could have been World War I, the fuss she made." Miss Lutyens found her "of great value . . . [but] the most frightful bully". Stevie gave a calmer account of this evening. The "dishy young man" was Rodney Phillips, who had been the financial backer of a journal called *Polemic*. "He brought me back from Oxford to Palmers Green in 1½ hours," Stevie said. Miss Lutyens's music she judged to be "just right".

Among the more unusual books Stevie reviewed in 1959 was a collection of letters by E. H. W. Meyerstein—scion of a wealthy Jewish family who travelled, wrote poems, loathed women (according to Stevie, as a result of the "possessive and capricious furies" of his mother) and engaged, when he could, in masochistic acts. Stevie had known "the old Turk", as she fondly called him, and wrote to congratulate a close friend of Meyerstein on the collection of letters he had edited. She recalled an incident when Meyerstein was her "dragooned escort at a Newnes Office Dance (help!)" which "seemed funny enough at the time but much funnier now". She thought him "a real hero to have done what he did with his life & talent among all those hair raising hates, loves & crochets [sic] of his needy old heart. There's nothing older, timeless really, than a 15 year old heart in an aging body, bossed by a scholar's mind." In these sentences Stevie catches not only Meyerstein, but also something of herself. "Pompey Casmilus in her teens", Professor Häusermann called her, in response to her success in passing off as "juvenilia" poems recently written.

Much of July Stevie spent with the George Lawrences at Brockham End, just outside Bath. In August she wrote to tell Hans Häusermann: "Have just had a heavenly day doing what I like best of all—walking & swimming turn & turn about in heavenly hot weather in the country." Her companion was Stanley Bate, the composer she had visited in Milan. Stevie was amused by Bate, a "feeble little ass, [who] wouldn't bathe, & when I said, 'We've got to get under this wire but mind you don't touch it because it's electrified,' nearly *fainted*!" She thought people who like the country "in the right way, i.e. being able to recognize a cow but not gushing", were rare, and she wondered if her companion thought farmers wanted to kill their cattle by electrifying fences. She was scornful, also, of his asking "if the river Lea (very much in its upper reaches) was 'salt water or fresh?' " Stevie could not have known, of course, that two months later Stanley Bate would commit suicide.

As the 1950s came to an end, she managed to publish a few poems and continued to review, though her pace was slackening. She also read her work in London as well as at the universities in Nottingham and Birmingham. She was guest of honour at a party in Birmingham, and was put up "in a sort of *palace*, dear", she told Helen Fowler. "Very different from the home life of the old poet, but I am back now shovelling the coals & washing up. Ha ha & I like it, you know. I am really getting awfully domesticated, & I really think it is nicer—easier

anyway—than writing." She had made three dresses for herself and
joked that "As I am the same back & front, it is rather simple." Aunt
now had a hearing aid which, Stevie said, rattled and whistled in an
extraordinary way. Although Aunt could now hear her, "there
doesn't seem much to say".

It was around this time that John Guest, an editor at Longmans, met
Stevie at a crowded party given by the philosopher, Maurice Cranston. Crushed against a wall next to her, Guest impulsively said he
loved her poetry. Stevie just as impulsively asked, "Would you like to
publish me?" So the plan to do a *Selected Poems* came about. A short
time later, Stevie confessed about her poems to Professor Häusermann
that she did not "feel too keen on any of them" and did not "feel at all
inspired to write". (Several months later she told another friend, "I
know they're *something*, but is it poems?") But the invitation from
Longmans must eventually have roused her, because before long her
tune changed, and she wrote to tell Rachel Marshall that she had
finished "a lot of new poems, almost enough for another book".

The Longmans book was to be a selection from old books with the
new poems added. Stevie would sort through old poems, then stop
and "write a new one for a change". Making this selection occupied
her through much of 1960. In her letter to Rachel Marshall, Stevie
added that she was sleeping "far too much, but feel better for it & have
put on 1/2 a stone, I am now 8 stone." The other literary project she
entertained in 1960 was compiling an anthology for Deutsch of
writings about hell. She started "a Hell File", but in the end did not
complete the book.   *Depression*

Throughout 1960 Stevie was hampered by an arthritic knee. She
went to a specialist who recommended an operation, which she
delayed, perhaps because the results could not be guaranteed. But a
deeper distress slowed Stevie's efforts. "If only I could get over this
neurotic attitude to work," she told Diana Athill, "being afraid of
getting it wrong, being afraid of everything really, and thinking how
much nicer it would be to be dead . . . I say it's laziness, but of course it
isnt, but it sounds less dotty." None the less Stevie continued to do X
some reviewing. Among the books she wrote about for *The Observer*
was a collection of Ronald Knox's sermons. Stevie said Knox had
"always been for me . . . a fascinating yet a painful figure". She loved
his sweetness, and his bidding everyone in a Christmas sermon to
creep as children again upstairs to the nursery appealed very much to
her. But she went on to say that what they learn there may "bring the
little children downstairs again looking rather pale". Once more
she condemned Catholicism for sowing seeds of cruelty and guilt in
its adherents. The review engaged Stevie for the third time in a

correspondence with the Oxford student who had protested to her about "The Necessity of Not Believing".

Although there was no further assurance from her physician that an operation would repair Stevie's knee, damaged originally in school hockey games, she decided to undergo surgery. At Wood Green & Southgate Hospital, in January of 1961, a surgeon removed her right patella. Several friends came to visit Stevie in hospital where she lay in "lazy comfort", feeling "spoilt" by presents she received. Aunt wrote long letters nearly every day, replying to Stevie and reporting on domestic details. Signing herself "Yours lovingly", Aunt expressed concern that Stevie was "spending a fortune on stamps". In hospital Stevie read a "wonderful" mid-Victorian edition of *The Arabian Nights*; *New Grub Street*, lent her by her dentist who was a Gissing fan; Henry James's *The Awkward Age*; Graves's *The Golden Fleece*; and "lots of tecs, A. Christie & so on". She remained in hospital until early February and so had to put off until March a short recording on books the BBC invited her to give. Meanwhile she spent several weeks recuperating in Buckinghamshire at the home of Nancy Hodgkin, who sculpted Stevie during her stay.

In March her knee was still "pretty feeble", but the BBC provided a car and a rug and Stevie went to the studio and recorded her talk on a poet's reading. When she heard the broadcast she became upset about the sound of her voice: "as if I had a code iddy head!" She wrote at once to Rachel Marshall for advice about the "awful nasalness of my voice. . . . Is there anything I could *do* about this?" she asked her old friend. What with Aunt's arthritic hip and her bad knee, she said, "we are a nice couple of lame dogs". Several weeks later she wrote again to Mrs Marshall with the news that her "poor horrid little knee has *got worse*" and she had difficulty getting about. She felt disinclined to do anything and considered herself "a deserter to ill health". "My poems always come home to roost," she wrote, thinking of "The Deserter". "*They* at least know who they were written for!"

In June Stevie had a brief contretemps with Victor Gollancz and the National Campaign for the Abolition of Capital Punishment, of which he was joint chairman. She wrote and asked that her name be removed from the "anti-death penalty list". Gollancz could not find her name on any list and wrote back to say so, lamenting the "ugly" sneer in her letter. "How you nineteenth century rationalists misunderstand us," he told Stevie. "We do not believe in 'heaven', nor in a merciful old man with a long beard. . . . Nor do we think for a moment that death is 'the greatest calamity that can befall a human creature'." Stevie seems to have supported the death penalty in her tiff with Victor Gollancz, and managed as well to air other favourite themes. And she went public with her quarrel in a nearly libellous

portrait of SS by Cubitt

way in an *Observer* review published a few weeks after this exchange. Speculating on how "serious" contemporary writers might refashion a story in which a cat saves the life of a dog in the Canadian wilds, Stevie wrote: "they would let them sit in a ditch and wait for God. Or perhaps, if it was contemporary history, one animal would have murdered all the animals of Canada and Mr. Gollancz would come forward and turn the other cheek in the least private way imaginable."

Some of Stevie's loathing of the "Lefty goodhearts" may have been operative in this. But one friend she saw often in the years 1960–65 was the architect and sculptor, James Cubitt, a passionate Marxist. He and his second wife, Anne Sitwell, frequently gave parties at their home in Regent's Park, and Stevie attended, along with Rosamond Lehmann, Anna and Michael Browne, and others. Afterwards the host usually drove Stevie home to Palmers Green. He felt even then that "Stevie was a sort of celebrity" and remembers that they both had "a common dislike of Christianity, to say the least", and that this subject (rather than politics) was one they frequently discussed. Drawing was another, although Cubitt thought her sketchbook a failure because of the distortion that occurred when her drawings were blown up. The drawings themselves he greatly admired, noting that Stevie touched in all of her work "the same nerve of uncertainty as the later Giacometti". "She drew a cat conceptually," he said, "not anatomically . . . a cat flows, and basically that was what Stevie caught." Cubitt found Stevie "attractive as a woman . . . nice figure, graceful" and thought she "had a genius for friendship and also a genius for the reverse". Although appearing to be vulnerable, Stevie was in her own way "very tough", he said, and particularly about her art. It was Stevie who directed him to Ladislav Horvat, whose client Cubitt briefly became.

Stevie's sister retired from her job as County Drama Organizer in the spring of 1961 and had a bungalow built in Buckfast, into which she moved that year. Stevie described her as "rather downcast" by her difficulties in securing a loan. The bungalow turned out to be more than Molly could afford, and Stevie eventually made two interest-free loans to her, wanting her "to be easy & happy, specially in case the poor darling doesnt as they say 'make old bones'". She went to visit Molly in Buckfast early in August, and travelled from there to Hartland (near Bideford in north Devon) where she was a guest of the Watneys—Antoinette, her husband John, and their son, Marcus.

The Watneys were full up with guests and first proposed to Stevie that she come later in the year. "No, no," Stevie said, intent on coming when it suited her. (Mrs Watney thought her "immensely self-centred" and "not all sweetness and light" though "terribly funny".) She came, and they put her up in the village pub partly because, as Stevie said, "the child of the people I am half-staying with . . . developed Chicken Pox." Once installed there, Stevie was adamant about doing *what* she wanted, *when* she wanted. Largely the Watneys devoted themselves to their son who was on school holiday and, at least part of the time, unwell. When he felt up to it, his parents often took him and his young friends off to the beach. Before they got to their picnic lunch, perhaps exercising again her competitiveness with children, Stevie would insist on going back to her room. "But you can't go back now," Mrs Watney would insist. "That was a side of her I didn't know before," said Mrs Watney, who found Stevie "totally unbending, inflexible". ("I can vouch for the 'inflexibility' [Antoinette Watney] speaks of," recalls Sally Chilver. "Nothing really came up to the cossetting Aunt gave [Stevie], and also called for.")

The weather was unusually wet during Stevie's visit and she spent part of her days in bed. She did like the coast and the bathing which was

> wonderful, but fierce beyond words, on this coast of granite, the rocks & cliffs like awful teeth & knives, black ones, but the waves are wonderfully green & enormous, only one is afraid of being too tired to crawl out again, & of course gets rolled about rather.

The people in the pub were nice, Stevie told a friend, but "There is also a Voice I hear everlastingly here on the telephone reading out Racing Prices (if that's what it's called?) No Good for Us Higher Types." Aunt was sorry Stevie's holiday with the Watneys was not going well, and wrote to say, loyally:

> Why cant people be honest? Why if they did not intend to treat you more as their guest did they keep on asking you to that cottage & why tell "a needless lie" (as someone once said to me) must they say the distance to the sea was only half a mile. Abominably rude I call it to suggest you went off by yourself & they would pick you up from anywhere you liked to go. I am glad you are comfortable at the Inn but you dont want to stay in bed all the day when on holiday—At the least if you don't want to go to the beach they might suggest you brought your lunch & sat in their garden in the cottage if the weather was bad—.

By the third week in August Stevie had arrived at the Brownes' in Norfolk where, she told Aunt, her reception was far more cordial than at Hartland. Anna Browne recalls that "Stevie felt the Watneys were unduly pre-occupied with [their] child at her expense". But Stevie was being "unreasonable", Anna felt, and she sympathized "very much with the Watneys—including the child".

"Your experience there," Aunt said (referring to the Watneys'), "would form a groundwork for a novel or short story & if they recognized themselves serve them right—." Such an idea had already occurred to Stevie, who, on 10 August, had sent to *The Observer* a poem called "The Holiday", which they published a month later:

> The time is passing now
> And will come soon
> When you will be able
> To go home.
>
> The malice and the misunderstanding
> The loneliness and pain
> Need not in this case, if you are careful,
> Come again.
>
> Say goodbye to the holiday, then,
> To the peace you did not know,
> And to the friends who had power over you,
> Say goodbye and go.

Mrs Watney saw the poem and presumed Stevie was writing about her family, although a number of people said to her, "Don't be silly, it's about a visit she had with *us*."

Stevie was in hospital again in October, this time for the removal of a tumor from the area of her right breast. "Isn't it absolutely ghastly," she said to a friend, but typically her wit was engaged. "I have something they call a BENIGN LUMP . . . sounds like a fairy godmother beginning to put on weight." The operation was successful, and eventually her knee also healed, for her physician recalls being called by his children, some time later, to see "Miss Smith" turning a cartwheel on his garden path.

In December Stevie broadcast on the BBC World of Books and she also acceded to the request of Peter Orr, then Deputy Director of the British Council Recorded Sound Department, that she do a reading. She arrived for lunch "in a costume that suited her—it wouldn't have suited anyone else", he said. After a substantial lunch, Stevie brought her manuscripts into the studio and "in a very unpretentious way spoke and sang her poems which were recorded. She seemed quite prepared for you to think her dotty and to believe you might be right. One didn't at first see the sharpness and soundness and practicality behind this."

It was around this time that David Wright organized a poetry reading at a London department store. Stevie and the Irish poet, Patrick Kavanagh, were to read; and Brian Higgins, a remarkable poet who published three books before his premature death at the age of thirty-five, was to compère the evening. The reading was a disaster. David Wright gives this account:

> Stevie and Kavanagh took an instant dislike to each other, & Higgins, himself an adept at creating chaos, proved totally incapable of dealing with the chaos of others' creating . . . at the end of the reading I escaped with Stevie Smith . . . Kavanagh & Higgins & the rest of the audience went to the George, the BBC pub; which proved an evening . . . of grief for Higgins; who got into a fight with somebody, & was thrown out with a black eye & a bitten ear.

Stevie's nights out became fewer. In 1962 Aunt was progressively more immobilized by her arthritic hip and eventually forced to stay upstairs in her room. As a consequence, Stevie's domestic routine became more strenuous. "*I* (rather late in life) now do all the cooking, etc., & take both our meals upstairs on a tray," she told a friend, "& there we camp, very cosily, but it doesn't seem to leave much energy for anything else." Dr Curley speaks of Stevie's efforts to look after Aunt as "brave", but it would only be human for Stevie to have been overwhelmed from time to time by her responsibilities. Once, a friend recalls, Stevie came in tears to tell her that she had struck Aunt and was horrified at what she'd done. And in a letter Molly, who had always been a bit less ardent about Aunt than Stevie, wrote to her sister: "I fancy things are not so easy. I think probably it is because you are far more 'anxious' of the old lady's possible difficulties & discomforts than I am . . . we both know how ungracious she can be without meaning it." But Molly confesses in this letter her concern for Aunt's failing sight and offers to bring a new lamp if it would prove useful. Molly wished Aunt would come and spend the colder months with her, but by then Aunt's immobility obviated any such arrangement. And late the following year Molly herself suffered a coronary that left her invalidish thereafter.

Molly did stay with Aunt in Palmers Green in August of 1962, when Stevie spent a fortnight in Venice with Sir John Lawrence and his wife Jacynth. "Stevie was perfectly awful on the Venice trip", they agreed. The Lawrences thought it would be ideal to sightsee in the morning, picnic at the Lido, and spend a leisurely evening. But Lady Lawrence found, as had Mrs Watney, that "with Stevie you had to do exactly what she wanted". Stevie disliked Abroad. She thought Italian food was "disgusting", and at a restaurant in the Piazza San Marco ordered rice pudding. The Lawrences thought the trip to Venice with Stevie was worth it—once—but "I wouldn't have done it again", said Sir

John. Stevie would suddenly say, after dinner, let's go sightseeing, and the Lawrences found this meant sitting on the edge of the canal where Stevie would say how pretty it all was. "Most exhausting", was their opinion about Stevie's behaviour. Nor could they persuade her to look at anything else. One wonders, of course, if the daytime heat wasn't disagreeable to Stevie. Pompey certainly would have disliked the summer heat in Venice: "a heat wave and all the fixed hard colours take it out of you", she tells us. So Stevie dodged the daylight in Venice and went about as much as she could at night. Sir John did persuade her one day, though, to abandon her search for safety pins and to go instead with him and his wife to see the Tintorettos. The picture Stevie liked best was Tintoretto's "The Creation", in which all the beasts are "streaming out from the hand of God" except for one which, having had a look at the world, "is going back". "That's me", Stevie said.

She was certainly not Mediterranean-minded. In one of her reviews she made this plain when speaking about "the feeling one has sometimes in Italy, that too many people have been living there too long, and with a fancy for something barer—Northumberland coming up to the Wall, say." A few years later she tried to persuade Lady Lawrence to travel with her to France, but her friend declined the invitation.

That autumn Stevie's *Selected Poems* was published in England by Longmans, and Lady Lawrence said in her *Frontier* review: "They speak of terrible things. Of doubt and death, of love and fear and loneliness and despair. They would be too painful without the laughter. Stevie Smith is an avowed agnostic. But these are religious poems." The editors had been back and forth with Stevie about which poems to include, and when Jonathan Williams complained about the absence in *Selected Poems* of "Tender Only to One", "Come", and "Do Take Muriel Out", she told him: "My poor publishers were trying to choose them all, and anything they chose I didn't like. And anything I chose, they thought that wasn't right either. I'm amazed it ever got published."

The reviews, though, were enthusiastic. Although many of the critics noted the usual and flattering resemblances—Blake, for instance, Emily Dickinson and Edward Lear, there was more emphasis on Stevie's singularity. Most of the reviewers came round to a judgment voiced by Kathleen Nott in *The Observer*: "There is simply nowhere to put her . . . she puts you in mind again and again of something to which you must on no account compare her." Both Anthony Thwaite and the *Spectator* reviewer mentioned witches: "The animus behind Miss Smith is . . . a witch", the latter wrote.

A reviewer in *The Queen* (for which Stevie herself began to write reviews in 1963) suggested that Stevie Smith had been "enormously under-valued", and it is true that with the appearance of *Selected Poems*, first in England, and two years later in America, her stock as a writer soared. After comparative neglect in the 1950s she was, in the 1960s, suddenly in demand, particularly for poetry readings.

The *TLS* reviewer who wrote that Stevie might be "one of those who attract addicts" was prophetic. Such readers appeared in increasing numbers in the 1960s, among them Sylvia Plath, who wrote in November of 1962, a few months before her tragic suicide, to tell Stevie, "I am an addict of your poetry, a desperate Smith-addict", and to ask where she could obtain a copy of *Novel on Yellow Paper*. Earlier in the year, *The London Magazine* had printed Plath's response to their question, "What living poets continue to influence you, English or American?" She had replied, "The poets I delight in are possessed by their poems as by the rhythms of their own breathing. Their finest poems seem born all-of-a-piece, not put together by hand: certain poems in Robert Lowell's *Life Studies*, for instance; Theodore Roethke's greenhouse poems; some of Elizabeth Bishop and a very great deal of Stevie Smith ('Art is wild as a cat and quite separate from civilization')." In her fan letter, Sylvia Plath asked if Stevie would come to tea or coffee at the London flat in which she planned to be living by the end of 1962, and she received "a deliciously Smithish letter" in reply:

Dear Sylvia Plath,

Thank you so much for your letter—I was glad to hear from you & glad you enjoyed the Harvard record.

I'm afraid I really dont know where you would find a copy of *Novel on Y.P.* now. It did go into a Penguins (in 1950, I think it was) but that sold out & they did not reprint. When I go downstairs—I camp upstairs most of the time with my aged Aunt, she is 90!)—I will look out the address of a man who sometimes manages to track down books for me. He lives in this neighbourhood oddly enough but is very *shy*—just sends the book & the a/c—wh. after all is what one wants.

I do hope your novel goes well & I do hope the move in the New Year goes well too—if only as you suggest, so that we can meet some time.

I feel awfully lazy most of the time, even the idea of writing a novel makes me feel rather faint! And as for poetry, I am a real humbug, just write it(?) sometimes but practically never read a word. That makes me feel pretty mean spirited when poets like you write such nice letters.

Yours ever,
Stevie Smith

Sunday Looks as if I'd been for *days* upstairs—but it's just the Oblomov in us all!

Another poet who appeared in print in 1962 praising Stevie's poetry was Philip Larkin. In his review of *Selected Poems* for the *New Statesman* he wrote, "I am not aware that Stevie Smith's poems have ever received serious critical assessment . . . her poems, to my mind, have two virtues: they are completely original, and now and again they are moving. These qualities alone set them above 95 per cent of present-day output."

James Laughlin of New Directions, the American publisher of *Selected Poems*, told Stevie how difficult it was to "'break through' with a new name, especially a poet", but he had sent her poems around to friends and they were met everywhere with enthusiasm. Among those who sent Laughlin praising sentences about Stevie's book were Brendan Gill and Howard Moss at *The New Yorker*; Muriel Rukeyser (who commented, "Stevie Smith is our acrobat of simplicity", a remark which left Stevie grateful but puzzled); Louis Untermeyer and Babette Deutsch. Ogden Nash's praise is still widely cited, at least in excerpted form. He wrote:

> Who and what is Stevie Smith?
> Is she woman? Is she myth?
> Winging ever out of grasp
> Like a butterfly or wasp,
> Slipping from her secret nook
> Like a goblin or a spook,
> Searching out her God to haunt him,
> Now to praise him, now to taunt him,
> Then to sing at Man's expense
> Songs of deadly innocence.
> Across the world I hear her still,
> Singing underneath the hill.

Robert Lowell also sent James Laughlin a puff for Stevie's book, saying: "I was wild about [Stevie Smith's poems] . . . so witty, charming, gruesome and unlike the usual poem trying so hard to be a poem."

Stevie thought being valued for quaintness a "sad fate", and fearful of being described as an Ogden Nash *manquée*, she wrote her own blurb for the book:

> The English Poet Miss Stevie Smith is unlike any other poet writing today. There may be echoes in her work of past poets—Lear, Poe, Byron, the gothic romantics and Hymns Ancient and Modern —but these are deceitful echoes, as her thoughts may also seem deceitful, at first simple, almost childlike, then cutting at depth with a sharp edge to the main business of her life—death, loneliness, God and the Devil. Her metric, with its inner rhymes and assonances and the throwaway line that can seem mischievous, is very subtle. "A daring and skilful technician" the *Times Literary Supplement* calls her. Another poet says: "I have heard her both read and sing her poems; the beat is formidable." She is certainly funny. But it is not a humour one would care to meet on a dark night. Her own drawings which illustrate the book add to the general nerviness. Elizabeth Jennings says she is a poet who appears to be on the brink of a nervous breakdown. I would say at worst, she might bolt . . . like the girl she draws on the cover—a bolter if ever there was one. But the control is very strong.

Stevie also quoted to Laughlin remarks she obviously approved by George Stonier, who said she could "drop into poetry", and John Betjeman's "She is a good girl and bites suddenly."

Laughlin also succeeded in placing some of Stevie's poems in *The New York Review of Books*, *The Nation*, *The New Yorker*, and *The Atlantic Monthly*. When *Selected Poems* appeared in America the reviewer in *The New York Times* wrote: "I unreservedly say: Admire Stevie Smith." A unique response to Stevie's book came from the poet Marianne Moore, who wrote to "Mr. Laughlin": "We all of us need her *Selected Poems* and drawings—to counteract the estranging determination of some writers of prose and verse to obtrude on us their wanton unnaturalness. I have thought so for a long time. You do send me good things." At the bottom of her note Miss Moore typed, "Mrs. James Laughlin", and Stevie wrote back to Laughlin: "Isn't it super about M. Moore. . . . Isn't the change of sex interesting too?. . .Mr. above, to Mrs. below. I hope you are feeling *all right*."

Stevie seemed sometimes in high spirits when she stayed with the Fowlers in Scotland in August 1963. The house was near a river, and close by grew some giant hemlocks. Having Stevie around was sometimes like having another, rather naughty child to look after, Helen Fowler recalls. Her son was then seventeen, and Stevie suggested to him how marvellous it would be to break off one of the large hemlock limbs and plant it some night in a neighbour's garden. He promptly did so and tied to the tree a little tag reading, "Hemlock and After". Helen immediately guessed that Stevie had put him up to it. She found Stevie to be more jealous of animals than of children: Stevie did not like people to pet cats, for instance. She thought her friends should be paying attention to her, not to their animals.

Robert Lowell met Stevie when he sat next to her at a poetry broadcast festival held at The Royal Court Theatre that summer. Day Lewis, Roy Fuller, Peter Redgrove and others took part. This was only one among many readings Stevie gave as the 1960s went along. Word of her performances had reached New York, and The Poetry Center at the 92nd Street YMHA invited her to appear there. The Academy of American Poets also hoped to persuade Stevie to make a "college reading circuit" in the next year. Stevie felt "as long as I don't come to America" the reviewers who confused her with the young girl depicted on the cover of *Selected Poems* wouldn't find out she really looked "a young 150 in a 99 year old who's worn badly".

In March 1964, Stevie read at a meeting of the Oxford University Poetry Society. She described the experience to her Cambridge friend, Helen Fowler, as "simply awful".

Stevie's drawing on the cover of the American edition of her 'Selected Poems' (1964).

The young man turned up & almost failed to meet me (meandering up & down the platform instead of planting himself firmly at the ticket barrier) & had to borrow the taxi fare & the cost of a few drinks from me. And only about 20 turned up, & the room wasn't heated & I did not get anything to eat until 9:30. And then I had to change my hotel because we had a row because they wouldn't bring us our dinner! Ha, ha. I think Oxford is a *bad place*. (Always have.) They promised to pay my expenses, so I let them off a fee. But I doubt if I shall ever see the £5 that they came to.

In the same letter Stevie describes her lethargy, "the sort of awful mood in which one sits on some draughty provincial station waiting for a beastly train that never somehow *comes*." Betty Miller, her friend in former years, was in hospital with a degenerative brain disease, and Francis Hemming dropped dead at a dinner party Phoebe Pool gave. "Oh what awful luck poor Phoebe does have", wrote Stevie, who wished "we could all go & sit on a beach somewhere—in the hot sunshine. A nice flat broad East Anglian beach. (I nearly wrote 'Anglican'—shows how theological I'm getting!)"

On a Sunday night in November Stevie took part in a poetry reading staged at the Aldwych Theatre. She described it as "mixed up with Jazz and several other poets . . . and some poems of poor Sylvia Plath said by an actress. The theatre absolutely crammed and hardly anybody over the age of twenty." Glenda Jackson, on the same programme, found Stevie to be a revelation:

She came up to me and said "hello," and I said "hello". . . . She was wearing one of her famous white Peter Pan collar blouses under a black schoolgirl jumper . . . She was very tiny and had this straight across fringe. I remember the absolutely straight way she stood before me. Most people meeting other people for the first time introduce some form of protective curve—there was no protective curve to her at all. It was not a stiff . . . soldierly stance. I mean it was just an absolutely direct very honest thing. And those eyes that were observing, read your whole life and dispensed with it in about 30 seconds. And then she leapt on to the stage and read "Not Waving but Drowning", and I could not believe that that tiny creature could encompass what that poem said.

This was the only occasion on which the actress and poet met, but it was enough to inspire Glenda Jackson, after Stevie's death, to read her poems for a phonograph recording, and to enact the role of Stevie Smith on stage and screen.

17

Feebleness and Fame

As Christmas 1964 approached, Stevie wrote a long letter to Rachel Marshall describing life at 1 Avondale Road. She typed it because she'd burned her fingers cooking ("I'm always doing this . . . very clumsy of me"). Visits to Cambridge were no longer easily arranged. Aunt, who still managed "to totter around" and kept pretty well, had not been downstairs for nearly three years.

> I camp with her in her room, which we have managed to rig up as quite a presentable little den for the Lion, and only come down to do my work, writing, I mean, and of course, the house work which doesn't get done too well, I fear. We have had a series of awful chars and now have no one, and I'm rather glad we haven't, tell you the truth. I am getting quite a dab at it all especially the cooking.

The rest of her energy Stevie kept for trips into London that involved such professional chores as being interviewed. She continued to read and sing her poems quite often, although she complained about a painful thickening up "of my m's and n's" and had plans to see a doctor at Guy's about it. Generally, though, she felt fit and attributed her good health to looking after Aunt. Her suburban life left her lonely, however: "The almost absolute dearth of companionship in Palmers Green does drive one into writing, and, like a good many other writers, I need driving." In general she saw herself as "slightly lost", although with the success of *Selected Poems* in America and her inclusion as one of the three poets in *Penguin Modern Poets No. 8*, Stevie admitted, "I am becoming quite famous in my old age, isn't it funny how things come round."

Stevie continued to worry about Molly, as she would for the rest of her life. "Poor Molly, I am afraid she will never really get over the coronary she had last Christmas." Her sister had recently come home, Stevie told Rachel Marshall, but she seemed far from well. At Buckfast her life was happy, Stevie thought, though it was "an R. C. Centre. Not quite in my line, I fear." Stevie did find everyone there friendly and seemed more mellow about her sister's religious predilections.

Although in these years Stevie's dedication to domesticity curtailed her expenses, and on the credit side her royalties were mounting, memories of economic perils in her family endured, and a letter she wrote to an editor early in 1965 shows her vigilant and very much on the alert about finances. The copyright for all of her books published by Cape had reverted to her, she wrote, although "Chapman & Hall, André Deutsch and Longmans, have prudently hung on to their copyright". In this instance the editor wanted First Rights publication and Stevie was surprised, as his "small magazine" paid no fee or honorarium. "I had experience once of a token fee publication in some small magazine which subsequently blocked a *New Yorker* offer for the same poem at a dollar fee working out at about £75. Those things are doubtless sent to try us. And how well they succeed!"

In January Stevie also wrote to thank Ladislav Horvat for his "splendid statement" of the income her investments had generated. "I do admire arithmetical understanding but, as you know, I am not awfully good at it." She slips foreign expressions into her letter to this multilingual Czech and says, "all these international touches are due to reading toshy novels in foreign languages, especially I like Agatha Christie in French." Another business correspondent had confused "shall" and "will" writing to her, and she told Horvat that "he must be Scotch—they always muddle their wills and shalls".

"I don't want to write particularly", Stevie had said two years earlier:

> I don't really like the thought of writing. I—that's why I'm living rather a domestic life. I'd rather do anything than write. I mean, I'd rather wash up. It's an excuse from writing, you see, it's a perpetual excuse for not getting on with other work really I suppose. It's much easier to cook and wash up than it is to write, isn't it. And I think women are very lucky in have [sic] that excuse. I was thinking only the other day, you see, I've taken to this domestic life rather late. . . . And it's creative and it satisfies one's creative and one's destructive impulses. I mean, could anything be more destructive than cleaning out a room and throwing everything away . . . and

quote this xx

> this wonderful relationship between preparing a nice, a tender young vegetable and eating . . . it. . . . This is complete life . . . I think really that women naturally don't want art, because they've got all this wonderful thing going on at home all the time.

Stevie concluded that "it was exile from domesticity that produced the poems", adding that there'd been conflicts in her life which brought along poems—"if you live a tranquil life you probably wouldn't write a word".

But writing to a friend in March of 1965, she demonstrates that in contradiction to the view she had expressed earlier, life at home was proving productive for her. "I have been writing quite a lot of poems but am very doubtful about them—as always." Aunt looked "absolutely blooming. I flatter myself it's all that heavenly food we have, ha ha, heavenlily cooked of course is what I mean." The great difficulty she faced was finding someone reliable who could stay with Aunt and allow Stevie a visit to Molly in Buckfast. A charwoman was in the offing, though Stevie said they were more a burden than a help and usually left after a short stay. Stevie had longstanding though comic suspicions about chars. During the war she had reviewed a book describing how the Germans penetrated Polish defences by parachuting behind Polish lines troops disguised as nuns and charwomen. Her advice to readers was, "keep an eye on the heavens—and do not, however short-handed you may be, take on a charlady who arrives by air."

Despite her quiet life, the media were paying considerable attention to Stevie and in the early months of 1965 she was the subject of several interviews, one of them a "Monitor" film shown on BBC TV. Jonathan Miller, the producer of the series, had proposed that the film be made at 1 Avondale Road, but Aunt demurred. Stevie told Miller: "It would upset my Aunt, and she's an old lady. I simply won't have it. You'll just have to find somewhere else." In her account of this to Kay Dick, she seems amused and irritated that the BBC got muddled and mistakenly chose a house in Parsons Green far to the west of Palmers Green, in Fulham.

> So I said, "Well it's rather a long way isn't it?" "Oh, we'll send a taxi to you," they said. "Well," I said, 'It's easier to get there by train, because taxis are never heated in the morning." The taxi arrived—it was in January—and by the time we got to Parsons Green I was practically frozen. Then we went in; a house slightly like this, larger; it was immediately opposite a school, so we had to keep knocking off, waiting for the children to finish their playtime and go in again, because the noise was so terrific that we couldn't do a

thing, and this went on and on, and then, if it wasn't that, the aeroplanes would be coming in. It was all right in the end as these things often are, but the worse catastrophe was when the old lady down in the basement—the housekeeper—thought she'd get on with her ironing, and of course every light in the place fused.

Stevie had no television at home so she went round to the vicar's to watch the programme. Everyone who commented on the film told her how terribly ill she looked, but Stevie thought there wasn't much difference between looking ill and looking old. "One doesn't really bother about it."

On the telecast Stevie made generalizations about her work: "People in rather odd circumstances are what most of my poems are about, mixed up with arguments, religious difficulties, ghosts, death, fairy stories, and a general feeling of guilt for not writing more." "The general feeling about love in the poems is nervous," she said. "Life bears hard upon the children in the poems, but sometimes the little children are odious." Her manner was exceedingly lively, especially her piercing dark eyes which darted about unceasingly.

Giles Gordon, "Boswell" in *The Scotsman* for a time, wrote to Stevie in 1965 and proposed an interview. She agreed, with the proviso that she could see proofs, and they met by Stevie's preference at Brown's for lunch. She was especially interested in the vanilla ice cream with hot sauce on it which "must be hot sauce, not chocolate blancmange half-melted". The interview was jolly, and in time the piece was printed in *The Scotsman*. Stevie told Kay Dick a few years later that Giles Gordon's was one of two interviews she really liked. The other had been conducted in February 1965 by John Horder, who was a friend of Stevie from the early 1960s. Years after her death Horder revealed, partly at Gordon's urging, that she rewrote their interviews. "The Stevie I knew never lost contact with her child self till the day she died", wrote John Horder. "Who else but an adult, tyrannised by the omnipotent child, would have dared to have completely rewritten my interview with her for *The Guardian* and Giles Gordon's for *The Scotsman*, and got away with it?"

In June of 1965 Stevie spent a weekend with the Clutton-Brocks and later sent her hostess a letter which mentioned a party she went on to after the weekend. "I twisted", wrote Stevie, sounding amazed at turning trendy in her sixties. "I never tried it before but now I want to go on doing it & am looking for honorary grandsons to do it with." Her friend Barbara Jones had also been in the weekend party at Chastleton House, and Stevie enjoyed the ride home "in that marvellous open car of Barbara's".

Clifford Doyle, the vicar of St John's, Palmers Green, and his wife, both friends whom Stevie dropped in on often during these years, remembered that in the mid-'sixties she was amazed at her growing fame and at the many invitations she had to read her poems. The revival of interest in her poetry, he said, really did take Stevie by surprise. Among other readings she gave in 1965 were those at The Hampstead Festival; Chipping Camden Weekend; the I.C.A.; the *Tribune*; the Theatre Royal, Stratford East; the Cheltenham Festival; Eton; the Stroud Festival; and many stops on a six-day tour through Devon and Somerset with Patric Dickinson. Stevie was less nervous before a reading than she was before a party, according to Anna Browne, and was always eager to read herself as the financial rewards were greater. She especially liked reading to the young, and Helen Fowler noticed that the success of the readings in the 'sixties pleased her immensely, particularly those she did at schools. Helen thought they "gave her a new lease on life".

Stevie generally felt herself on less friendly ground when she read at universities. At Sussex, Colin Amery, the president of the literary society in 1965, recalls that about fifty students came to Stevie's reading, but few if any faculty members. "I don't think she liked academics or bothered with them . . . she didn't go in for analysis", he said. Stevie stepped off the train "dressed in clothes that were thirty years too young—a red tunic, white stockings and red shoes—dressed like a school girl." He had the feeling that her very calculated performance, successful as it proved, was a "bit of a cover-up operation", based on uncertainty about her work. It was her evident vulnerability, he thought, that won students to her side, and the immediacy and availability of what she wrote. Also the contradictions she embodied: "an atheist from respectable Palmers Green; a middle-aged woman behaving like a child; not being able to sing, but singing. Very unresolved, almost adolescent, but she sustained it, and at a high level."

Stevie's grandest event that year was the Edinburgh Festival. She travelled there on a sleeper after a holiday with the Brownes in Norfolk where she enjoyed, as always, bathing in the sea. At the festival she was one of a company of poets that included George Barker and W. H. Auden. She followed W. D. Snodgrass, and her friend Jonathan Williams, hearing her on radio as he landed at Southampton, wrote to say, "You sounded marvellous, but it must have been weird to follow Snodgrass's mangling of poor Walt Whitman." To Williams Stevie replied, "I have been doing a lot of Poetry Readings, round & about. Only one walked out . . . A clergyman, it was." A film crew making a documentary of the festival happened

to enter a pub, in which they came upon and recorded the phenom-
enon of Stevie and Auden singing hymns together. "I don't think
Auden liked my poetry very much," Stevie later reflected, "he's very
Anglican."

An interesting glimpse into Stevie's home life at this time is
provided in a memoir written by Gertrud Häusermann-Häusermann.
Stevie had first heard of her in a letter from Professor Häusermann
written several years earlier. At that time his wife Siv was dying of
leukaemia, and after describing her condition, Häusermann said he
wished Stevie could read German because a niece of his wrote "novels
for girls, and her latest, *Die Geschichte mit Leonie* . . . is decidedly more
than just a children's book. Her name is Gertrud Häusermann; her
married name is G. Voegeli." In 1965 Hans Häusermann and his niece
visited Stevie in Palmers Green, and less than two years later, he wrote
to tell Stevie of their marriage. Mrs Häusermann-Häusermann recalls
that it was on a Sunday when she and her uncle visited Stevie, who
looked drawn and tired. She had come to the door in a red miniskirt
with a grey pullover, very animated, but perhaps more animated than
she actually felt.

Her house struck Gertrud as joyless: old wallpaper, doors of faded
colour, everything old and poor. Stevie seemed unaware that she
sighed at the end of every sentence, a kind of moaning. She took her
guests upstairs to greet Aunt, who sat motionless with a blanket over
her knees, and spoke clearly. Aunt said she was happy Stevie could go
out with the professor and his niece, and that a neighbour was coming
to stay with her while they were out.

Downstairs Stevie prepared a meal, her actions punctuated by the
half-subdued moaning, as if a mountain were pressing her down.
Cooking seemed to be a terrible effort, and she had nothing to serve
but a few potatoes and a can of peas. They ate with tin-plate flatware
and from very old worn plates. In an animated way Stevie talked about
the Edinburgh Festival, and how she had lost her eyeglasses and asked
the audience, while on stage, if anyone had glasses which she might
borrow. "I am farsighted," she said, and it began to rain glasses from
the public. In the end she did find a pair corresponding to her
prescription. She went on to say she had begun to sing her poems, and
she sang a few lines for her guests. Her singing struck Gertrud as
touching, half-speaking half-singing, as if coming from her unconscious, the source of her suffering. They spoke of authors and books,
and Hans Häusermann tried to persuade Stevie to travel to Geneva.
Although the thought of travel pleased her, the trip itself would be too
much effort, she said, and she did not want to abandon Aunt, who was
always there for her when she was a child.

Later they drove to the country and Stevie led her guests over the Roman road. She and the professor walked in front, talking seriously. She seemed to be opening up, and Häusermann spoke to her emotionally. Gertrud followed slowly, feeling she should not take part in the conversation. She felt embarrassed about her "youthfulness", being about twenty years younger than her uncle and Stevie. Stevie, for her part, seemed embarrassed by the professor's charm.

There was no joy in Stevie's laughter, Mrs Häusermann-Häusermann recalls. In her heart Stevie was sad, even when she laughed. She was hard towards herself because she had to be hard in order to maintain herself without affection. Everything was a burden to her: the aunt, the house, life in general. She rarely accepted invitations, because she had come to feel less and less relaxed among people. And everything was so complicated: the long road from Palmers Green to central London, finding someone to stay with Aunt. Aunt's imminent death occupied her constantly, as though everything would lead to that point, and then?

Stevie was never free for long from the melancholy that continued to overtake her. In a note accompanying some books he sent her for review in *The Observer*, Terence Kilmartin said, "Sorry you are feeling so gloomy." But despite her sadness and fatigue, she continued to write reviews—several in 1965 for *The Listener* and *The Observer*, and one for *Encounter*, an evaluation of a symposium called *Contraception and Holiness*. Indeed the subject of nearly all the reviews she did that year is in one way or another religion: Antonia White's account of her return to Catholicism, for instance, about which Stevie makes a point raised in a number of the other reviews. "Faith's best friend", she says, is "Desire" and, as she earlier said in an interview, "it is because man is so lonely that he must invent God . . . to have somebody interested in one, even if only to damn one as against this, this huge . . . indifferent universe. Man must have religion. . . . And yet he must learn to get on without it." She found, in reviews of their work, much to admire about both Simone Weil and Calvin, though the latter was not "'all of a piece throughout'; but of course few people are". In writing about Isherwood's book on Ramakrishna, Stevie comments on the glumness of the reincarnation theory: "I often wonder why people, in these matters where nothing is known, do not imagine more agreeable things."

Death and the Afterlife, never far from Stevie's thoughts, were perhaps more urgently present to her as 1966 commenced and Aunt celebrated her ninety-fourth birthday. Molly was able to come home (as Stevie always put it) for this, which delighted Aunt and her. Aunt seemed well, though Stevie complained herself of "getting remarkably feeble" and in an awful state of inertia. She was also finding it more and more difficult to locate someone to stay with Aunt: Molly couldn't be given the task because her heart condition kept her from climbing stairs.

Stevie thought Aunt's highest praise came after she heard her niece had won the Cholmondeley Award for Poetry. Aunt said, "I wish your mother was alive and could have known about this dear." Stevie received the Award in May of 1966, and the next month she gave a reading at Cley in Norfolk. The holiday which preceded her reading was "too marvellous. I do nothing at all but sleep, eat & go on short trots over the saltings & soon the boat will be all right . . . so we shall do some sailing. I am becoming fat & agreeable—& quite incapable of reading Fr. D'Arcy" (whose book, *Facing God*, *The Observer* had asked her to review). While Stevie was in Norfolk, Aunt spent two weeks in a nursing home which specialized in short-term stays for the elderly.

Seventy to eighty people attended the Cley reading in the Methodist church hall, and after Stevie read "The Best Beast", her poignant poem about the suffering animal who has won the Fat Stock Show competition at Earl's Court and is taken off in the lift to an unknown and perhaps lethal destination, someone rose in the audience and said, "My husband went up in the lift with the Best Beast."

The Cley reading was followed by the one Stevie gave in the Albert Hall, which Christopher Logue has described (page 2). It was sponsored by a wealthy friend of the arts, "The Boulting Boy", as Stevie called him. She found him unpleasant and was rude to him at a preliminary meeting, so expected to be dropped from the reading. But in the end Stevie went on. The event began on a sunny Sunday at about 6 p.m. and continued until midnight. (Stevie had told Terence Kilmartin the arrangements sounded dotty.) Many fine poets read, Robert Graves among them. "By 7 p.m. on the following day," Christopher Logue recalls, "8,000 people had turned up. Everyone was in a genial mood." He found two performances particularly memorable: the Austrian poet, Ernst Jandl, along with Pete Brown and Michael Horovitz, gave their version of Kurt Schwitters' (onomatopoeic) "Fury of Sneezing", which concludes with "Happapeppaisch" repeated four times, followed by "Happa peppe", and then "TSCHAA!" It "brought the house down", according to Logue. Stevie was the other surprise attraction and proved a huge success.

Stevie wrote to Professor Häusermann in August of 1966 that "Aunt seems very well at present, though the poor darling gets terribly bored, not being able to see or hear properly, and worrying that she 'can't do anything to help.'" Stevie recounted in this letter some of the readings she continued to give—one, for instance, at

Cranborne Chase, Alice Browne's school, where Stevie had for the first time an audience only of girls and very much loved the "heavenly castle" [Wardour Castle in Wiltshire] where the school was located. The *Listener* and *The Observer* had published new poems, and the *TLS* had recently printed "A House of Mercy". The Marvell Press had just issued an LP recording of Stevie reading her poems. But the big news was that Longmans were planning to publish that autumn Stevie's next collection, *The Frog Prince and Other Poems*. Stevie was ambivalent: "Why it gives me a feeling of nausea thinking about my poems, I do not know. I only really seem to enjoy actually reading them aloud. And that's chiefly to forestall other people doing it. Other people, especially actresses, do make the most terrible things out of my poems, when they read them. They sort of get up to tricks with them . . . you know."

The *Frog Prince* contained seventy-two poems which had not appeared previously in book form, but in addition the volume reprinted ten poems from *Not Waving but Drowning*, fourteen from *Harold's Leap*, twenty-two from *Mother, What Is Man?*, twenty-one from *Tender Only to One*, and sixteen from *A Good Time Was Had By All*. The book is in part a retrospective of the poems Stevie had written over three decades, and suggests she probably thought readers had forgotten her early poems.

When *The Frog Prince* appeared, John Horder hailed Stevie as "the most original poet writing in English today. She is most completely herself. She hides nothing, submits herself to a complete exposure which is the hallmark of the great poets." Stevie's openness, which Horder emphasizes, in part accounts for the power of her poems. But as Horder indicates, that openness did not prevent loneliness, a feeling which shades many of the poems in *The Frog Prince*. "The word 'frozen'", he noticed,

> occurs throughout the book . . . and one can only guess that the something frozen deep down in the poet's soul prevents her from having any real relationship with other people. This is hell, to be completely shut inside oneself . . . Stevie Smith seems to have stoically accepted that there is no way out. Except, perhaps, by the writing of these poems. It is out of anguish, despair and a deep desperation that the poems are wrought.

The *TLS* reviewer listed three or four pervasive themes in the book: the pitiable condition of man along with his "luminous or alarming oddity"; the significance of humans as a race; man's relation to a God who very likely doesn't exist; and death. The reviewer also commented on the poet's direct confrontation with the world, and added that "the mannered mode of speech is the clue to her work: a convention enabling her both to compass the pity and terror of her themes and to respond to them with rueful courage and humour."

Next to "Not Waving but Drowning", "The Frog Prince" may be Stevie's most admired poem.

> I am a frog
> I live under a spell
> I live at the bottom
> Of a green well
>
> And here I must wait
> Until a maiden places me
> On her royal pillow
> And kisses me
> In her father's palace.

On one galley of the poem, Stevie noted the "Frog's fear of death". The Frog Prince recounts to himself what he finds good about his life and seems not at all eager to be translated to the palace and a princely existence. But then another thought comes to convince him that it is the spell which tricks him into embracing his frog existence and into fear of disenchantment. So he realizes

Stevie's drawing which she placed by her poem "The Frog Prince"

> It will be *heavenly*
> To be set free,
> Cries, *Heavenly* the girl who disenchants
> And the royal times, *heavenly*,
> And I think it will be.

"'The Frog Prince' is a religious poem", Stevie said, "because he got too contented with being a frog and was nervous of being changed back into his proper shape and going to heaven. So he nearly missed the chance at that great happiness, but, as you will see, he grew strong in time." The poem concludes:

> Come then, royal girl and royal times,
> Come quickly,
> I can be happy until you come
> But I cannot be heavenly,
> Only disenchanted people
> Can be heavenly.

Enchantment was an allure to which Stevie was susceptible and about which she was, as well, wary. In her poem "How do you see?", she used the word several times, typically in this way: "we must put away the beautiful fairy stories / And learn to be good in a dull way without enchantment." And in the poem "Pretty", published in *The Frog Prince*, Stevie announces again the motif of this collection, saying that to "be delivered entirely from humanity / This is prettiest of all, it is very pretty." (The speculation about where this deliverance will lead the soul is the burden especially of "Animula, vagula, blandula", Stevie's translation of the Emperor Hadrian's deathbed poem.) Throughout this book Stevie's need to be delivered from "The nerviness [in an earlier draft 'The loneliness'] and the great pain" is insistent. It is "Sweet Death, kind Death" which "Throws it [pain] on the fresh fresh air / And now it is nowhere."

Anna Browne remembers that Stevie liked her poem "Tenuous and Precarious" very much and said about the speaker, "You know she killed them all." Stevie called the poem "cheerful as it is a word-play poem (on Latin endings to adjectives). But there is murder hidden in it. I think she murdered everybody except the cat."

> Tenuous and precarious
> Were my guardians,
> Precarious and Tenuous,
> Two Romans.
>
> My father was Hazardous,
> Hazardous,
> Dear old man,
> Three Romans.
>
> There was my brother Spurious,
> Spurious Posthumous,
> Spurious was spurious
> Was four Romans.
>
> . . .

According to Lady Lawrence, Stevie wrote "Tenuous and Precarious" while visiting at her home. Sir John suggested some Latin words and "Stevie just made it up as she went along, almost like a conversation. It was easy but erudite." This confirms the impression that Sally Chilver, Elisabeth Lutyens, and other friends had that in some cases Stevie did toss the poems off the top of her head, though perhaps with much forethought. But the existence of multiple drafts demonstrates that she carefully revised some poems.

"The Best Beast of the Fat-Stock Show at Earls Court" is one of many poems in this volume which features an animal as its subject. Others are about fish or fowl. If Stevie felt drawn towards the preternatural—towards ghosts and witches, angels and divinities for instance—she felt as well a link to the orders of creation that dwell below man on the chain of being. (In 1969 "The Best Beast" became the title poem of a volume of Stevie's work published in America by Knopf.) Stevie seems to empathize (if not identify) as much with the Best Beast as she does with the Frog Prince, and her depiction of the Beast's suffering is poignant:

> When he lay in the straw
> His heart beat so fast
> His sides heaved, I touched his side
> As I walked past.
>
> I touched his side,
> I touched the root of his horns;
> The breath of the Beast
> Came in low moans.

Stevie once told Anna Browne's daughter, Alice, that only after she'd written the poem did she notice she'd done it all in monosyllables.

In "Exeat" Stevie tells again the story of one of the cruellest of the Roman emperors who visited his prisoners in dungeons and when they'd beg for death would reply

> Oh no, oh no, we are not yet friends enough.
> He meant they were not yet friends enough for him to give them
>   death.
> So I fancy my Muse says, when I wish to die:
> Oh no, Oh no, we are not yet friends enough,
> And Virtue also says:
> We are not yet friends enough.
>
> How can a poet commit suicide
> When he is still not listening properly to his Muse,
> Or a lover of Virtue when
> He is always putting her off until tomorrow?

The last stanza of the poem is immensely personal:

> Yet a time may come when a poet or any person
> Having a long life behind him, pleasure and sorrow,
> But feeble now and expensive to his country
> And on the point of no longer being able to make a decision
> May fancy Life comes to him with love and says:
> We are friends enough now for me to give you death;
> Then he may commit suicide, then
> He may go.

Stevie's drawing which she placed by her poem "The Best Beast of the Fat-Stock Show at Earls Court". In her sketchbook Some Are More Human Than Others, *she placed under this drawing the caption "I was alive and am dead."*

In presenting this poem to an audience Stevie used this introduction: "The next poem is about committing suicide before you get trapped up in an Eventide Home; if that happens to be how you feel about it." She deleted her first, more expansive, and autobiographical lead-in:

> I am haunted by a lot of fears, and one of them is the fear of being an old helpless person in an Even-tide Home, as they call these retreats for the agèd. I would rather be dead, and I think a lot of the old people would rather be dead, too. (Judging by the number of letters I received, after this poem was published, asking me "how to do it", I think certainly a lot of them would rather be dead.) But by the time you get into a Home, you have lost the power of decision, and on the practical side, it is difficult to find the opportunity, and the means. I do not think you can expect the doctors to help you, you

must help yourself while you are still able to do so. I will read you a poem about it—it is called "Exeat".

On a holograph of her poem, "Longing for Death because of Feebleness"—

> Oh would that I were a reliable spirit careering around
> Congenially employed and no longer by *feebleness* bound
> Oh who would not leave the flesh to become a reliable spirit
> Possibly travelling far and acquiring merit.

—Stevie jotted down "The Frog Prince strikes the same note." But even more does "Exeat".

In introducing her poem "Phèdre", Stevie once called it, "a literary criticism, but couched in rather school-girl language. It might be called 'From the Classical Sixth Form at St. Agatha's'." She imagines herself rewriting Racine's play, a great favourite of hers.

> Yes, I should like poor honourable simple sweet prim Phèdre
> To be happy. One would have to be pretty simple
> To be happy with a prig like Hippolytus,
> But she was simple.
> I think it might have been a go,
> If I were writing the story
> I should have made it a go.

And along the way Stevie vents her objections to actresses again. Marie Bell was legendary as Phèdre, but Stevie complained that the French actress portrayed the young woman "In awful ancient agonizing."

In many of the poems in *The Frog Prince* Stevie exercises the gift for story she'd put aside nearly a decade earlier. "The Last Turn of the Screw" is based, of course, on Henry James's story, and Stevie told her audience on at least one occasion that though Miles, the boy possessed by the ghost of a dead servant, dies in James's story, he does not do so in her poem. Stevie had years before sent a copy of "The Last Turn of the Screw" to Professor Häusermann, and it provoked an interesting exchange between them. Stevie wrote:

> The poem deals with the price of intelligence & whether in some cases, this price is worth paying, in fact, in the ultimate account, whether morally speaking, it should be paid. *I* really think it should and must be paid. But there are moral, religious (of course!) moods when one thinks *not*. Roughly Christ's words—"What shall it profit a man if he gain the whole world and lose his soul?" is what it is.

Stevie continues at length to assess the relation between Miles and the governess in Henry James's story. The governess, in Stevie's view, does not know she is in love with Miles. He recognizes this, although he does not love her, only likes the warmth of sitting on her knee and being out of the cold. Stevie says that Miles, like herself, slightly despises people "who find evil 'romantic' and 'interesting'. Because it shows they don't know what evil is & that means (and this is what Miles feels) that they are 'silly little things'—more cruelly, say, stupid, and of course also vulgar." She then describes the process of evil as Miles comes to understand it through the tutelage of Quint: "from silliness to stupidity, through envy, malice & vulgarity, through all false values and greed in them, to ultimate absolute cruelty. In my poem I have him going too far, or thinking he has gone too far, to draw back without cutting off . . . his intelligence—which 'first let sin in'."

Stevie also writes interestingly in this letter about the "peculiarly Anglo-American thing to romanticise the upper classes. We are still doing so with our Nancy Mitfords & Evelyn Waughs & even Anthony Powell." She calls Ford Madox Ford "one of the earlier sickening examples of this sort of leaping middle-class writing about high class nonsense." Even good writing, "like Meredith's", she finds full of romantic views of the upper class. Stevie concludes with her thoughts on the relation between vulgarity and evil, rooted in falseness. "The Devil is the father of lies," she writes, and then asks Professor Häusermann to "forgive the sermon from your agnostic friend!"

Five "Voice from the Tomb" poems appear in this volume. Stevie said "the melancholy note is struck rather often in the poems. It is like the poet who did not want to write and then he died: and this is what he thought was going to happen to him . . . I had this nightmare once, after reading the Parable of the Talents in the New Testament."

> Here lies a poet who would not write
> His soul runs screaming through the night
> Oh give me paper, give me pen,
> And I will very soon begin.
>
> Poor Soul, keep silent; in Death's clime
> There's no pen, paper, notion,
> And no Time.

"Voice from the Tomb (2)" reads like autobiography:

> I trod a foreign path, dears,
> The silence was extreme
> And so it came about, dears,
> That I fell into dream,

> That I fell into dream, my dear,
> And feelings beyond cause,
> And tears without a reason
> And so was lost.

"Is it Happy?" Stevie called "a poem that is really a novel, only cut down to a page and a half." She said it included everything that was "popular in novels of a few years back, when you still had to be upper middle class, unhappy at school—preferably Eton—(I have left that bit out), sensitive, of course sensitive, and at odds with one's parents."

"Fairy Story" is an account in five rhyming couplets of a girl who wandered into the wood and encountered a little creature:

> He said if I would sing a song
> The time would not be very long
>
> . . .
>
> I sang a song, he let me go
> But now I am home again there is nobody I know.

The shift from iambic tetrameter to an irregular alexandrine perfectly conveys the *dépaysement* of the girl who has returned from the dark forest.

In "Dear Child of God" Stevie turns again in a perturbed way to the ramifications of Darwin's theory:

> In the beginning, Father,
> You made the terms of our survival
> That we should use our intelligence
> To kill every rival.
>
> The poison of this ferocity
> Runs in our nature.

She audaciously says to her "Father in heaven": "It is not often we remember / You put this poison in us", and sees humans in their natures as "Faithful but unfortunate".

Although characters are often flawed (those in "A House of Mercy" are exceptions) the poems are stirring, admirably crafted and witty. In a consoling way Stevie, who earlier argued that "Nobody writes or wishes to / Who is one with their desire", reminds the reader (and herself) in her closing poem:

> Some are born to peace and joy
> And some are born to sorrow
> But only for a day as we
> Shall not be here tomorrow.

Stevie envisioned one speaker of that poem as "a little child who has found a defence against the inequalities of fortune". The echo of Blake can be heard in the verses.

Near the end of the volume is a lovely poem called "Nodding" which evokes the atmosphere of a room lit by firelight on a dark, wintry night. The speaker watches her beautiful cat asleep by the fire. The uneven ticking of the clock is audible, as is the owl hunting in the Old Park.

> One laughs on a night like this
> In a room half firclight half dark
> With a great lump of a cat
> Moving on the hearth,
> And the twigs tapping quick,
> And the owl in an absolute fit
> One laughs supposing creation
> Pays for its long plodding
> Simply by coming to this—
> Cat, night, fire—and a girl nodding.

On 6 December Stevie celebrated her new book in a public way when she read a selection from *The Frog Prince* on the BBC.

18

The Death of the Lion Aunt

Just after *The Frog Prince* was published, Stevie wrote to Rachel Marshall, thanking her for purchasing a copy of the book. Stevie was particularly pleased that her LP record of poems, which had been made in the home of her childhood friend, Doreen Diamant, appeared just as the book came out. Stevie hadn't heard the recording as she did not own a record player, nor did she like listening to her own voice. She was grateful to *The Observer* for having reviewed both the record and the book favourably and in time to attract Christmas shoppers. But Bernard Bergonzi's criticism, published in the *Guardian*, of her "Little-girl persona" and slyness, angered her. She was fearful that more such criticism was imminent, as Cape were about to reissue *Novel on Yellow Paper*, "a book I now detest".

Stevie was settling more than ever into domestic life and particularly liked cooking. She would prepare meals and then take them upstairs to Aunt's room where they ate together. She rarely went out. "So often when evening comes I simply cannot face the prospect of going into London for dinner parties", she told Rachel Marshall. Writing to another friend early in 1967 she said that "for the last five years I have been getting more and more domesticated and am now quite a good cook." Aunt, she reported, was still wonderfully alert mentally, "but alas her old hips are arthritic and she can only walk with two sticks and is even so apt to fall over. She is also very deaf and getting a bit blind too. So it's a bit triste for her, eh?" Stevie described herself as still in a love-hate way about poetry, "and now I've so taken against my own that I am saying 'no' to readings most of the time. I don't know why writing sort of gets me in such a state, peeling potatoes by contrast is so nice."

Stevie confessed the same doubts about her work to James Laughlin, who had written to say he did not plan to publish *The Frog Prince* in America. She wondered if her sort of poems were right for America—or her prose either, given the slim sales of *Novel on Yellow Paper*. But she half-heartedly asked Laughlin for the names of American publications that might be interested in printing poems that had first appeared in English papers. In 1967 she seems mostly to have declined the invitations she did receive to read her poems, telling David Wright, who wanted her to come to Leeds University while he was a Fellow in Poetry there, that it was too difficult to arrange "aunt-sitters and, worse still . . . I am getting awfully sick of reading my poems . . . I was just brooding over some Feb. and March 'dates' . . . and wondering if I could cancel the lot . . . and surmising —whether I could or not—I would." She shared David Wright's enthusiasm for Yorkshire but found in her present languid mood that even going to central London was a tiring trip. "I just stay at home and get absolutely fascinated by doing the same thing at the same time over and over again every day. I should be quite lost without it. Tonight —for a *wild* sort of change, I have been boiling a jersey purple. . . . You have no idea how sinister the brew looked." Aunt missed being able to sew or read, but Stevie tells David Wright, whose book *Deafness: A Personal Account* she later reviewed, that by some miracle Aunt could hear the radio, especially when it was turned low. Probably she had Aunt in mind when she wrote in her review of Wright's book, "it is of course the elderly deaf ones who are truly cut off and unhappy."

Stevie had once said, by way of introducing "The Frog Prince" to an audience, that "the poem is a parable of life, that we may get to be afraid of thinking of what may lie beyond life and so lose the chance of heaven. We must remember, as Chaucer says, 'Here is no home.' " Little danger of Stevie forgetting this as she watched Aunt loosen her hold on life, and felt herself grow feebler. But conflicts persisted in her and, ironically, made new poems possible. In "Oh grateful colours, bright looks!", published that summer, Stevie rejoices in the texture and colour of life:

> The grass is green
> The tulip is red
> A ginger cat walks over
> The pink almond petals on the flower bed.
> Enough has been said to show
> It is life we are talking about.

Then she celebrates common, man-made things such as "Bricks, slates, paving stones". (Sally Chilver remembers that once Stevie entered a building, inhaled with pleasure, and said, "Oh, my favourite smell—coke.") The poem ends with an exhortation reminiscent of Hopkins:

> Men!
> Seize colours quick, heap them up while you can,
> But perhaps it is a false tale that says
> The landscape of the dead
> Is colourless.

In June 1967 Stevie did take part in a reading sponsored by the Poetry Society in the Purcell Room at The Royal Festival Hall. The Chairman of the Society visited her dressing room before the reading and presented her with a bouquet. Five minutes after her performance ended he arrived at her dressing room, but she had vanished. Instead he found a trail of flowers. It led out of the building and towards the Thames. Next day, when he rang up, he was relieved to find Stevie at the other end of the line, and sprightly. "Possibly she didn't like flowers", he theorized, "or possibly she didn't like the Poetry Society."

The following month Stevie attended the Royal Garden Party, although she complained about having been sent a single invitation. "Only one thing to do at such a party", she said, outlining her strategy. "You make straight for the Tea Tent and straight for a table at which only one person is sitting, preferably a man, and you say: 'do you mind if I sit down here,' and then he feels he must go and get you a cup of tea."

The last time Elisabeth Lutyens met Stevie was two months after the Royal Garden Party. They were going to stay in Newcastle-on-Tyne with Miss Lutyens' sister, Viscountess Ridley, who had arranged a concert of theatre music in a gallery there. Miss Lutyens was astonished when Stevie arrived at Kings Cross railway station with badly dyed hair. She had a large handbag into which she had crammed an evening dress, also badly dyed. At the gallery Stevie read poems as part of the evening's entertainment, and two of her poems were sung by professional singers. But neither the male nor the female singer who had been engaged suited her, and she said she'd envisioned one poem as sung by a small, dark man, and the other by a tall, fair man. (Miss Lutyens found Stevie impossibly demanding.) Lady Ridley eventually committed suicide, and her sister felt Stevie had not been a good influence, with her insistence that "after the gin and tears come the poems". She also found Stevie to be a bully, and said she could be devious in trying to have her own way. All of this she attributed to an egotism rooted in Stevie's appalling lack of energy. She agreed with Olivia Manning's view that Stevie made a good acquaintance but a bad friend.

The peculiar flavour of their friendship was captured by a reporter who interviewed them after the programme of poetry and music. Stevie, wondering why her poems appealed to the young instead of her contemporaries, quoted a boy who told her, "It's because we can see you've been through the mangle." When she then commented she would call her next book *Through the Mangle*, Elisabeth Lutyens drily noted, "Mangles are out of date." As the conversation proceeded, Stevie observed in a "vague and polite" way: "'Yes, Liz has very kindly set some of my poems to music. The music's wonderful. I'm delighted . . .' She tailed off vaguely, then impishly stifled a giggle with her handkerchief and added confidentially: 'As a matter of fact, I can't remember what she has done!' Surrealism took over at this point," the reporter noted. The interview concluded with Elisabeth Lutyens saying she was once "short-listed for a job at the BBC" because, when asked whether she could do the job and look after four children, she replied: "That's none of your business." Stevie then asked, innocently, "Did they mean four of their children?"

In October 1967 Stevie found someone to stay with Aunt and, *mirabile dictu*, flew off on the 6th with several younger poets to read at the Palais des Beaux Arts in Brussels. The troupe of poets had been organized by Michael Horovitz. Stevie rode about in a charabanc and got on famously with the "poet-toughs", as Alasdair Clayre called them, who were reading with her. The Belgian government financed the reading, billed as a "Psychedelic Feast", and a large, very enthusiastic audience turned out. The government provided an unlimited supply of wine and, according to Clayre and Horovitz, Stevie kept up with the younger poets as they drank it. The evening started off promisingly. As the audience entered they were handed, each one, a fresh rose and biographical sketches of the twelve poets. Next day, a reviewer praised the readings of Frances Horovitz and of Stevie, the flute playing of Neil Oram, the humour of Adrian Mitchell, and the songs of Mr Clayre. But "the whole thing quickly disintegrated into a general free-for-all that apparently had to go on until all the bottles . . . on the speakers' table were emptied. And since there were about as many bottles as there were speakers, the thing lasted a long time."

The paper described the poets as rambling on and becoming increasingly drunk. Eventually the audience took to scolding them. Michael Horovitz recalls that it was when "all hope had gone" that Stevie went to the microphone and was greeted by "riotous hoots of applause . . . she was convinced she was getting the bird & begged leave to 'stand down'". He led her back to the microphone and "she stood defiant", waited for the clapping and whistling to stop, and then read her poems. "As far as Stevie was concerned, she was bravely outfacing a derisive mob: which might just have been what they were applauding, I suppose, unless—with more justice still—it was the astonishing unimpeachable fact of her." The European papers were not complimentary about the "Psychedelic Feast," and they were ironic about the frequency with which "flowers, love, and peace" were mentioned. "After Stevie Smith, a grand lady (what's in a name?) . . . the evening clearly fell apart. . . . What we got out of it was enough to encourage us to buy Stevie Smith's works (Penguin)." One reviewer's impression was that the "British happeners" had no feel at all for the conservatively dressed, conformist Belgian audience. One of the poets drank so much he fell asleep on stage, another made rude noises in the microphone ("sound-poems", according to Michael Horovitz). One was so drunk and noisy that the airport officials refused to let him board the flight home.

When Stevie returned to England late on the night of 7 October, she went to the home of Anna Browne, who recalls that she quickly drank several sherries. She told Anna the experience in Brussels had exhausted her. Stevie rested at the Brownes for a day and then travelled to Chastleton House for a ten-day visit with the Clutton-Brocks.

The experience of her Brussels trip hardly encouraged Stevie to accept a grant that year from the Society of Authors, who offered her £250 to travel abroad. Nor, of course, would Aunt's health have allowed Stevie to go. On 2 November Miss Spear made a will, leaving everything to Stevie. The following month Stevie wrote to James Laughlin that Aunt was remarkably well. "The heart-block she had for some years has somehow taken itself off, and we all, including the doctor, are very pleased if a little mystified."

Christmas was deliciously quiet. Stevie slept a lot, she told Mrs Clutton-Brock, "just getting up to cook and eat. Hope you had a heavenly time too." Stevie sent her a copy of "The Poet Hin" who wonders why other poets are so much more honoured than he. "I am much condescended to, said the poet Hin, / By my inferiors." The speaker consoles the poet by reminding him that he knows the correct use of shall and will: "something we may think about":

> But ah me, ah me,
> So much vanity, said he, is in my heart.
> Yet not light always is the pain
> That roots in levity. Or without fruit wholly

As from this levity's
Flowering pang of melancholy
May grow what is weighty,
May come beauty.

In many ways Stevie was offering here an assessment of her poetry —its origins, its attainments.

During these years she continued, and often in brilliant form, to assess the work of others too—in the reviews she wrote for *The Listener* and *The Observer*. Editors continued to send her mostly books about religion. In the reviews, Stevie's preoccupations remained the same.

> The contradictions in the New Testament do not trouble these writers . . . the doctrine of eternal hell which looms so hideously clear . . . is ignored. Every age has its own Jesus. It is curious that ours, so violent . . . should see Christ only as "kind", but of course it is also a sentimental age. Nor do the Christian contributors find difficulty in the doctrine that God in Jesus took our manhood upon him, yet could not sin as we can.

All of these themes had for long preoccupied her in poems and in other writings. Stevie found a biographical study of Ste. Thérèse of Lisieux "fascinating" and to some extent seemed to identify with this young French woman, obsessed with God and religion and bereft at an early age of her mother.

Otherwise, in 1967, Stevie reviewed books about Thurber and Vicky, and four books about cats, one of them by Doris Lessing, and another by Olivia Manning. Stevie parries a bit with her old friend, saying a few such provocative things as "Always, on controversial subjects, Miss Manning is apt to be more agitated than original." But she is generous with praise for her, too.

As we have seen, people tended to think of Stevie, especially after the publication of *Cats in Colour*, as a cat enthusiast. But, as she wrote to Ann Thwaite:

> I don't know . . . about cats, I mean. I quite like the animals but I've never had one (bar an ancient tom we had when I was a child, called, most unsuitably "Fluff"). I seem to have a name as a "Cat Lover" but I fear I do not deserve it . . . Cat poems I have in plenty, but they are often a shade sharp in tone, e.g. "Monsieur Pussy-Cat, Blackmailer".

Stevie may have had in mind as well her most famous cat poem, "The Galloping Cat", which begins

Oh I am a cat that likes to
Gallop about doing good

Stevie's drawing placed by her poem "The
Galloping Cat"

(much *accelerando* in Stevie's reading of the second line)—and comically reveals the hypocritical nature of this cat who considers himself "a
martyr to doing good". In her *Observer* review of Doris Lessing's
book on cats, Stevie gives this view: "our poor dumb-wits of pets are
in the same boat with us and it's a very cold night outside. True
enough, but I would still rather stroke a cat than read about it."

In January 1968 Stevie sent her poem "Exeat" and a new poem,
"Like This", to *The Observer*. Although "Exeat" had already appeared
in book form, she thought that poems should be reprinted if they were
"to the point—I mean, some point that is always cropping up in
people's minds? Like suicide & Old People's Homes, etc. Oh, & mad
people." "Like This" is the soliloquy first of a young man, then of a
young girl, in an asylum. The lonely young man declares: "nobody
like this *likes* this, / Or likes another like this." The lonely young girl
.says: "What love means is, To speak to me / Not leave me in the cold."

Worrying over the illness of Aunt, Stevie could scarcely have been
more sympathetically disposed to appreciate the loss Rosamond

Lehmann experienced when her newly married daughter caught polio and died at the age of twenty-four. Miss Lehmann wrote of this in *The Swan in the Evening* which *The Listener* sent Stevie for review late in 1967. The review appeared on 4 January, Stevie saying at the outset that "It is . . . a difficult book to write about." It was Miss Lehmann's belief in spiritualism that disturbed Stevie, who said it "lays open a world of anarchy". Miss Lehmann describes in her book "returning presences and other supernatural experiences, . . . traffic with departed spirits, at seances and in solitude." Finally she describes a vision of Christ in Gethsemane and thinks of becoming a Christian. Stevie assumes rather the role of apologist for the Judaeo-Christian teaching, reminding Miss Lehmann that according to this teaching "the souls of the departed are in the hands of God and no harm shall touch them. Nor can we help but feel that among the harms that might touch them, were they not protected by God (or by non-existence, if you are an atheist), is the harm of earthly lovers trying to get in touch with them."

> Miss Lehmann thinks that people who do not wish to traffic with the dead are cold, careless or timid. They may be more full of love than she thinks. For if you believe in God, you will let the dying go, glad that the pain of loss is ours, not theirs. They have finished with the imperfections of human love, its dark places of egoism, greed and idolatry. And, as even the new-born baby, with a full span of life ahead, cannot really be said to have to live very long, is it asking too much that we should love our dead and leave them alone, waiting for our own deaths to know what it is all about? Or to know nothing ever again.

Miss Lehmann thought the review "prim orthodox—churchy . . . almost as if she was shocked—and I wrote to her then." In an epilogue to a recent edition of *The Swan in the Evening* Miss Lehmann writes that Stevie's criticism "with its raw note of censure, hurt me, particularly coming from Stevie whom I thought of as a friend." She quotes Stevie's letter to her:

> I do not see the dead as a "frightening and hostile society," but as being . . . either non-existent, or, in the hands of a loving God . . . I love the thought of "departing" and of the human patterns being broken forever, either in nothingness or in love that is beyond our thought . . . I am sure we cannot touch the dead. All of which goes to show how right you are when you so gently imply . . . how very imperfect an agnostic I am. I love to think of the dead as in the hands of God and safe from us.

Perhaps Stevie's sorrow at the imminent loss of Aunt was present to her as she wrote this letter to Rosamond Lehmann in early 1968. Aunt was more than ever her preoccupation that year. In a review of an anthology of women poets published in May, Stevie goes out of her way to mention "'Michael Field', that odd amalgam of Aunt and Niece." (Her only other review that year was of books about religion for children, a subject which in various ways suited her moods and concerns in 1968.) On 6 February Aunt had her ninety-sixth birthday, and Molly came for a visit. Nina Woodcock brought daffodils: "I'm going downhill Nina", said Aunt. Four days later Stevie was at lunch with Anna Browne and Father Gerard Irvine when a call came from Molly that Aunt had had a stroke and been taken to North Middlesex Hospital. Stevie was "shattered" by the news. Father Irvine rushed to her side and put his arm around her. "Very little hope", wrote Stevie in her diary. "Gerard brought me back from Anna's & down to the hospital. Poor Molly had had to cope with everything."

On the 11th Aunt began to regain consciousness. Writing to Hans Häusermann that day, Stevie told him:

> We saw Aunt again this morning & the Vicar came with us & gave her the holy oils & I could see she was looking quite different from what she looked yesterday. She is conscious & can recognise us. . . . I don't know whether we shall ever have her at home here again, but I do hope we shall. But it was getting rather difficult as she is so heavy to lift. Isn't it wonderful, though, that at the age of 96 she can rally round after a "severe cerebral haemorhage" [sic]?

Marghanita Laski sent flowers to Aunt, and Stevie wrote on the 24th to thank her and say that Aunt was in a good deal of pain. As soon as it could be managed, she was to be moved to St Christopher's Hospice at Sydenham "before the poor darling gets quite desperate".

From 16 to 18 February Stevie managed to get away for a weekend in Kent with Norah Smallwood. She loved the little churchyard in the village and on one winter visit was ravished by the snowdrops growing there. They entered the church and talked with the parson. The whole thing made a big impression on Stevie: the little country church, the crumpled parson, the snowdrops, so she pressed a pound into his hand. On the journey home Stevie misdirected Norah Smallwood, who felt she'd done so deliberately to prolong the weekend. When finally they reached Palmers Green, Mrs Smallwood said, "Do come again, Stevie." And Stevie replied very earnestly, "Those are the words one wants to hear when a weekend is over."

Aunt's recovery proved to be chimerical. On 5 March Stevie told Marghanita Laski that "Aunt is slipping away, she is much weaker &

does not really now recognise us. So the plans for St. Christopher's Hospice may come to nothing . . . she is in a good deal of pain, & one cannot want *that* to go on long." She added, "Molly—my sister—is staying on with me. I am thankful to have her."

In the midst of so much sorrow, Stevie did not neglect either her friendships or her muse. While Aunt was in hospital Stevie sometimes dined with friends. And she allowed Granada TV to come round and film her on 7 March. Four days later, 11 March, on which "Darling Aunt died in North Middlesex Hospital at 5.30 p.m.", Stevie sent some poems to *The Observer*, including her "Sapphic (in mixed speech)":

> The mune ha gien her loicht an' gan
> The stardies eek are flee
> Upon ma bett in durchet nich'
> Ah lane ah lee.

[handwritten: mixed dialects]

"It's got a mixture of dialects", Stevie explained to Terence Kilmartin, "English (plain, archaic and cockney), American (negro: ah for I), German and of course Scotch. In fact it was after reading some old Lallans poems I much admire that I wrote it. I think 'alane ah lee' is beautiful and 'are flee' very desolate."

Aunt's funeral took place on 15 March at St John's, Palmers Green. The Vicar was ill but Gerard Irvine read the lesson at the house and another priest conducted the service. Stevie noted in her diary that the Brownes, Armide Oppé, Rosemary Cooper, Doreen Diamant, Bobby Woodcock, and others attended. All of her friends rallied round Stevie, concerned how she would manage after Aunt's death. Most of them felt, with Norah Smallwood, that "Stevie was content to lie on the bosom of Aunt", and loved to be cherished and cosseted by her. Once, when a friend told Stevie she'd be much freer after Aunt died, Stevie said "don't be silly, it's Aunt who looks after me". Two days after the funeral, the Woodcocks, living nearby, had Stevie and Molly to lunch, and a few nights later, on the 21st, the Cleverdons gave a dinner party that included Stevie, Sir John Betjeman, Lady Elizabeth Cavendish and John Wells. Stevie brought four of Aunt's long Edwardian petticoats as a present to their daughter Julia. All through dinner Stevie sang her poems which the others found hilarious and, at the same time, strangely moving.

Shortly thereafter, Stevie went to Norfolk for nearly a month's holiday with the Buxtons and the Brownes. While there she wrote to thank Doreen Diamant who had forwarded her mail. Of course just then there would have been many condolence letters, but one neighbour reported that his son had worked part-time delivering mail and never ceased to be astonished at the volume and variety of the letters sent to Stevie.

In late April Stevie returned to Palmers Green. Bobby and Nina Woodcock had gone over the house and Stevie found all in order and fresh flowers in the vases. The Brownes had driven her home and on the journey their dog, Pug, sat on her lap. This inspired the poem "O Pug"—about the Brownes' dog, but even more about Stevie who just then was surrounded by loving friends, and yet frightened and lonely too:

> But at heart you are frightened, you always have been.
> O Pug, obstinate old nervous breakdown,
> In the midst of *so* much love,
> *And* such comfort,
> Still to feel unsafe and be afraid,
>
> How one's heart goes out to you!

When Stevie returned to 1 Avondale Road she seemed to the Woodcocks to be lonely and sometimes desperately so. She often rang them up sounding agitated or, alternately, depressed, and asked if they would come over to see her, or could she come to visit them. Once she sounded hysterical and began to cry. She kept repeating, "I wish Aunt were here." Helen Fowler remembers sitting with Stevie at about this time, and Stevie said how much she missed Aunt and then began quietly to weep. Anna Browne believes that after Aunt's death Stevie drew nearer to Molly, her last family link—that from then on the sisters were very much closer.

Early in June Stevie went to Buckfast to spend two weeks with Molly. On the eve of her departure she attended a party at Gerard Irvine's and subsequently described this Whitsunday event to Anna Browne: "I went to church—Gerard's—as after evensong he had arranged a little poetry reading & singing. . . . When that was over we had Benediction (in English) & somehow you know, with the best will in the world, it does seem what Aunt always called 'odd'." She commented on the wonderful Bavarian rococo madonna in the church. The C. P. Snows attended the party and told Stevie that at a Fourth of June celebration at Eton, the Captain of the School had recited her poem "Do Take Muriel Out". "What glory", Stevie said in her letter to Anna.

In August Stevie published her poem "So to fatness come":

> Poor human race that must
> Feed on pain, or choose another dish
> And hunger worse.

There is also a cup of pain, for
You to drink all up, or,
Setting it aside for sweeter drink,
Thirst evermore.

I am thy friend. I wish
You to sup full of the dish
I give you and the drink,
And so to fatness come more than you think
In health of opened heart, and know peace.

Grief spake these words to me in a dream. I thought
He spoke no more than grace allowed
And no less than truth.

Stevie told Hans Häusermann she had been trying to write since
March to inform him of Aunt's death. What resolved her to do so was
the fear he would read "So to fatness come" in *The Sunday Times* and
guess what had happened. After her several visits to friends and to her
sister, she was now, in mid-August, back at 1 Avondale Road and
confronting her loneliness and grief. Of Aunt she said: "I think in her
mind I always remained the rather feeble child I was when she first
came to take charge of us all. But I always told her—and how
truthfully—that I depended on her just as much as she did on me."
Aunt was irreplaceable and Stevie accepted that. "One either lives
with someone one loves VERY much", she told Professor Häusermann,
"or one lives alone." She did not want, she said, to live with someone
whom, after a while, she would begin to hate.

It was around this time Stevie wrote a brief and moving essay which
was returned to her by *The Observer* in September, and seems never to
have been published. It is obvious, she began, that those who believe
in a God of love are best fitted to bear the death of loved ones, for "the
most comforting of all beliefs" is that "the souls of the departed are in
the hands of God and no harm shall touch them":

> My own *feelings* incline me to this belief but . . . it is a feeling only,
> or perhaps I might say it is what I should like to be true, it is Hope
> rather than Faith that speaks. But I have had an upbringing in a
> church I loved . . . and grief turns one back to childhood. The
> prayers and the love of God one knew then run in again like a spring
> tide to cover the rocks of objection. But the rocks are there and the
> tide will ebb.

As she could deny neither rocks nor tide, Stevie went on, she must
embrace contradiction. It is common for humans to live comfortably

with such dichotomies, she said, but she did not know whether the ability stemmed from "mental laziness or a modest estimate of human reasoning":

> When people one loves die, one's love for them fans into a great flame (my thoughts seem to run on fire and water) and all love seeks its source and its destiny in Love, in the idea of some great Love, that is beyond the human pattern. But that perfect love may be what I have often written of as the greatest of all blessings: Death as a scattering of the human pattern altogether, as an End.

Stevie then quoted her poems "Why do I . . ." and "My Heart Goes Out". One's own physical health has a lot to do with "what one thinks about life and death", she supposed, and detachment from life might be less a virtue than the result of "some physical or chemical deficiency":

> I have found myself that the testing period comes some months after the death one mourns. Then, when the elation of love passes to the fruitless melancholy of loneliness, a dreadful restlessness oppresses the mind . . . friends are too much absent, or, if present, they seem like phantoms. . . It is best to pray then, whether your intellect allows it or not, pray, to have love or at least think every night and every morning . . . about love, remembering that self is the great enemy of love and of peace. Remember too that mourning . . . can be selfish. One thing I am sure of. It is shameful to believe evil of a future life that may or may not exist, shameful to believe in eternal hell, shameful to try and bewitch the dead back into communication . . . there is one thought that is neither comfortless nor subject to doubts or questioning: You do not mourn if you do not love.

"It is this I was thinking", Stevie added, "when I wrote 'So to fatness come'."

Sally Chilver thought that after Aunt died, Stevie was lonely in Palmers Green with her friends so far away, but that she couldn't bring herself to move. One of Stevie's neighbours also noticed how lonely Stevie was after Aunt's death, and how kind she was to a gardener they shared. He was called "Old Charley" and lived entirely out of doors sleeping in parks, after a brother had cheated him of the home his parents had left to both sons. He was put in a Home once, but left after a bad experience there. Besides paying him some small sum for looking after her patch of garden, Stevie would give him a sandwich and fill his thermos flask on Sundays when restaurants were shut. One Sunday Stevie was in a terrible hurry and crossly banged the door on him. He went round the corner, sat down on a step, and began to cry.

Stevie's neighbour saw him and asked what had happened. "I lost my best friend," he said. "Stevie banged the door on me." The neighbour fed Charley and when next she saw Stevie told her how hurt he was. "I'm sorry I did that," said Stevie. "Here's a pound. Will you give it to him and tell him to come back soon?" "There is an Old Man", a poem found among Stevie's papers after her death, tells the story of Old Charley who slept in parks. It ends:

> Oh living like this is much jollier for me
> Than anything I've found for the Elderly.

In 1967 Ladislav Horvat had found a house for Stevie near Kensington High Street—a five-flat house near a home for gentlefolk where Aunt would have stayed. He thought of it as a splendid investment and Stevie first told him to go ahead and purchase the house. Two days later she called and cancelled her plans. At the time Horvat was angry and refused from then on to act as Stevie's financial adviser. But later he thought that she just couldn't have sent Aunt to a nursing home. Stevie had once told him "I don't want to be well off", and he thought she rather preferred "playing the role of a churchmouse". In late summer of 1968, after Aunt's death, Stevie wrote:

> I have decided to stay on living here, I really love it, it is so quiet and sweet in a way and I dont feel I want . . . to have anybody else living with me. At least not just yet. Marghanita Laski asked me to go and spend a few weeks with them in the south of France. . . . And there was some talk also of my going to Malta with some friends. But I just dont want to go anywhere—except dear old Norfolk, where as a matter of fact I am going, yet again, next week for a fortnight to stay with the same long-suffering friends I always do go to. I might feel more adventurous next year.

Stevie's late August holiday in Norfolk was "heavenly", and when she reached home Nina Woodcock had again filled the house with flowers. Stevie said sadly, though, that judging by the mail that had accumulated, "the only people who want to talk to me are always the newspapers", and added that lack of companionship drove her to make an ass of herself in interviews.

That autumn Stevie gave poetry readings in Hampstead and Oxford, and on 7 December she joined Eddie Linden, Brian Patten, and other poets at the Lamb & Flag in Bow Street. Eddie Linden remembers the animosity between these poets and those who got published in the *New Statesman* and the *TLS*. But Stevie fitted right in with them somehow, he recalls. She was fearful that the reading wouldn't go well, so Linden met her earlier at the Arts Club, Leicester Square, in an effort to calm her. The audience numbered about 130, and once Stevie got to the platform she read wonderfully. Afterwards she told everyone she had earned a fiver for her reading, and invited a number of poets to drinks. When John Heath-Stubbs put Stevie's name on a list of people who could be helpful to Linden in starting a new magazine he planned to call *Python*, she wrote back (no doubt motivated by her horror of snakes) to say, "You cannot possibly call it that. That's a terribly cold name."

Two nights later Eddie Linden went with John Heath-Stubbs to hear Stevie's talk to the St Ann's Society. Among those attending were Antonia White and an unidentified Roman Catholic priest. In the discussion that followed Eddie Linden (born an Irish Catholic) attacked Irish Catholicism, and Stevie rose to support his views. Her talk was titled "Some Impediments to Christian Commitment", a later version of "The Necessity of Not Believing". Rosamond Lehmann refers to the talk as an "impassioned Apologia for agnosticism" and indeed it was a curious, but in its own way deeply spiritual, address for the Advent meeting of a church society.

Stevie's Christmas letter to Rachel Marshall told her friend the news of Aunt's death. "We miss her *awfully*. Molly was the greatest possible help & lots of friends came to the funeral to 'honour her'. All that was really wonderful. But, well, it is still *awful*." She confessed that she rather liked "how old everybody is now getting . . . it's so *restful &* *sleepy* somehow. Or perhaps that is just these dark days before Christmas that I always love so much."

Not reading at the Palais Voor Shone Kunste, Brussels 1967. Sketch by Michael Horovitz.

A Breath of Fresh Air

Early in 1969 negotiations were proceeding between Longmans and Knopf for the publication of *The Best Beast* in America. The collection contains forty-four of Stevie's poems, some that had appeared in *The Frog Prince*, and some that were to appear, posthumously, in *Scorpion and Other Poems*. James Laughlin wrote that New Directions was considering reprinting *Selected Poems*, and Stevie said again how much she admired the look of that edition and regretted that the book hadn't made a fortune for him. She also mentioned poems in the *Sunday Times* ("So to fatness come") and in the *New Statesman* she thought might have alerted him to Aunt's death. The *New Statesman* poem may have been "Grave by a Holm-Oak", which was published a few weeks before Stevie sent her letter:

> You lie there, Anna
> In your grave now,
> Under a snow-sky
> You lie there now.
>
> Where have the dead gone?
> Where do they live now?
> Not in the grave, they say,
> Then where now?
>
> Tell me, tell me
> Is it where I may go?
> Ask not, cries the holm-oak
> Weep, says snow.

A few weeks after Stevie had told Laughlin of Aunt's death, she received a condolence letter from him upon whom Aunt had made "a most profound impression . . . we [he and his wife] had the sense that we had been privileged to know and share, even a little, in the life force of a very extraordinary person." He was pleased that Stevie had no reservations about the continuance of her *Selected Poems* in the New Directions paperback series. "Despite the rather modest sales, it is one of my real favorites on our list, and a book which gives me the greatest satisfaction." Prophetically he added, "And I have confidence that it is going to do better in the future." He wondered, now that Aunt had died, whether Stevie might give some thought to a reading tour in America.

Rosemary Cooper remembers how Stevie repeatedly wondered aloud to her after Aunt's burial how her decomposing body might look as time passed. This seemed a morbid preoccupation to Rosemary, and she finally told Stevie to stop talking about it. But Stevie's essential wonderment must have been, as "Grave by a Holm-Oak" suggests, "Where have the dead gone?" As she wrote in the poem "If I lie down":

> If I lie down upon my bed I must be here,
> But if I lie down in my grave I may be elsewhere.

Stevie was still grieving deeply for Aunt. "I wasn't able somehow to get it out before, and feel ghastly doing so now", she told James Laughlin. But Stevie reports with pride the fight left in the Lion towards the end, and repeats with good humour the question the hospital Sister put to her: "Has Miss Spear been always in a Position of Authority?" "I would never wish her back", said Stevie, describing Aunt's last year of deafness and increasing blindness, "but It is Still Awful, and I don't think I am really making much of a job of living alone . . . Aunt was 96. I only hope I never am. But as I have more or less said in practically every poem I have ever written . . . It Lies with Us."

None the less Stevie often preferred to be by herself. In "What Poems Are Made Of", which *Vogue* published in late winter, she writes of watching with pleasure birds, animals and children in the park and thinking how fortunate she is that they aren't hers. "I do not know how people can manage to have animals, wives and children and also write. Of course isolation can be very painful. Many of my poems are about the pains of isolation, but once the poem is written, the happiness of being alone comes flooding back." In this mood the loneliness Stevie had felt during wartime, and which she condemned in "The Failed Spirit", she saw in a different light. The speaker of that poem "was so foolish as not to recognise his nature and its solace", she said. "Instead he tried to do some war work." But loneliness "is not a stony pasture", she now declared.

Stevie had some comic relief early in 1969 with two pieces of writing she did after reading a biography of Swinburne. One is a review of the book, published in *The Listener*, entitled "Swishing and Swinburne". She seems fascinated by the love relationship, perverse as it became, between Swinburne and his beautiful cousin, Mary Gordon, whose accomplishments she lists, adding "I daresay she thought *un peu de vice* on top of it all did not signify. I do not say she was right", Stevie wittily adds. Soon after completing her review of the Swinburne biography, Stevie composed "Seymour and Chantelle or Un peu de vice, *In memory of A. Swinburne and Mary Gordon*", about two children who give each other pleasure by inflicting pain. As she wrote to Terence Kilmartin:

> That book about Swinburne I had to review has made such an impression on me that I have now written a *vicious* poem. (I think the reason I am writing such a lot of poems is because I am supposed to be doing something else . . . i.e. collect an anthology for children of other people's poems, not mine of course, and it's such a bore it makes one write). Anyway, this poem shows it's basically childish whereas all poor Swinburne's didn't. Mind you *Suppress* it if it is even a shade of what darling Aunt called Unnecessary! Lots of vicious old love.

It was a curious subject for Stevie, who was exceptionally horrified by cruelty, but a small closet drama achieved with psychological accuracy:

> Seymour, when you hold me so tight it hurts
> I feel my ribs break and the blood spurt,
> Oh what heaven, what bliss,
> Will you kiss me, if I give you this
> Kiss, and this and this? Like this?

A review of *Against All Reason* that appeared in an April number of the *New Statesman* is another of Stevie's meditations on religion in which she is again concerned with cruelty mixed into goodness. She seems especially intrigued with the desolation felt by a Carmelite nun sixty-six years old—just Stevie's age then, so perhaps she saw her with the writer's eye as a discarded self. Certainly Stevie finds much that is noble, if illusory, in the lives of those religious *dévots* described in this book.

In the late winter and spring Stevie kept busy with proof correcting for *The Best Beast*, collecting, as she said, pieces for *The Batsford Book of Children's Verse* (published in America as *The Poet's Garden*), visiting friends, and giving an increasing number of poetry readings. On 5 May, accompanied by Bobby and Nina Woodcock, Stevie and Molly went to St John's, Palmers Green, where Father Doyle dedicated a reading desk given by the sisters in honour of their mother and Auntie Maggie. The lectern, inscribed with their names, is still in use. Then, on 24 May, another crushing blow fell on Stevie. She wrote to tell Doreen Diamant that

> poor old Molly had a stroke last Saturday. . . . At first it seemed only a minor one, but then she had another. Anyway, I came down here [Buckfast, south Devon] on Sunday & got to the hospital (at Ashburton) about 6 p.m. Matron said it was very unlikely she would live the week out . . . But by now I think she is a little better, but it is *awful* to see her, really, because she is paralysed all down her left side & her speech is affected . . . what wd. be really awful is if she "got better" as they say, but remained paralysed.

Stevie stayed on in Buckfast for some weeks, visiting Molly twice a day and "looking after that rather detestable cat", then spent much of the summer shuttling back and forth between London and Buckfast. While she was away from Palmers Green, Doreen Diamant and the Woodcocks looked in from time to time at 1 Avondale Road. One request she made of Mrs Diamant was that she put the "two 'Busy Lizzies' out in the back garden so they cd. get the rain . . . I think I put them in rather too sunny a position. Perhaps a northern aspect . . . might be better. Silly thing to think of at this rather awful time—but there it is! I really *cant* think of the real awfulness much of the time, it is almost too much."

A few days later she wrote of the news of Molly's stroke to Lucinda Fowler and said that if her family were still at "Frog Cottage (heavenly name) do ring up & come over". The Fowlers did visit Stevie in south Devon that summer. Stevie chose a particularly gloomy part of the Dart River as a picnic spot. She discussed a television interview she had taped earlier in the year with Ludovic Kennedy and said she'd been chatting briskly with him when he unexpectedly asked about how she had felt when her mother died. At once she had burst into tears. The day was hot but the water was uninvitingly dark. Stevie donned a regulation school bathing dress she'd borrowed from Helen's daughter Catherine, and plunged into the river. She came up gaily, reciting a poem about the Welsh (it was the summer Prince Charles was invested as Prince of Wales and Stevie and Molly had watched the ceremony on the hospital television). Then Stevie's mood shifted again and she emerged from the water speaking about death and the race of water that would take one over the weir into the still vale.

In June Stevie gave a reading with Phoebe Hesketh in Skegness and learned that, with an ill husband, Hesketh was financially hard pressed. Stevie wrote to Rupert Hart-Davis asking if the Arts Council could assist Phoebe Hesketh. Hart-Davis replied, saying he would see what could be done. The news of Aunt's death had been a blow to him: "the very thought of her takes me back to Bedford Square and Hamish Miles and the days when we first met."

That same month Stevie delivered the manuscript of *The Batsford Book of Children's Verse* to Sam Carr, her editor. Carr had put in a few of the selections (Wordsworth's "The Sun has long been Set", and the ones from Pope and Melville), but on the whole the volume reflected Stevie's tastes "as they are now and as they were in my childhood". Not surprisingly, she included five poems of her own among the eighty-six entries from the Bible, Shakespeare, the "Romantics", and others. When the book was published, a few months later, Hardy's "The Rejected Member's Wife" was missing because of "last minute pressure", although Stevie's reference to it had been left in her Preface. "I struggled & struggled to get them to spell Catullian Hendecasyllables with the 'i' (as Coleridge does)", she also noted, "but, no: they will have Catullan." She thought it "a v. pretty book", however, despite the fact her choices constituted a "rather *sad*" collection.

In June of 1969 Stevie also took part in a BBC panel with Maurice Edelman, Cyril Connolly, Antonia Byatt and Edward Lucie-Smith. Stevie was more and more a hit on the poetry-reading circuit and invited everywhere. In July she read at Stratford and that allowed her a visit to Chastleton House nearby. She managed, despite so many trips back and forth to Buckfast and to poetry readings, not to neglect her writing. At this time the *New Statesman* was her principal outlet for poems and she sent Anthony Thwaite, its poetry editor, her poem "The Donkey"—in two versions because she found it hard to judge which was better and "sometimes get a bit nervous as to too jog-trotty rhythms". She celebrated in this poem "the hilarity that goes with age" and "the sweet prairies of anarchy" for "After a life-time of working / Between the shafts of regular employment / It was now free to go merrymaking":

> And the thought that keeps my heart up
> That at last, in Death's odder anarchy,
> Our pattern will be broken all up.
> Though precious we are momentarily, donkey,
> I aspire to be broken up.

Stevie wrote to tell Anna Browne that Molly was weeping with boredom and having to remain motionless. She had treatments three times a week and though making progress, had to be in a wheel chair all the time.

In July Anthony Powell wrote Stevie a letter congratulating her on her reviews in *The Listener* of Malcolm Muggeridge's *Jesus Rediscovered* and of the biography of Swinburne. In her criticism of Muggeridge's book, Stevie is sharp about the limitations, as she sees them, of the Christianity Muggeridge embraced and of his hatred and contempt for those in power. Muggeridge and Powell had, of course, been colleagues when they edited *Punch* and sometimes published Stevie. So, many years later, Stevie wondered in her reply to Powell "how it [Muggeridge's religious passion] all began? But I gather you have not been on terms recently." These reviews for *The Listener* were nearly the last she wrote, although she had offers to do others. Terence de Vere White evidently asked her to review some poetry but she declined saying, "I always have refused to review poetry and I think I really must be in the peculiar position of only liking writing it, bar some of the old tried (and dead) favourites of course." She recommends non-writers of poetry as the best critics, as "they have more interest more patience and more outward going thoughts ha ha."

Most of the summer Stevie spent at Buckfast, managing in late August to have a week's holiday with the Lawrences at Brockham End before returning to London on the 26th. A bit earlier she had written to tell a friend that Molly would not be able to drive nor to live alone again. In mid-July Molly was still in hospital. This friend had asked for advice about publishing her poems and Stevie suggested publishers but said she dare not forward the manuscript as she was "well known now for *not* being able to judge one end of a poem from another, which is why I never review it (them?)." She also inquired of this friend if she knew any nice, preferably RC, person who might share Molly's bungalow at Buckfast. Stevie praised the beautiful Dartmoor scenery and said Molly had recovered her speech, her mind was good as ever, "so some retired but hearty lady might quite like the job if job it is." Molly was still paralysed but power seemed to be returning to the right arm and hand.

Writing a few days later from Palmers Green to the same friend Stevie says she was "full of doubts this morning, mostly about the love of life wh. some people are said to feel!" Stevie had plans to return to Buckfast in a few days. Of Molly she wrote:

> I am so desperately sorry for her but the poor darling has a romantic temperament, seeing things rather as it would suit her . . . & now

has built up quite a vivid picture of going down with little sister to the grave hand in hand, after many a long if ageing year close closetted à deux! I'm afraid in this race to the grave little sister wd. have won before Molly lifts her starting foot! But you can imagine how *wicked* it makes me feel & how (I hear from a Buckfast spy) feeling in that tight little catholic-masochistic society is veering strongly against me, *unless* I "do my duty". Only my own doctor furiously writing to Molly's doctor says "not on any terms".

In August Stevie heard from "an old fan of the thirties", Denis Johnston. After the war he had gone to America where he spent twenty years as a university professor. Back in Britain and watching television he saw Stevie and said, "it sent me to my shelves to read Yellow Paper again. It contained an inscription, a small photograph signed Stephanotis . . . I cannot resist writing to you again to say that it is a lovely book—far more up to date than ever it was, that you [are] a bloody good televiser, and that I am feeling greatly stuck up over having been an admirer from the very start." Stevie wrote right back to say "you know you were the first person to write to me when the book was published in 1936". She hadn't seen herself on television as in Devon "BBC 2 cometh not".

The problem of Molly's future was not easily solved. Stevie had been to see her own physician in Palmers Green and he had entered Molly's name on the waiting list of the Home for Distressed Gentlefolks but learned that it would be a long wait. Meanwhile an acquaintance of Molly called Aylet Hyams had entered the picture. Stevie had met her on a number of occasions before Mrs Hyams began to write Stevie letters in August, describing Molly as mentally ill and vicious, plotting to have Stevie come to live with her. "You must not dream of living with her. All your gifts will be wasted and you will be eaten up alive." She said further that Molly was complaining Stevie was larking about in Norfolk while she (Molly) was about to be "incarcerated in a sort of Belsen". These letters must have been disturbing to Stevie until she realized the bias of their author. Meanwhile Molly engaged a Mrs McCleod as housekeeper, and shortly after was discharged from hospital. Stevie stayed in Norfolk from 28 August to 11 September when the Brownes drove her back to Palmers Green.

That autumn she did more poetry readings, among them one in London in October where she read to a group of American students. After the reading Stevie and the professor who directed the group went to the Museum Tavern for drinks and lunch. He remembers very well that, when the topic got on to suicide, she said she would probably commit suicide herself. "She said it casually, not out of any perceptible feeling of despair, but as if suicide would be the normal, logical end for her. The remark stuck with me . . . because of the *way* in which she said it. . . . It struck me, in its curious way, as somehow an expression of strength—as if the awareness of that choice enabled her to cope." That same month Stevie read at Roedean and Bristol.

News of her being awarded the Gold Medal for Poetry brought letters from many friends, among them Suzannah Jacobson. Stevie told Suzannah of Aunt's death and said, "I have been living here on my own . . . getting quite to like it in a sad little way, but miss the old lady awfully really." She also described Molly's situation and the temperamental housekeeper who "at least 3 times a day says: 'I cannot stay one more minute'." Everything depended, as Stevie told another friend, on the housekeeper staying. "Poor old Pug is now dead", she tells her friend, "aged 13½." For some days Stevie was rung up by interviewers for newspapers and by photographers, a number of whom called on her. One of these was a reporter for the *Evening Standard* which published a few days later an account of Stevie emphasizing her dislike of change and her love of the familiar.

Christmas came and Stevie spent it with Molly in Buckfast. The housekeeper was still threatening to leave, but Stevie thought Molly and she liked each other, fundamentally, and so did her best to keep Mrs McCleod on. Stevie was growing fond of village life in Buckfast, and "being able to *walk* in to see people". She expected to find life back in Palmers Green "awfully isolated after the to-ing & fro-ing that we have here". Molly seemed much better. They drove over Dartmoor and watched television. "I don't think I should ever do any writing at all down here, the days rather fritter themselves away."

In January of 1970 Stevie was back in London for a reading. She was also photographed at the request of the National Portrait Gallery which wished "to include a photograph portrait of you in the National Record". This request was followed a few weeks later by one from William Plomer who congratulated Stevie on the Gold Medal and her upcoming appearance at the Aldeburgh Festival. He then asked Stevie if she would agree to sit for the painter Robert Buhler, a member of the Royal Academy, who was doing a series of pastel portraits of contemporary writers. Stevie replied that she'd be delighted to have a portrait painted by Buhler, whose work she admired when she saw it in the Leicester Square Gallery. The portrait was painted during the spring, and in May she wrote to the painter thanking him for a photograph he'd sent of the portrait but adding, "I do not think it is like me. It's the mouth and perhaps the lower part of the face that seem to belong to somebody else . . . to Disraeli, I think . . . for whom of course I have a great admiration, but whom I think I do not resemble."

In March the *New Statesman* published "Smudgers and Meddlers", Stevie's review of the New English Bible. Her review is devastating:

> Everything that was bright is dulled, what was sharp, blunted. Does their great labour, then, serve only to smudge and betray? . . . To smudge, to weaken, to blunt, to make pallid, every beautiful word and thought it carries—was this worth 24 years' work, with the alterations so trivial, nothing to make even the attempt worth while? But let not the friends of mediocrity be cast down. For nowadays we do not often have the AV or the RV or great Cranmer's Prayer Book. The churches and chapels have chosen the new translations. They think they are right for the times we live in. But nothing that is second-best is right.

Sir John Lawrence said that Stevie had strong views about the translation of the New Testament, published some years before the complete New English Bible which, of course, was predominantly the Old Testament. When Stevie was sent the New English Bible she wrote her review largely out of her reactions over the years to the translation of the New Testament. Soon after she rang up Sir John and told him that she did the review, and he pointed out to her that she could not apply her reactions to the translation of the Old Testament, which had been done by completely different people and in a very different style. Stevie ignored this and, said Sir John, "as she died so soon after, I didn't want to blacken her memory". He called it "a pretty unscrupulous bit of reviewing". Although the review attracted attention of a most admiring kind, there were readers who had strong reservations. In a letter to the *New Statesman* one of them noted areas where contemporary scholars would disagree with Stevie's assumptions. She did not write in response, but told Anthony Thwaite her errors were the result of "sheer ignorance".

In April Stevie "skidded on a new pair of shoes just after giving a reading at Dulwich", cracked three ribs, and damaged her knee, "the one that had the kneecap removed from it a few years ago", she told Kay Dick. She was preparing to read as part of the May "Funeral Festival" organized by the director of the Brighton Pavilion on the theme "Death, Heaven and the Victorians". She joked that if Olivia Manning turned up in the audience "I shall have to cut out all the singing. I suppose it is ridiculous but friends do make one feel nervous." Stevie was also disapproving of Olivia's having received a grant of £1,000 when her efforts on behalf of Phoebe Hesketh only produced £500. Stevie told Kay half jokingly she thought she might apply as "now I have to pay for other people to do all the things one generally does for oneself, it does cost quite a lot."

In April she heard from Eric White at the Arts Council, with whom she had deposited manuscripts to receive estimates of their worth. Evidently she was trying to sell them, as Eric White advised her to name her asking price so he could find out whether the British Museum or other libraries wished to acquire the materials. (*Novel on Yellow Paper* was valued at £500 and subsequently purchased by the library of the University of Hull.)

It was probably in these weeks, too, that Stevie visited the Kensington Church Street Book Shop to inquire if they had sold any of the manuscripts she had left with them. The proprietor remembers her saying her sister was very ill. Ralph Steadman, the painter, was there when Stevie came in, and she was pleased that he gave her a drawing. Stevie asked him to wait a minute, rushed next door to a sweet shop, bought a candy bar, and autographed the wrapper for Steadman, who now has it framed and displayed in his home. The Kensington Church Street Book Shop had not sold any of Stevie's manuscripts, nor were her coffers greatly enriched by the reading she gave in Brighton, for which she received £15. There was a large audience, however, and Stevie's poems seemed just the thing to launch an exhibition on "Death, Heaven and the Victorians". Everything on display had to do, of course, with the theme. Mauves and blacks, recorded Victorian hymn tunes, and Gothic arches contributed to the lugubrious atmosphere. "Oh, it's grue, it's very grue," Stevie exclaimed as she walked around. "I think she was slightly shaken by the gruesomeness," recalls John Morley, the director of the Pavilion, "as quite a few people were."

The day after her reading Stevie had a picnic lunch with Kay Dick, the Morleys, and a friend of the Morleys, Dora Evans. According to John Morley, Dora had a powerful personality, was plain spoken, and quite perceptive. Years after the picnic her impression of Stevie stuck in his mind. Dora found her "a poor thing", and predicted she wouldn't last another year.

In June Stevie sent some poems to Anthony Thwaite, uncertain of their worth. But "I do claim all the same that with Scorpion I have found a *new grievance*, at least I dont remember anyone else feeling aggrieved on just this point." "Scorpion" was published in the *New Statesman* and became the title poem of Stevie's posthumously published collection. Soon after it appeared she wrote to ask Kay Dick, "Did you see Scorpion in the N.S. . . . My poor beloved unspeakable (just like my tum) Scorpion? It is just what I feel like." The speaker begins with a complaint:

"This night shall thy soul be required of thee"
My soul is never required of *me*
It always has to be somebody else of course
Will my soul be required of me tonight perhaps?

(I often wonder what it will be like
To have one's soul required of one
But all I can think of is the Out-Patients' Department—
"Are you Mrs Briggs, dear?"
No, I am Scorpion.)

I should like my soul to be required of me, so as
To waft over grass till it comes to the blue sea
I am very fond of grass, I always have been, but there must
Be no cow, person or house to be seen.

Sea and *grass* must be quite empty
Other souls can find somewhere *else*.

O Lord God please come
And require the soul of Scorpion

Scorpion so wishes to be gone.

The other poem she sent to Anthony Thwaite and which also appeared in the *New Statesman* was the eerily titled (as it turned out) "Black March":

I have a friend
At the end
Of the world.
His name is a breath

Of fresh air.
He is dressed in
Grey chiffon. At least
I think it is chiffon.
It has a
Peculiar look, like smoke.

The speaker calls him "Black March" because his eyes are "like March raindrops / On black twigs." The poem ends with Black March saying:

Whatever names you give me
I am
A breath of fresh air,
A change for you.

On 21 June Stevie read at the Aldeburgh Festival and then, for nearly three months, she stayed with her sister in Buckfast. In a letter to Jean MacGibbon, she described Molly's condition:

> She is so sweet, poor darling, & *good* but so very sad. I'm afraid she is now as well as she ever will be—i.e., she has permanently lost the use of her left arm & can only walk with great difficulty & a 4-pronged sort of stick. I feel quite sick really all the time I am away in case something awful happens. You can imagine. But all through October I have readings to do. . . . I crammed them all into one month so as to get down to Molly as soon as possible.

Stevie did manage, during that summer, to stay with the Fowlers for several days. The visit ended a bit tetchily when Helen, feeling bullied by Stevie and exhausted after the ninety-mile round trip to collect her from Buckfast, refused to take her all the way back and instead drove Stevie to Exeter where she could board a bus. During the journey Stevie lectured Helen about how poets need to be cherished and looked after and how they have to husband their strength. Helen steeled herself and in the end did not yield to Stevie's importunings.

In October Stevie read her poems in Cambridge, at Stockwell College in Bromley, St Mary's College in Newcastle, the Stroud Festival, the Brighton Poetry Society, and her old school, The North London Collegiate. At Stroud she stayed with Lord Moyne (the poet and novelist, Bryan Guinness), the director of the Festival. In the midst of this busy schedule she found time to write to friends. To the son of her friends the Russell-Cobbs she sent congratulations on the essay he'd written at Westminster School about her work, and apologized for its "deathward trend" but said the feeling was "*very* strong" in her that "if there were no death, one would indeed be trapped". She also sent two poems to *The Scotsman*, "Sapphic" and "Francesca in Winter". In March "Francesca . . ." had been the "Poem of the Month Club" selection and *The Scotsman* did publish it late in November 1970. It is a beautiful poem which, at one point, echoes Dante's imagery (of souls tossed lightly on the winds of hell) when he tells the Paolo and Francesca story in the *Inferno*:

> O love sweet love
> I feel this love
> It burns me so
> It comes not from above
>
> It burns me so
> The flames run close
> Can you not see
> How the flames toss

Our souls like paper
On the air?
Our souls are white
As ashes are

O love sweet love
Will our love burn
Love till our love
To ashes turn?

I wish hellfire
Played fire's part
And burnt to end
Flesh soul and heart

Then we could sit beside our fire
With quiet love
Not fear to look in flames and see
A shadow move.

Ah me, only
In heaven's permission
Are creatures quiet
In their condition.

On 6 November Stevie recorded for "Dial A Poem" in a "Poet of the Week" series. She did the "Old Sweet Dove of Wiveton", "Tenuous and Precarious", and a short poem about Lucrezia Borgia ("Yes, I know"). She enjoyed doing it, she told Lord Moyne, and because of a technical fault and the telephone breaking down, she was continued for a second week. Lord Moyne thought Stevie read the poems too swiftly and she attributed this to the stop watch used by the producer which prompted in her "the *unsuitable* aspiration . . . of beating the record". The next day Kay Dick came to 1 Avondale Road and taped the splendid interview which was published in *Ivy & Stevie*. A few days later, doubtful that she'd said enough of the right things, Stevie sent Kay the John Horder and Giles Gordon interviews which were in effect Stevie's interviews of herself.

Stevie went to stay with her sister in Buckfast again in November and seemed cosily ensconced as she wrote to Lord Moyne about the ceaseless rain in south Devon that year. But she had begun to feel ill. Wild dashes of energy were followed by periods of collapse. Two days later, Stevie wrote to John Guest that she was "*not well*" and nobody knew the cause. She had recently read at Plymouth and had more readings scheduled in the next week. But she began to have fits of nearly fainting, she told her London physician, and losing control over words. When she was really tired she heard ringing sounds and had to sit down. She said the physician in Buckfast was wrong about her loss of word-control being "in the category of 'er-er' & a slight stammer before being able to go on. It (in my case) is almost a mental black out . . . I tried to say to . . . someone that I could not find the word I wanted. But instead of 'word' I said 'milk' first & then 'snow'."

Stevie was planning to stay at Molly's over New Year and then to go to Leicester, Edinburgh and London to read her poems. But on Boxing Day she wrote again to say she was worse off, unable to travel, and had cancelled all her readings. "Doctors in command", she told Robert Nye, *Scotsman* editor. In a letter to Lord Moyne she said that, between peeling potatoes and passing out, she had managed to write two poems, one of which was "The Stroke" (with its dedication, *For M.*):

> I was a beautiful plant
> I stood in the garden supreme
> Till there came a blight that fell on each leaf
> How I wish this had not been
> Oh I wish this had not been.
>
> I can feel the sun, and my blighted leaves
> In an elderly way grow glad
> But oh in my depths I bleed, I bleed,
> From a heart that is youthful and sad
> From a heart that is piercèd and sad.

Donald and Molly Everett, neighbours and friends of Molly Ward-Smith, recall that when "The Stroke" was published in *The Observer* Stevie rushed to them in a panic asking, "how am I going to stop Molly from seeing this?" Mrs Everett replied, "You can't, someone will give it to her." Far from being upset when she saw the poem, however, Molly declared, proud as Punch, "She's written it for *me!*"

Stevie was still at her sister's, struggling to set her records straight with the Inland Revenue, when, on the night of 6 January, her illness became worse and she had to go to Torbay Hospital in Torquay. In between the medical tests she tried to conduct her affairs as best she could. She was especially concerned to put right the other poem she had recently written, mentioned in her letter to Lord Moyne. Molly Everett visited, with a revised version typed out. Looking like a little girl in one of Mrs Everett's nighties, she saw it and approved. Although her speech had not been good for some time (she had been missing words), Stevie read the poem aloud perfectly. It was a dramatic moment. Visitors, who were on their way out, stopped:

I feel ill. What can the matter be?
I'd ask God to have pity on me,
But I turn to the one I know, and say:
Come, Death, and carry me away.

Ah me, sweet Death, you are the only god
Who comes as a servant when he is called, you know,
Listen then to this sound I make, it is sharp,
Come Death. Do not be slow.

Although Stevie had misgivings about this poem, "Come, Death
(2)", she very much wanted it to appear in her forthcoming book,
Scorpion.

From hospital Stevie wrote again to Robert Nye:

I'm afraid I've been snatched away into a hospital & dash & dash it all
I have been touched & dished all over with glass & steel & all those
swords & things. Anyway they cant quite make out what the odd
thing is (i.e. I can't quite often speak anything but the sort of
scrambled Eggs important offices, especially the War Place, loves to
speak. I alas can't speak anything else. Perhaps perhaps *they* cant
either ha ha). At any rate I was snatched half dead into the hospital
but as they may have finished checking me up I *may* be with my
poor sister but I feel I ought *not* to find anyone to make them look
after me but just stay in a hospital. Ugh! but often I must say very
nice warm & lazey with lots of nice things to eat &, with luck, found
[?] & friends to come & have tea!

Friends did come to visit: Anna Browne, Helen Fowler, the Bray-
brookes, Everetts and MacGibbons. When Helen Fowler told Stevie
about her husband's promotion, she replied, "I'm so glad Laurence
has got his brigade." Jean MacGibbon rang Olivia Manning to tell her
how gravely ill Stevie was, and Olivia said, "Well, if she's really
dying, send her my love." When Stevie was told this she laughed and
laughed.

Throughout the last weeks of December and into January John
Guest had been corresponding with Stevie in an effort to arrange the
poems for *Scorpion* in a way that suited her. By January her letters had
become "untidy" and sometimes difficult to puzzle out. Guest kept a
log of her correspondence:

Dear John, I'm afraid I'm a sad nuisance for you now as I can't really
do much as I keep having *sort* of fits mixed with some poisoning
down below, ha ha. *When* the fits come I am almost unconsaince

[sic] with something coming upon me [two deletions] cannot speak proper words or read them it is a nightmare of muddle of words & frightful bells ringing & unknown crowds crying out advice, warnings etc. that I cannot understand, or when sanity returns remember. I am now with 2 doctors & have to go at once to a 3rd who is a specialist as soon as possible & may then—awful—have to go into some hospital. I have bored you with all this and will now try to be sensible with my poems and your *two* [? letter damaged] kind lists. [Here follows lists of poems etc., difficult to follow as instructions. The letter ends:] As a matter of fact I like all the *others* (poems) pretty well. And here is one I should like you to add, please. I have just written it, in this *awful* medical case that to me is rather like Hampton Ct. maze. I'll write it on a separate sheet. [This poem . . . . was "Come, Death (2)" . . . .]

My dear sister is better & all, really pretty well *all*, the inhabitants of Buckfast etc treat me forbearingly & the doctors are very efficient & are fun *too*, & Molly *looks* much better as she is fatter for one thing.

I have a thought dear John that all that say 4 years of enteritis we both had has now in me come out in these *Fits* from a Poison [Prince? or Pounce? I think *Prince*, a reference to the "Frog Prince"] Lots of love Stevie. [She then adds a long P.S. reverting to various poems. At the end she adds:] I expect I shall be better soon I hope so & I dare say you hope so too! [In a final P.P.S., attached to the copy of "Come, Death", she writes:] punctuation: wd you think some commas better? [these I subsequently added.] Perhaps it isn't v. good, it *speaks* very well because of the slightly ironic accent of [? on] the last line. Because Death, god though he is, has to come when he is called, it is suicidal of course.

[She then forwarded to me two letters to deal with that had been sent to her, one from the Midlands Arts Centre for Young People, and another from Julian Gardiner, dated 14 January, who had set some of her poems to music. Stevie had scribbled, with *many* erasions in the margins of both letters:] Dear John, I'm afraid I dont quite know about *any* of these letters. though Julian said once ages ago when he wanted to music my poems. I am not *very* well I'm afraid! Could you be an angel and tell Julian Gardiner *what* or what *not* to do, *please*. Love from Stevie . . . [On the other letter, at top and bottom she had written:] When the doctors have finished here (awfully nice here, in Plymouth somewhere) when they've spent ages & ages & mornings & mornings & afternoons & afternoons, *dear*, stick with sticks & glass tubes & being rather sharp with steel rods, *then* they'll know what's the matter! I'm not *sure* I'll be *very* bright ha ha as so often I cannot speech [? k] properly I scramble velly velly well. Do forgive me dear John if I've been already over & over & over all this again & again. Love Stevie. [. . . a scrap of paper attached to the two above letters . . . referred to two neighbours (the Everetts) who had shown Stevie and her sister much kindness, and who had indeed written to me informing me of her illness and her whereabouts when she was taken to hospital. The note simply said:] The Everetts have helped me so much. John, could you please send them a copy of my *Selected Poems please* (and I will pay for it) and sent [sic] it to them here *please*. [She then added their address and telephone number].

Stevie's head had been shaved so the doctors could perform exploratory surgery. Friends who saw her in hospital remember how striking she looked, rather like the death mask of John Donne in St Paul's, said one, with her head in white bandages. When the bandages were removed she wore a "shocking pink" scarf, which she would tug at continually, distressed that it sometimes slipped and revealed her white dome and grizzled stubble. "The gamine child she never ceased to be", recalls Helen Fowler, "coexisted with the witch, the black fringe had done for both. Now, under the scarf, the face and brow took on the nobility that had always been there, disguised." The George Lawrences went to see Stevie and she said to Mrs Lawrence, "Olga, this is terribly important, please listen carefully", and then lost the power to say what she meant, summoning only dissociated words. Mrs Lawrence remembers Stevie with the pink scarf around her head and with a kind of translucent inner beauty. Lady Lawrence found Stevie quite cheerful in hospital and presumed that was because she really did want to die. Inez Holden had accompanied Lady Lawrence, and Stevie said quietly to Inez, "I love you." Then she signalled they should have tea.

Stevie had been moved from Torbay Hospital to Freedom Fields Hospital in Plymouth, and finally to Ashburton Hospital. Her illness had been diagnosed as a brain tumour—malignant and inoperable. James MacGibbon recalls that she showed him "Come, Death (2)" with the word "Death" encircled. He couldn't understand what she was saying, but was certain she wanted him to fetch her sleeping pills from Palmers Green so she could take them all and end her life.

Helen Fowler remembers that in those last weeks Stevie was always beginning sentences and

then leaving them trailing, to begin another inconsequential one straightaway. They were only the neutral beginnings of communication: "Of course I always said"; "But then again, if one thinks"; "If one doesn't know about it"; "I was so anxious"; "They told me". . . . These beginnings were so unstressed, so casual, it was as if, on the surface of her mind, the detritus of a hundred faltering conversations lay fragmented. The nearest to the old Stevie was once when she said "heigh-ho" in a little, mocking, sighing way as she always did.

Stevie's last letter was one she wrote to Anthony Thwaite on 17 January. Delayed by a lengthy postal strike, it arrived the day after he had heard the news of her death:

Dear Anthony,

I had a very peculiar toss down. I wasn't *very* well for about a month & even then I did often find myself using all but quite extraordinarily odd wrong words. It is like the telephones scrambling their eggs—but I never managed for about 10 minutes at least to spike any proper eggs. Then I'm afraid on Jan. 5th (I was of course & had since Nov. 10th been living with Molly, my sister, in Buckfast) then on Jan. 5th I fell almost dead, I mean ridiculous, etc. The doctor at 10 P.M. arrived & at once shipped me off to hospital, not this one I've had to come to for endlessly *more* odd takings (?) *with* can't remember what. (Anyway I've had now to sign glad pleas for happy gladdings of more endless pops & goes. I have signed my name *but* now I feel awfully frightened. Am I saying wisely in these solemn legalnesses) that any little murder, caused by here & there, has been quite let off by my names having been signed so beautifully!??

How does happens in life come for you & Ann. I do hope you & the children are wonderfully well. I wish I saw you all sometimes in Devon but lor! it's far away. How are you finding Crossman's *New Statesman*'s new ah *quite* possibly *too* new life! Isn't it nice for me I found Terry (Kilmartin) did not like the extra verses I wasn't really any surer about than either of us! I did send one called "Come, Death" to Jack Lambert but if he doesn't like it I shd. love to send it to you. It's being *fond* of *Death* old Crossman (*is* that his name?) would blast his bussynuss [with?] & would *hate*. Love to you (& thanks for sending your home address) & Ann & *les enfants*

During one of their visits to Stevie in Ashburton Hospital, Helen Fowler and James MacGibbon held her hands, each side of the bed, and talked to her. "James was wonderful," Helen recalls, "kind, loving, patient—telling her what a wonderful smile she had and that was what people remembered about her." Stevie's eyes were still dark and lively, Helen remembers, but talking to her was like talking through a telephone and having to leave a recorded message. "Wouldn't it be fun," [Stevie] had said once when staying with me and the phone kept ringing, "to have an Answerphone that said, 'they have all gone into the world of light?'"

She was restless and fidgety suddenly: then racked with a paroxysm of shaking. . . . I saw her several more times, only a shape in the little room, with the pallor of a wintry March on the hills outside.

Stevie died at Ashburton Hospital on 7 March 1971. Years earlier, commenting on the poet Gerard Manley Hopkins, she had written, "it is only on his deathbed that he could truly say 'I am happy', as only on his deathbed any . . . sensitive poet can say these words—or for that matter, any human being at all, conscious of exile."

20

Memorials

Five days after Stevie's death her funeral service was conducted in the Anglican Church of the Holy Trinity at Buckfastleigh. Father Sebastian Wolfe, a monk at Buckfast Abbey, played the organ, and James MacGibbon read "Come, Death (2)" as part of his oration. Many residents of Buckfast attended in addition to Stevie's sister, who sat in the front pew with the Everetts. The Fowlers walked behind the coffin on the way out, and Helen remembers seeing, among the long rows of flowers outside the church from such famous people as Flora Robson and Anthony Powell, a small bunch with the note, "We did not know you, but we loved your poems." "We drove James to the crematorium in Torquay of all places," Helen recalls, "how Stevie would have laughed—the Shangri-la of the Midlands, she called it once."

Molly Ward-Smith was not happy that her sister's body was to be cremated. Towards the end Stevie had decided against cremation, and Molly had got the monks to agree to bury her sister in the grounds of Buckfast Abbey. But Stevie's will had not been changed, and it stipulated cremation, so cremation it was. As the coffin slid into the flames, Molly shrieked in a theatrical way, "Stevie", and Helen found it "embarrassing, like all grief".

Two weeks later the *Evening Standard* ran a story that Stevie would be remembered by her friends in an unusual way. As she enjoyed giving her friends tea at the Ritz, a tea-party was to be held there in her memory. Helen Fowler had suggested the idea to James MacGibbon, whom the *Evening Standard* quoted as saying that, because Stevie was a non-believer, "it would be wrong to hold the usual kind of memorial service". None the less, on the last day of March, a memorial service according to the liturgy of the Church of England was held in London, at Saint Matthew's, Westminster.

The Venerable Edward Carpenter, now Dean of Westminster Abbey, took part in the service, as did Stevie's friend, Father Gerard Irvine, who gave the eulogy. In it he evoked the "apparent antinomies" of Stevie's life and character: her innocence and sophistication, her compassion and familiarity with the glee of cruelty—which caused her to fulfil, at one and the same time, "the roles of child and witch". He reminded the congregation of Stevie's "light-hearted gaiety and her low reserves of nervous energy", her enjoyment of the Ritz tea-lounge opulence and her equal enjoyment of the unchanged Edwardiana of her house in Avondale Road. Stevie was ambivalent about religion, said Father Irvine, as she was about life:

> In her at any given moment love and hate, faith and rejection coexisted—not, as with most of us, in the moods of successive moments, but at one and the same time. One might say that she was a believer who did not much like the God revealed by her faith. Still less did she like the God others claimed was revealed to them by theirs. She was acquainted with academic theology: she was offended by many of its formulations. The God of whom it seemed to speak appeared to her to be (to quote one of her poems) a person "in a fairy wood": a Nobodaddy to be rejected at once by reason and by moral sensibility. In this, as in so many other respects—more indeed than are at first apparent—she has much in common with that deeply religious poet, William Blake.

Father Clifford Doyle thought "Gerard won over some people" who had gone to Saint Matthew's "wondering if they'd done the right thing". Among those present were Sir John Betjeman, Tom Driberg, M.P., Inez Holden, Lady Lawrence, Naomi Lewis, Armide Oppé, Ian Parsons, William Plomer, Norah Smallwood, Eric White, Nina and Doreen Woodcock, and the Brownes, Cleverdons and Russell-Cobbs.

Some of the people at the memorial service, such as the Brownes and Woodcocks, also attended the small tea-party hosted by James MacGibbon and Norah Smallwood almost a month later. Held in the Marie Antoinette Room of the Ritz, it made for "a rather comic, strangely assorted gathering", according to Elisabeth Lutyens. Patric Dickinson gave a talk about Stevie, remarking on her "Blake-like innocence", whereupon Miss Lutyens looked at Olivia Manning "and almost winked, for we knew our Stevie," she said, "and anyone less 'innocent'—Blake-like or not—I have yet to meet." When Molly

Ward-Smith learned the cost of this tea-party, charged to her sister's estate, she was furious.

In her will Stevie named James MacGibbon as her executor and bequeathed to him her copyrights and royalties. She expressed the desire, if the income he so received were to be "more than he wishes to keep for himself", that he would "devote such money towards the relief of some needy writer or artist". The rest of her estate she left to her sister, hoping that Molly would give or bequeath some of it to the National Society for the Prevention of Cruelty to Children. In August of 1971, *The Daily Telegraph* printed a notice that the estate came to £23,442 net. The financial dealings of Ladislav Horvat, and Aunt's will, which had bequeathed to Stevie the house in Palmers Green, accounted for much of this.

Molly lived on until June of 1975, staying for less than a year, at the end, in a Torquay nursing home. One of the letters she received, after her sister's death, was from the man who used to wash Stevie's windows in Palmers Green. He never cleaned the windows of anyone who had entertained him so much, he told Molly, and he would miss Stevie terribly. Some of those who visited Molly from time to time after Stevie's death found her feelings about her sister to be an odd mixture of pride and envy of Stevie's fame. A monk at Buckfast Abbey remembers that Molly would "enthusiastically recite" her sister's poems, and Armide Oppé recalls the pathetic way Molly used to say about Stevie, "she loved me at the end, she really loved me". "Molly was difficult before her stroke," recalls Helen Fowler, "and quite impossible after that." Helen was appalled that, a month after Stevie's death, Molly sold her sister's books for about £40. In her will Molly left her freehold property in Buckfast to the Everetts, and a large sum to Buckfast Abbey. In accordance with her sister's wishes, she bequeathed the Gold Medal for Poetry to Stevie's godchild, Lucinda Fowler, and a substantial sum to the NSPCC.

Stevie's greatest legacy, of course, is her writing, and it is also her best memorial. Since her death a *Collected Poems* has been published (1975), all three of her novels have come back into print, and a volume of her previously uncollected writings and drawings has appeared (1981). Also, a popular stage play and film have been based on her life, and her reputation among readers, poets and critics has continued to grow.

When a Borough of Enfield committee considered putting a commemorative plaque outside Stevie's house in Avondale Road, newspapers reported the "row" that ensued, with one councillor asking, "Should we do this for every Tom, Dick or Harry?" But the plaque was there by the summer of 1978 when two Americans, admirers of

Stevie's poetry and at work on her biography, visited Palmers Green. The *Palmers Green and Southgate Gazette* published an account of their visit, which provided further evidence of Stevie's growing fame. When their biography was published in 1985, its readers learned, from a prefatory note by Stevie's executor, that a second biography was in preparation. Two biographies! How Stevie, who had enjoyed her late fame so very much, would have laughed.

Acknowledgments

No one can know the whole truth about another person, and even self-knowledge is limited, for the whole truth about anyone is the truth as it exists in the mind of an omniscient God. Biographers, therefore, have reason to be humble, and even more reason in that even their partial truths would be impossible to come by without the cooperation of many. It is inevitable that not all testimony coincides. As any biographer would, we have tried to weigh evidence in a context as informed as we could make it, and we have cited our sources. Biographers must suppress their egotism when it would close their minds to evidence which goes against their presuppositions, and they must resist the egotism of their subject's friends when it does not allow the friends to admit that their knowledge, however intimate, is partial. In writing this book we have found most of those who knew Stevie to be understanding of our efforts, and cooperative beyond our hopes. It is a pleasure to acknowledge their generosity, and that of the many others whose help has made the biography possible.

One acquaintance of Stevie, an English publisher, told us he supposed our biography would be in the American tradition of the beaver, as opposed to the English tradition of the butterfly. Other terms for this distinction are those used by Bernard Crick when he contrasts the writing of a life with the English tradition of character study. He once preferred the novelist to the historian as biographer he says, in a wise essay published in the *London Review of Books* (7–20 October 1982), but he came to regard that view as "romantic". We have been beavers, we admit, but our book is not completely lacking lightness of touch—how could it be, when we quote so much from Stevie Smith!

Our debt is greatest to Stevie's closest friends: Anna Browne, Helen Fowler, Sally Chilver, and Nina and Bobby Woodcock. They all consented to multiple interviews, allowed us to copy Stevie's letters to them, answered numerous letters of inquiry, read and commented on drafts of this book, and encouraged us by their friendship.

Helen Fowler's contribution is especially notable. After Stevie's death she searched the house in 1 Avondale Road, rescuing many valuable papers which she allowed us to see. They were then turned over to Stevie's executor, and some have now been published in the book, *Me Again*. In fact, all of the material cited in the Acknowledgments to *Me Again* as being in the possession of James MacGibbon is from that batch of papers found by Helen Fowler in Stevie's house. We have, accordingly, referred to the collection as the Palmers Green Papers, or PGP, in our Notes. Mrs Fowler also shared with us the copious notes she made about the interior of Stevie's house, and her notes on her interview after Stevie's death with Molly Ward-Smith. These notes include a key to the people on whom characters in *Novel on Yellow Paper* are based. Every case in which we were able to cross-check Molly's memories and identifications resulted in confirmation, thereby strengthening the credibility of her testimony as a whole. Finally, Helen Fowler allowed us to quote freely from her unpublished account of Stevie's death. In our Notes we refer to Mrs Fowler's notes and unpublished memoir as the Helen Fowler Papers, or HFP.

Other friends of Stevie who consented to interviews, answered inquiries, shared papers, or otherwise helped are: Diana Athill (and the firm of André Deutsch); Patricia Beer; Neville and June Braybrooke; Michael, Maria and Alice Browne; Horatia Buxton; Sam Carr; Richard Chilver; Mr and Mrs Douglas Cleverdon; Barbara Clutton-Brock; Rosemary Cooper; Doreen Diamant; Kay Dick; Mr and Mrs Patric Dickinson; Molly and Donald Everett; Kathleen Farrell; Laurence Fowler; Margaret Gardiner; John Guest; Sir Rupert Hart-Davis; Dr Polly Hill; John Horder; Michael Horovitz; Ladislav Horvat; Dr Audrey Insch; Prebendary Gerard Irvine; Suzannah Jacobson; Terence Kilmartin; Marghanita Laski; George and Olga Lawrence; Sir John and Lady Lawrence; Naomi Mitchison; Lord Moyne; Armide Oppé; Ronald Orr-Ewing; Olive Pain; Anthony Powell; Joan Prideaux; Sheila Raynor; Naomi Replansky; Trevor and Piers Russell-Cobb; Mary Lee Settle; Jane Stockwood; Francesca Themerson; Anthony Thwaite; Dame Veronica Wedgwood; Eric W. White; Gertrude Wirth; Jonathan Williams; David Wright; and Francis Wyndham. James Laughlin was especially hospitable when we visited his home in Connecticut, one snowy day, to interview him and to study his important collection of letters from Stevie.

Some of those who helped and whom we hoped to please have died: Mungo Buxton; Lord Clark; Alasdair Clayre; James Cubitt; Reverend Clifford Doyle; Stevie's dentist, Wallace Finkel; Rosalinde Fuller; T. R. Fyvel; David Garnett; Gertrude Hermes; Denis Johnston; Elisabeth Lutyens; Hester Marsden-Smedley; Ian Parsons; Professor Joan Robinson; Dame Flora Robson; Norah Smallwood; and Stevie's friend and charwoman, Mrs Wright.

Others who were acquainted with Stevie and provided information about her are: Ian Angus; her accountant, J. W. Arnold; Colin Amery; Mrs Marshall Baynes; Professor Bruce Berlind; Dr Marjorie Boulton; Mr and Mrs Bradford; Miss K. Brett; Robert Buhler, R.A.; Lady Georgina Coleridge; Lettice Cooper; her physician, Dr James Curley; her solicitor, A. J. Davey; Mrs C. Day-Lewis; Nigel Dennis; D. J. Enright; Gavin Ewart; Dr James Gibson; Giles Gordon; Geoffrey Handley-Taylor; Glenda Jackson; Margaret K. Ralph; Dr Elisabeth Schnack; Dr Shelley Jacobson; Francis King; Dr Ruth Landes; Professor Sir Edmund Leach; Rosamond Lehmann; Naomi Lewis; Eddie Linden; Christopher Logue; George MacBeth; Margaret Macbeth; Derwent May; Jonathan Mayne; Dr Jonathan Miller; John Morley; Peter Orr; Mrs Peacock; Peter Porter; Phebe Snow; Paul Stephenson; Bernard Stone; Michael Thomas; Antoinette Watney; and Terence de Vere White.

Further assistance or information was provided by: The Chief Accountant of the Privy Purse, Buckingham Palace (concerning the Gold Medal); A. A. Balkema of Cape Town, and the City Librarian of Durban, South Africa (about Major Raven-Hart); Kay Boyle; Reverend Benet Conlon, Prior, and the monks of Buckfast Abbey; Professor Bernard Crick; Reverend Ian Douglas; Margaret Evett, Secretary of Westonbirt School; John Gardner; Richard Garnett; Celia Goodman (cousin of Inez Holden); Patricia Hann (Broadstairs Library); Judith Hemming; Geoffrey and Kitty Hermges; George Lawson; Brenda Lee (Phoebe Pool's sister); Marian Mainwaring; T. C. Marsden-Smedley; Professor Hamish Miles; Mrs June Nethercut (niece of Eric Armitage); Robert Nye; Diana Oakeley (sister of John Hayward); Derek Parker; Lady Pearson; Josephine Pullein-Thompson and the PEN English Centre; Alan Russell; B. E. Scutt of the Archives of the Ministry of Defence (concerning the naval career of Charles W. Smith); Miriam E. Shillito, Archivist of The North London Collegiate School; Sara Smith for The Bodley Head; Basil Toole-Stott (cousin of Stevie Smith); P. N. Turner, the London Borough of Enfield Librarian and Cultural Officer, David Pam, Local History and Museums Officer, and his assistant, Graham Dalling; Jonathan Vickers; L. A. Wallrich of the Ontario book shop, About

Books (owner of Stevie's letters to Robert Nye); Janet Watts; Hugh Whitemore.

In the course of our research we obtained copies of material from several American libraries and we would like to thank the staffs of: Special Collections, Mugar Memorial Library, Boston University; Manuscripts Department, Lilly Library, Indiana University; Rare Books and Special Collections, the Fred Lewis Pattee Library, Pennsylvania State University; The Poetry Collection of the University Libraries, State University of New York at Buffalo; the Rare Book Room, Smith College Library; the Modern Literature Collection, Rare Books and Special Collections, Washington University Libraries, St Louis; and the Harry Ransom Humanities Research Center, The University of Texas at Austin. The most extensive collection of Stevie Smith material is in the Department of Special Collections, McFarlin Library, The University of Tulsa, and we are especially grateful to the former director, David Farmer, and other members of the staff: Bonnie Mitchell, who organized the collection, Caroline Swinson, and Denise Kidd.

We thank too the staffs of the several British libraries and archives which we consulted: The BBC Written Archives, Caversham Park, Reading; The General Register Office, London; The British Library and its newspaper branch in Colindale; Somerset House, London; Special Collections, University Library, The University of Birmingham; Special Collections, University Library, Durham; The Department of Manuscripts of the National Library of Scotland, Edinburgh; The John Rylands University Library of Manchester; and The Library, University College, London. Special thanks are owed to Dr Michael Halls of King's College Library, Cambridge, and Dr P. A. Larkin, Librarian, and N. Higsen, Archivist, The Brynmor Jones Library, The University of Hull.

It was a long time before we were able to discover what had become of Karl Eckinger. In Germany we are grateful to Frau Ursula Matthiesen of Matthiesen Verlag, and I. Ritter of the Staatsbibliothek, Preussischer Kulturbesitz (which supplied the *vita* and title page of Eckinger's dissertation). Those in Switzerland we must thank are: Mrs J. Wenger, Documentalist, Centre de Documentation et Archives, Université de Genève; the Kanzlei der Universität Basel; La Bibliothèque Nationale Suisse; and Peter Wirz, editor of *Der Zürcher Oberländer*, Wetzikon.

The firm of Sauerländer Verlag forwarded to Gertrud Häusermann-Häusermann our inquiry about a letter to Stevie signed "Hans". (Mrs Häusermann-Häusermann is mentioned in the letter, along with the title of one of her books published by Sauerländer Verlag.) She then got in touch with us, and arranged for Professor Georges Denis Zimmermann, Séminaire d'anglais, Université de Neuchâtel, to organize and make available Stevie's letters to her husband and his papers concerning Stevie. In our Notes this collection is referred to as the Hans Häusermann Papers, or HHP. We are grateful to Mrs Häusermann-Häusermann for making the collection available, for sending us a moving account of her visit to Stevie Smith, and for contacting the family of Karl Eckinger and obtaining photos of him which are now part of the Hans Häusermann Papers.

Those with slight, if any, connection to our research, but who read and criticized our drafts, assisted in the translation of documents and correspondence, or helped in other ways are: Helen Bajan; Lewis Barbera; Evonne Burdison; Roland Clarke; Helga Doblin; P. N. Furbank; Dr Joan Hall; Professor Carolyn Heilbrun (not to mention Amanda Cross); James Hickman; Richard Hoffman; Eric Hope; Lord Horder; Dr Andrew Kappel; Oleg Kerensky; John Lahr; Lady Mander; Regula Noetzli; Dr Charles E. Noyes; Dr Thomas O'Sullivan; Tino Perutz; Wendy Rintoul; Jack Sarch; Eileen Simpson; George Staempfli; Professor Thomas Staley; Raleigh Trevelyan; and Professor Aileen Ward. We owe special thanks to our editor Roger Smith.

We thank the MacDowell Colony, Alfred University Summer Place, the Ossabaw Island Project, and Yaddo for providing refuge while we undertook the actual writing of this book. We are also grateful to The American Council of Learned Societies, The National Endowment for the Humanities, Hofstra University, and The University of Mississippi for providing financial support. Parts of this book have been published in different form in *The Powys Review* and *The Journal of Modern Literature*.

The Ogden Nash poem, "Who and What is Stevie Smith?" on p. 243, is reprinted by permission of Curtis Brown Ltd, NY, © 1962 by Ogden Nash.

For permission to quote from unpublished letters by Rose Macaulay we thank Constance Babington Smith, and for permission to quote from their letters to Stevie we are grateful to Sir Rupert Hart-Davis, Terence Kilmartin, James Laughlin, Rosamond Lehmann, Naomi Mitchison, Sir Victor Pritchett, Naomi Replansky and the late David Garnett and Ian Parsons.

All but one of the drawings in this book are by Stevie Smith. The drawing on the Contents page has not been published previously, and is from the Department of Special Collections, McFarlin Library, The University of Tulsa. The other drawings have been published in *The Collected Poems of Stevie Smith*.

Notes

The following abbreviations are used in these Notes: int—interview with; l—letter; pc—postcard; por—in possession of the recipient.

People

A	The Authors	TK	Terence Kilmartin
DA	Diana Athill	JL	James Laughlin
AB	Anna Browne	RL	Rosamond Lehmann
JC	Jonathan Cape	EL	Elisabeth Lutyens
SC	Sally Chilver	RM	Rachel Marshall
DD	Doreen Diamant	NM	Naomi Mitchison
KD	Kay Dick	AO	Armide Oppé
HF	Helen Fowler	OP	Olive Pain (née Cooper)
KF	Kathleen Farrell	ES	Ethel Smith
RHD	Rupert Hart-Davis	MS	Molly Ward-Smith
HH	Hans Häusermann	SS	Stevie Smith
JH	John Hayward	MAS	Margaret Annie Spear
SJ	Suzannah Jacobson		(SS's Aunt)
DJ	Denis Johnston	Woodcocks	Nina and/or Doreen
AK	Anna Kallin		(Bobby) Woodcock

Collections of Papers

Austin	The Harry Ransom Humanities Research Center, The University of Texas at Austin.
Birmingham	Special Collections, University Library, The University of Birmingham.

BBCa	BBC Written Archives, Caversham Park, Reading.
Deutsch	Files of André Deutsch Ltd, London.
Edinburgh	Department of Manuscripts, National Library of Scotland, Edinburgh.
GRO	General Register Office, London.
HFP	HF Papers, in possession of HF.
HHP	HH Papers, Séminaire d'anglais, Université de Neuchâtel.
Hull	SS Collection, The Brynmor Jones Library, The University of Hull.
King's	King's College Library, Cambridge.
PEN	Files of the International PEN English Centre, London.
PGP	Palmers Green Papers, in possession of James MacGibbon.
RMP	RM Papers, in possession of Kitty Hermges (*née* Marshall).
St L	The Modern Literature Collection, Rare Books and Special Collections, Washington University Libraries, St Louis.
Tulsa	Department of Special Collections, McFarlin Library, The University of Tulsa.

Periodicals

BPWN	*Bowes Park Weekly News*	*O*	*Observer*
DT	*Daily Telegraph*	*PGSG*	*Palmers Green and Southgate*
ES	*Evening Standard*		*Gazette*
JO'L	*John O'London's Weekly*	*S*	*Spectator*
L	*Listener*	*T*	*Tribune*
MW	*Modern Woman*	*TLS*	*Times Literary Supplement*
NS	*New Statesman*	*WR*	*World Review*

Books

CP	SS. *The Collected Poems of Stevie Smith*. Ed. James MacGibbon. London: Allen Lane, 1975.
H	SS. *The Holiday*. Introd. Janet Watts. London: Virago, 1979. (1st pub. London: Chapman & Hall, 1949.)
I&S	KD. *Ivy & Stevie: Ivy Compton-Burnett and Stevie Smith*. London: Allison & Busby, 1983. (1st pub. London: Duckworth, 1971.)
MA	SS. *Me Again: Uncollected Writings of Stevie Smith*. Ed. A. London: Virago, 1981.
NYP	SS. *Novel on Yellow Paper; or, Work It Out for Yourself*. Introd. Janet Watts. London: Virago, 1980. (1st pub. London: Cape, 1936.)
OTF	SS. *Over the Frontier*. Introd. Janet Watts. London: Virago, 1980. (1st pub. London: Cape, 1938.)
PGHS	*Palmers Green High School: 1905–1955*. Ed. Edith Gale. Published in 1955 by the school to celebrate its 50th jubilee.

Essays and Interviews

Boulton	Completed questionnaire returned by SS to Dr Marjorie Boulton, 1 June 1951, por.
Horder	John Horder. "Poet on thin ice." *Guardian*, 7 June 1965, p. 5.
Horwell	Veronica Horwell. "'Like Samuel I keep an open ear.'" *The Times*, 28 Aug. 1968, p. 5A.
"Not Believing"	SS. "The Necessity of Not Believing." *Gemini*, 2 (Spring 1958), 19–32.
Orr	"Stevie Smith." *The Poet Speaks*. Ed. Peter Orr. London: Routledge & Kegan Paul, 1966, pp. 225–31.
"Read Books"	SS. "How to Read Books." In *Discovery and Romance for Girls and Boys*. Vol. 2. London: Cape, 1947, pp. 267–72.
"Same Place"	SS. "The Same Place." In *Allsorts: 2*. Ed. Ann Thwaite. London: Macmillan, 1969, pp. 97–100.
Snow	Phebe Snow. "Stevie Smith." Unpublished biographical sketch in possession of Phebe Snow, based on her int SS, 1969.
Thomas	Yvonne Thomas. "A polite murder, cookery, and melancholy." *ES*, 3 Dec. 1969, p. 22.
"Too Tired"	SS. "Too Tired For Words." Third Programme BBC radio broadcast of 4 March 1957. Script at Tulsa.
Williams	Jonathan Williams. "Much Further Out Than You Thought." *Parnassus: Poetry in Review*, 2 (Spring/Summer 1974), 105–27.

We cite sources by tag phrase in the order of their occurrence on each page of the biography. We do not provide citations for poems by SS or lines from her poems which are published in *CP* or *MA*, and which are identified by title in our text. Citations of letters, essays or other writings in *MA* are by page number, without mention of dates or titles. All of our interviews with friends or acquaintances of SS took place between 1978 and 1984. We cite such interviews as sources, but see no point in specifying their dates and places.

Chapter One

page
1 **"Dear Eric":** SS, 1 to Eric White, 22 Nov. 1969, por.
"The thing about Stevie": Isobel English [June Braybrooke] in KD, *Friends and Friendship* (London: Sidgwick & Jackson, 1974), p. 96.
"What I especially liked": Karl Miller, "London Diary", *NS*, 66 (26 July 1963), 102.
"I say to them": SS in *I&S*, p. 74.
2 **"The audience were quite young":** Christopher Logue, 1 to A, 27 Oct. 1980, por.

"**The mis-firing**": Julian Jebb, "Poetry gala", *The Financial Times*, 4 Feb. 1969, p. 3.

the papers printed: See the *Sun*, 4 Nov. 1969, p. 5; *Guardian*, 4 Nov. 1969, p. 5; *PGSG*, 7 Nov. 1969, p. 1; and *O*, 9 Nov. 1969, p. 5.

"**Darling, I've been offered**": conversation as recalled by Eric White, int A. White's other recollections and speculations later in this chapter are from the same interview and are not separately cited.

3 **gave interviews to**: SS appointment book, PGP.

"**a couple of sherries**": SS to John Gale, "Death is a poem to Stevie Smith", *O*, 9 Nov. 1969, p. 21.

"**stick at the curtsey**": SS, 1 to Dr Audrey Insch, 12 Nov. 1969, por.

"**I'm sure H.M.**": SS, l to KD, 8 Nov. 1969, St L.

"**I shall now be terrified**": *MA*, p. 319.

"**everybody thought it was**": Reverend Clifford Doyle, int A.

"**She'd actually got**": Mrs Peacock, int A. Mrs Peacock is also the neighbour cited at the end of this paragraph: "**She didn't mind**".

stuff of legend: see Hugh Whitemore's *Stevie: A Play* (London: Samuel French, 1977), p. 40.

4 "**dumping ground**": SC, int A. The account of SS's sink, stove and larder are from HFP.

wrote one reporter: Thomas.

austere, clean, and light: SC, int A.

Little squares . . . on the walls: HFP.

nervous about being late: according to SS, *I&S*, p. 77.

Her appointment was for: the account in this and the next paragraph is based on SS's interview in *I&S*, pp. 77–78, supplemented by details from Thomas, and *MA*, p. 320. Also, the time of the appointment is given in a l to SS from Buckingham Palace, 3 Nov. 1969, PGP, and the anecdote about living in a house with nine rooms is recounted by Celia Goodman in letters to A, 22 and 23 Oct. 1980.

5 **Its obverse bears**: the medal is described in a l to A from the Chief Accountant of the Privy Purse, 16 Aug. 1983. See also Colin White, *Edmund Dulac* (New York: Scribner's, 1976), p. 157.

"**Nothing came and nothing came**": Norah Smallwood, int A.

an editor friend recalls: John Guest, int A.

6 "**She was really angry**": AB, int A.

Chapter Two

Two important documents for this chapter are HF's notes of her int MS, 14 April 1971, and a diary kept by SS's mother. We cite the interview as "MS to HF, HFP". The diary has the title, *E. Pauer's Birthday Book of Musicians & Composers*, and is inscribed, "Ethel Spear 25/6/95". We cite it as "mother's diary, Tulsa".

7 **wanted to join the Navy**: this and other details of Charles Smith's youth are from MS to HF, HFP, and *I&S*, pp. 64–65. The details are echoed in SS's fiction, *NYP*, p. 75.

coal exporting business: Charles Smith listed his father's profession as "Coal Exporter" on his certificate of marriage to his second wife in 1920, GRO. He listed his occupation as "Forwarding Agent" on SS's birth certificate in 1902, GRO.

1 September, 1898: the date and place are noted in SS's mother's diary, p. 226, Tulsa.

when her mother died: for death dates of SS's grandparents see her mother's diary, Tulsa.

one of Stevie's novels: *H*, pp. 37–40.

"Dear Daddy passed away": entry next to 22 Sept. on p. 245 of mother's diary, Tulsa.

honorary chief engineer and **she once remarked:** *I&S*, p. 64.

8 **the words of her aunt:** *NYP*, p. 74.

sardonic poem: "Papa Love Baby".

more sympathy, a "marker", and **yearned to join British troops:** MS to HF, HFP.

"There was, chaps": *NYP*, p. 76.

24 January, 1901 and **20 September, 1902:** pp. 19 and 243, mother's diary, Tulsa.

Stevie almost died: *I&S*, p. 64. SS's autobiographical poem, "A House of Mercy", refers to "two feeble babes", and Pompey says she was born with a broken arm, *NYP*, p. 73.

Pompey says it is: *NYP*, p. 162.

9 **at the Smiths' house** and **"The doctor had given":** entry next to 11 Oct. on p. 267 of mother's diary, Tulsa. The Smiths' address is recorded on SS's birth certificate, GRO.

went bankrupt: MS to HF, HFP, and *I&S*, p. 65.

"I sat upright": from the poem, "Papa Love Baby".

the White Star Line: MS to HF, HFP. MS said her father started as a pantry boy and rose to assistant purser.

A photo: PGP

substantial four bedroom house which still stands: MS to HF, HFP. It still stood, that is, when we visited Hull in 1980.

a good education: Marghanita Laski, int A.

changing her entry in *Who's Who*: SS was listed in *Who's Who* for the first time in 1955, and her entry noted: "b. Hull". She told Giles Gordon in 1965 that she added her early departure from Hull to her entry to tease regionalists, but the change did not appear until the 1971 *Who's Who* which read: "b. Hull; has lived in London since the age of three." See Boswell [Giles Gordon], "A Singer of Songs", *The Scotsman*, 24 April 1965, p. 5.

very poor: MS to HF, HFP, and *I&S*, p. 66.

"sheltered ladies": *I&S*, p. 65.

10 **according to Molly:** MS to HF, HFP.

an early draft: holograph in Tulsa.

£8,557 7s: will of John Spear, Somerset House, London.

Stevie reflected: *I&S*, p. 66.

11 **The move to London:** this anecdote is from, "Same Place," p. 97.
her first glimpse and **illumined:** *MA*, pp. 83–84.
semi-detached villa: the description of the house is based on *I&S*, pp. 61 and 66, and Frank Entwisle, "'Such a Nervy Business . . . ,'" *ES*, 24 Nov. 1960, p. 5. The house is semi-detached because it is a corner row house.
a foreign package: *MA*, p. 83.
she joked decades later: SS, l to Ann Thwaite, 18 July 1968, por. See "Same Place," p. 98.
a description of childhood in India: "This ape incident happened even earlier in my life: being an Indian childhood memory from the age of about four." Script of SS, "Swinging on a Star", Woman's Hour, BBC Radio 2, 4 Dec. 1969, BBCa.
having books read aloud: "Read Books", p. 270.
into a novel: *H*, pp. 99–100.

12 **common to have relations:** AB, int A.
"a golden age": *MA*, p. 84.
"Missum's School": this epithet and details about Osborne Road and Green Lanes in those years come from *PGHS*, pp. 26, 28, and 31.
attended the kindergarten, once described hearing, and **she also recalled:** *MA*, pp. 89 and 93. In *OTF* Pompey also recalls hymn singing in kindergarten, pp. 258–260. SS listed "Palmers Green Kindergarten & High School" next to "Education" on a questionnaire, Boulton.

13 **cure the disease in a few weeks:** according to SS, *I&S*, p. 66. The specific disease is named by MS to HF, HFP. Both Ms and SS mention that SS's stay at Broadstairs was for three years. See also *MA*, p. 300.
"rotten inside": SS, l to Dr Woodcock, in possession of the Wood-cocks. Our information about Dr Woodcock comes from his daughters.
"to the utmost limit" and **"children of educated people":** *Alfred Yarrow: His Life and Work*, compiled by Eleanor C. Barnes [Lady Yarrow] (London: Edward Arnold, 1923), p. 199.
"five-year-olds do not": SS, "Ghastly Child", *O*, 19 Feb. 1961, p. 29.
privileged and grand: *MA*, p. 300.
considered to count for less: Snow.
she told a friend: Maria Browne, int A.
"ghastly" matron: *MA*, p. 300.
who frightens Pompey: *NYP*, p. 155.

14 **they were forbidden:** Snow, and Barbara Clutton-Brock, int A.
"off to Valparaiso": *NYP*, pp. 76 and 164.
a postcard . . . dated 23 May: PGP.
Another card, mailed . . . 17 August: Tulsa.
one, sent . . . on 22 June, 1909: PGP
the first Teddy bears: MS to HF, HFP.
friends of Stevie recall . . . a parrot: Neville and June Braybrooke, int A.
"unrespected": see SS's poem, "Papa Love Baby". Pompey says

about her father's postcards, "a very profound impression of transiency they left upon me", *NYP*, p. 76.

prize for attendance and **"splendid system"**: *The Recorder*, 22 Dec. 1910, pp. 146 and 144.

The form-mistress . . . Stevie wrote: *NYP*, pp. 25–26 and *MA*, p. 89.

This instructress: i.e., Miss G. F. Hillier, identified in MS's key to characters in *NYP*, HFP, and in *PGHS*, p. 13.

15 **parrot Joey, "that died"**: *OTF*, p. 166. See also pp. 127, 149, and 222. There is a photograph of a parrot, labelled Joey, in a Smith family album, Tulsa.

drew all the time as a child: DD, int A.

local papers carried notices: See *The Recorder*, 1 Aug. 1912, p. 429; 14 Aug. 1913, p. 454; 24 Sept. 1914, p. 411. MS also received Honours Certificates for drawing in these years.

sentimental pictures in the spirit of Landseer: HF's inventory of SS's house, HFP; *I&S*, pp. 62 and 66; and *NYP*, pp. 92–94.

"for a short story on 'Strife' ": inscription in the volume, Tulsa.

"the life struggle": *The Sentinel*, 8 April 1910, p. [7].

an expurgated edition of Grimm's: The book is now in Tulsa. HF saw it by SS's Palmers Green bedside after SS's death (HF, int A). In a script for a BBC radio broadcast SS wrote: "I read a lot of Grimm when I was a child—and still do. I have just enough German to read them in German" ("Poems with a Conversation", Third Programme, 3 Sept. 1968, Tulsa and PGP).

16 **church records show**: see *St John the Evangelist: Report and Parochial Accounts for 1908*, St John's, Palmers Green.

the elimination of the debt: See *BPWN*, 20 April 1912, p. [7], and *The Sentinel*, 23 April 1920, p. [2], by which time it was noted that St John's holds "the premier financial position in the Deanery".

"Humphrey d'Aurevilly Cole": see *MA*, pp. 76 and 95, and *H*, p. 76.

[Footnote] **Barbey's writings fascinated Stevie**: In 1924 SS made notes about Barbey in a reading diary (Tulsa). The names "Travers de Mautravers" and "Majeur Ydow" from his story, "A Dinner of Atheists", probably inspired her use of the names "Morbid Maltravers" (in an early unpublished poem) and "Majeur Ydow" (in the poem, "Le Majeur Ydow"). Also, a phrase from "A Dinner of Atheists" became the title of SS's poem, "Après la Politique, la Haine des Bourbons," and she wrote Barbey's name by that poem in her copy of *Mother, What is Man?* (now in Tulsa). In the same book she wrote Barbey's name by French lines in her poem, "The Bottle of Aspirins". In her third novel the protagonist says: "I thought of the Barbey d'Aurevilly story I read" (*H*, p. 35). Later, SS wrote a short reviewer's note on Barbey's *Les Diaboliques* for *MW*, April 1948, p. 114.

paid for the Italian holiday: *MA*, p. 95, and Woodcocks, int A.

"He really is a trying individual": *MA*, p. 95. See also Williams, p. 108.

17 **"I liked to see the great banners":** *MA*, p. 154. Compare with *MA*, p. 95, where the name is "Edwin Alton Crumbles" instead of "Anna Maria Livermore".

Trouble began: see *The Recorder*, 11 May 1911, p. 382; *BPWN*, 6 May 1911, p. 10, and 20 April 1912, p. [7].

"Falling about in the pulpit": OP, int A.

"to be a little 'higher' ": "Not Believing", 19.

church bazaar wool stall: *BPWN*, 9 July 1910, p. 7.

The following July . . . fish pond: *The Recorder*, 20 July 1911, p. 498.

Miss Spear . . . holes in the surplices: MS to HF, HFP.

decide beforehand which hymn: *MA*, p. 95.

"Old London": these lectures are mentioned in *BPWN*, 18 Feb. 1911; 27 Mar. 1911; and *The Sentinel*, 13 Jan. 1911.

"fascinating lecture on the Moon": *MA*, p. 94.

18 **George Bernard Shaw spoke:** *BPWN*, 8 Mar. 1913, p. 1. See also *The Recorder*, 3 July 1918, for mention of Shaw again lecturing at St John's Hall in the Spring of 1918.

"intense interest": *The Sentinel*, 11 Nov. 1910, p. [3].

"delightfully serious": *The Sentinel*, 29 Apr. 1910, p. [3]. Another member of the cast was Bobbie Smith, presumably SS's cousin Bob Smith, who, MS told HF, lived at 1 Avondale Road for part of his childhood. "Bobby" Smith is also listed in St John's parochial accounts as having contributed to the Farthing's Fund in May of 1908.

"It is now quite the fashion": *BPWN*, 14 Jan. 1911, p. 11.

"delighted in stamping" and **"one of the hits":** *The Recorder*, 19 Jan. 1911, p. 192, and *The Sentinel*, 13 Jan. 1911, p. [5]. OP and the Woodcocks also acted in this production.

at the age of eight: She became eight years old in September of 1910, but that was the autumn she won a prize for attendance in her Transition Class.

Stevie and Pompey found so comforting: *NYP*, p. 155, and SS, Preface, *Batsford Book of Children's Verse* (London: Batsford, 1970), p. 3.

why especially now?: Horder.

Psychoanalysts charge: *NYP*, p. 15.

19 **an Australian ballad:** SS, "Ruthless Rhymes", *O*, 16 May 1965, p. 26.

she admitted it is not; *MA*, p. 112.

mentioned it in essays, and spoke: *MA*, pp. 110, 112, 129.

"delicious": *MA*, p. 113.

several of her poems: e.g., "What is the Time? or St Hugh of Lincoln", "Eulenspiegelei", "The Warden", "The River God", and "I rode with my darling . . .". Perhaps SS's exile at the Yarrow Home also lurks behind her poem, "A Dream of Nourishment," which recounts a dream wherein a breast is violently withdrawn from the speaker's infant mouth.

she drew the wry moral: SS, "Mr Vulliamy the Rebel," *JO'L*, 44, 8 Nov. 1940, 132.

She admired D. H. Lawrence's: AB, int A.

which Pompey says: *NYP*, p. 155.

she would explain: *MA*, p. 129.

20 **passing first class:** notices were printed in *The Sentinel*, 28 July 1911; *The Recorder*, 20 July 1911, p. 490; and *BPWN*, 23 Dec. 1911, p. 3.

free tuition: for MS's scholarship see *The Sentinel*, 28 July 1911, and *The Recorder*, 1 Aug. 1912, p. 429.

an especially hot summer: SS, l to Racy Buxton, 28 July 1968, por.

Pakefield . . . Clacton: the locations were mentioned by MS to HF, HFP.

Stevie confessed: *I&S*, p. 66.

any doubts . . . swallowed up: Snow.

in her second novel: *OTF*, p. 124.

21 **"the good poems and the bad-good poems":** *MA*, p. 119.

true poets when young are inclined: *MA*, p. 121.

[Footnote] **she recalled being thrilled:** *I&S*, p. 66.; **not a dog of Stevie's childhood:** SS, l to Racy Buxton, 28 July 1968, por.

22 **until she was in her twenties:** Orr, p. 225

One of her essays contains her first: *MA*, p. 94.

sent to her in 1920 and signed "F.G.": "F.G.", l to SS, 27 May 1920, PGP. F.G. may be Florina Gibbons who, we know from a diary (now in Tulsa) kept by MS, taught at SS's second school, which SS attended from 1917–20. The character "F. Caudle", in *NYP*, is based on Florina Gibbons, according to MS's key to the novel (HFP). Stevie may have resented the advice she received about her poetry from F.G. The autobiographical protagonist of *NYP* complains that F. Caudle, "the girl that taught me English at school", had "no sense nor judgment in that remark of hers that was quite uncalled for", and adds, "You certainly want to think before you go making remarks about people's poems" (*NYP*, pp. 107–8).

preferred adventure stories: SS names them in "Read Books", p. 267–8. The anecdote set in a school charabanc is told on p. 268.

"This of course made it": *MA*, p. 83.

Olive Pain (*née* Cooper) remembers: OP, int A.

"catching tadpoles": "Same Place", p. 98.

23 **"I once left a copy":** *MA*, p. 85. See also *OTF*, p. 96, where Pompey also recalls losing this book in a rabbit hole.

shooting to the top of a holly tree: OP and DD, int A.

"secret letters to each other": "Same Place", p. 98.

"I should have been a boy": Woodcocks, int A.

sprained her knee: Dr James Curley, int A.

eighty-yard race and an egg-and-spoon race: *The Recorder*, 16 July 1914, p. 340.

According to Olive Pain: OP, int A.

Molly once reminisced: MS to HF, HFP.

used to be taken to the "Ally Pally": SS, l to DJ, 20 Oct. 1938, por.

"The most exciting thing that ever happened to me": script of SS,

"Swinging on a Star", Woman's Hour, BBC Radio 2, 4 Dec. 1969, BBCa.
24 **"crawl along the top of it":** *MA*, p. 94.
   Beowulf, for example: *MA*, p. 89.
   The Priors: identified by OP, int A.
   "If a lady comes up to you": *MA*, pp. 86–87.
   her first stage appearance: SS, l to Roland Watson, 5 July 1962, Birmingham.
   "My mother was a great" and **"She would use the stories":** OP, int A.
25 **Living Whist:** see *BPWN*, 21 June 1913, p. 5, and *The Sentinel*, 20 June 1913, p. [3]. Mrs Cooper also directed *Aladdin in Japan* in the winter of 1916. Its cast included her daughter Olive, DD (*née* Coke-Smythe), MS as Aladdin, and SS as "a graceful and obedient slave" whose "good acrobatic entrances" were much admired. See *The Sentinel*, 22 Jan. 1916, p. [2]; and *BPWN*, 15 Jan. 1916, p. 1.
26 **praised for being "pathetic":** *The Recorder*, 16 July 1914, p. 314.
   "She always envied children": HF, int A.
   "split with war": SS's phrase in Orr, p. 227.

Chapter Three

27 **the cancellation of their . . . holiday:** MS to HF, HFP.
   Royal Naval Reserve . . . (French): B. E. Scutt, Ministry of Defence, l to A, 28 Aug. 1980.
   North Sea Patrol: *I&S*, pp. 64–65.
   to collect it . . . "couldn't pass a pub": MS to HF, HFP.
   "immensely loyal" . . . to the Baltic: *I&S*, p. 65. Cf. *NYP*, p. 164.
   "When I saw the suffering": *NYP*, p. 162.
28 **"Daddy is in Glasgow":** ES, l to SS, 25 April 1916, PGP.
   "Are your Nerves Upset": *BPWN*, 16 Oct. 1915, p. 9.
   "Everybody seemed souvenir mad": *BPWN*, 9 Sept. 1916, p. [7].
   shocking account of charred bodies: Woodcocks, l to A, 26 Mar. 1984, por.
   "I see the Zeps": SS, pc to ES, 25 Apr. 1916, Tulsa.
   "We were roused": ES, undated l to SS, PGP.
   "Just a line to send with your fare": ES, undated l to SS, PGP. We date it "1 May" on evidence of another PGP l.
   "in his motor" and **"Aunty Grandma":** SS, pc to ES, 3 May 1916, Tulsa.
29 **"fizzle"** and **"rudely interrupted":** *BPWN*, 6 Oct. 1917, p. [7].
   changes in routine: most of the information in this paragraph comes from *PGHS*, pp. 33 and 38. The information about Flora Robson comes from that booklet (p. 22) and from Dame Flora Robson, int A.
   "Miss Hum was a very wise": "Same Place", p. 99. See also *MA*, pp. 88–89.
   "had the strength of her": *OTF*, p. 259.

Ragged School Union and the Poor Children's Boot Fund: see *The Recorder*, 18 Dec. 1913, p. 104.

"a Belgian refugee girl": *PGHS*, p. 33. According to DD, int A, this girl's name was "Angèle," and she became a close friend of SS. SS may have based Josephine, in *NYP* (pp. 193–96), on her. In the typescript of *NYP* (Hull), Josephine had originally been called "Angela," and MS's key to characters in *NYP* (HFP) identifies Josephine as Angela.

"But she was very simple too": *OTF*, pp. 259–60.

30 **take her to High Church:** Much of our information about Sidney Basil Sheckell comes from SJ, int A. For the detail about his taking SS to High Church services compare the content of SS's poem, "A Soldier Dear to Us", with a remark by SS in an int John Horder: "we had a friend who was very . . . Anglican, and he was coming home from the war . . . he used to take me—I was about ten I suppose—. . . to a service at All Saints, Margaret Street" (unedited script at Tulsa of BBC Radio "World of Books", broadcast 2 Feb. 1963).

character based on him: MS key to characters in *NYP*, HFP.

"very full of incense": *NYP*, p. 177.

Pompey recalls that her "mother and": *NYP*, pp. 169–70.

his nurse: The nurse, Miss Champagne, is a character based on Elsie Porter—identified by MS's key to *NYP*, HFP, and Woodcocks, int A. SS's typescript of *NYP* (Hull) also calls her "Porter", p. 170.

Captain Joey Porteous, did Pompey's Latin: *NYP*, p. 181. Identified by MS's key to *NYP*, HFP.

Tommy Meldrum, sent Stevie a letter: *NYP*, pp. 227–28. The original is part of the PGP. Meldrum's surname is supplied by MS's key to *NYP*, HFP.

"those times with their unquietness": *NYP*, p. 228.

[Footnote] **Stevie's typescript:** now in Hull. The architect and dates cited in this note were identified for us by Graham Dalling of the London Borough of Enfield History Office.

in 1968 her "long long *long* poem": SS, l to SJ, por. The poem was published in 1968 in both *The Honest Ulsterman* and *Ambit*.

31 **"Tommy gave me his book":** now in Tulsa, inscribed "To Peggy from E.T.M./Feby 1919."

"That took them out of themselves": *NYP*, p. 175.

Miss Hum awarded: an inscribed volume of Carlyle, Tulsa.

"privileged," she recalled: *MA*, p. 121.

32 **As a superior student . . . did not win a scholarship:** Dame Flora Robson, int A.

Mrs Smith thought she should: MS to HF, HFP.

Stevie once repeated: *MA*, p. 292.

reviewing a biography: *MA*, p. 179.

"would rather have been thought naughty": Horwell.

"it was certainly at school": *H*, p. 118.

Molly once recalled a scene: MS to HF, HFP.

a single prize . . . (for Scripture): Horwell.

"put the other little girls off": *I&S*, p. 70.

33 "either you were clever": MS to HF, HFP.

"were both a bit bored . . . discovered them for ourselves": Margaret Macbeth, l to A, 23 June [1980], por.

"prayers and best wishes": inscription in the volume, Tulsa. Another book in SS's library at Tulsa is inscribed, "to my dear godchild Peggy Smith on her confirmation, 1918. F.W."

"really dashing air": for reviews of the play see *BPWN*, 4 May 1918, p. [5], and 1 June 1918, p. [5]; and *The Sentinel*, 19 Apr. 1918, p. [3], and 3 May 1918, p. [3]. The production was again directed by Mrs Cooper, whose daughter Olive starred as Ali Baba. The reviewers did not automatically praise the children, as the mixed notices Olive received make clear: she was praised as "very good" in one paper, but criticized in another for lacking dash.

a witness to Mrs Smith's will: will of ES, 18 July 1918, records at Somerset House, London.

Shortly after the Armistice: all information in this and the following paragraph comes from a diary of MS, Tulsa, except for the claim that SS

34 "revelled" in the role of Shylock, which was made by Margaret Macbeth, l to A, 23 June [1980], por.

who recalls hating trams: *NYP*, pp. 157–58.

"She was very white" and "What can you do?": *NYP*, pp. 158, 225–26.

"fearfully cut up" and visited home on several occasions: diary of MS, Tulsa.

demobilization . . . 1919: B. E. Scutt, Ministry of Defence, l to A, 28 Aug. 1980, por.

"retired to the country . . . high and dry": MS to HF, HFP.

raising poultry with a new wife: Mr Smith listed his profession as "Poultry Farmer" on his certificate of marriage (GRO) to Hylda Lingen, 22 July 1920, Kidderminster, Worcester.

Molly used to stay with them: MS to HF, HFP.

"fearful bore" and "Tootles": *I&S*, pp. 65–66.

During the period of her mother's final illness: All facts and quotations in this and the following paragraph, except for the last sentence, are from MS's diary, Tulsa.

36 "to general and spoken astonishment": MS to HF, HFP.

English scholarship and Scarborough in August: MS's diary, Tulsa.

was free to pursue: Snow.

"a keen member" and "did not learn Greek": Miriam Shillitoe, Archivist of The North London Collegiate School, l to A, 14 Sept. 1979, por.

"it needed a married woman": SS, l to HH, 17 Oct. 1957, HHP.

"at the same time I was having to listen": *MA*, p. 254.

"There was once a woman called Miss Hogmanimy": *NYP*, pp. 124–25 and 137–38. The character, Miss Hogmanimy, was based on a woman named "Mrs Bolton", according to MS's key to *NYP*, HFP.

37 **in 1920 her own departure:** Miriam Shillitoe, l to A, 14 Sept. 1979, por.

Building had been effectively limited . . . also contributing to the development: see David Pam, *The New Enfield* (London Borough of Enfield, 1977), pp. 17, 25, 15. The character Lord Rumbelow in *NYP*, pp. 166–68, is based on Captain Taylor according to MS's key to *NYP*, HFP. In *H* he is referred to as Captain Taylor, p. 107.

"the Armistice already saw": *NYP*, p. 167.

she once recorded her name: *MA*, p. 96.

"after the holy-man kind of palmer": "Londoner's Diary", *ES*, 25 Sept. 1969, p. 10.

38 **swinging arm in arm:** see *MA*, p. 103, and *H*, pp. 106–7.

"about lakes and people getting bewitched": *I&S*, p. 79.

"An artist, like a child, is happier with": Jacynth Ellerton [Lady Lawrence], *The Tablet*, 10 Apr. 1971, p. 360.

"He is not physically strong": "Plomer's Own Good," *T*, 17 Dec. 1943, p. 18.

"only intensification of her inner pride": Marie Peel, " 'Half in love with easeful Death,' " *Books & Bookmen*, June 1971, p. 6.

once expressed her admiration of Blake's lines: in fact, more than once. See *NYP*, p. 250, and SS, "A Happy Indian," *JO'l*, 60 (2 Mar 1951), 118.

"being alive is like being": *I&S*, p. 71.

she attributed to her stay: MS to HF, HFP.

wiggling and twisting in a chair: Woodcocks, int A.

[Footnote] **she joked:** *MA*, p. 277.

Chapter Four

Tulsa has journals in which SS kept notes about the books she read from 1919–30, and a collection of SS's unpublished poems from the 1920s. These journals and poems are quoted extensively in this chapter but, as it is clear from our text when they are being quoted, we do not provide a citation for each quotation.

40 **Mrs Hoster's:** KD, *I&S*, p. 81; MS to HF, HFP; and Snow.

the London School of Journalism: Margaret Macbeth, l to A, 23 June [1980], por.

an unsuitable candidate: Snow.

"I do not think I should have": SS, l to HH, 17 Oct. 1957, HHP.

several of her friends: e.g., AB and HF, int A.

pay for Molly's expenses: Snow, and Inez Holden, "Stevie Smith" [obituary notice], *Royal Society of Literature Report*, 1971–72, p. 52.

often absented herself: Snow.

sometimes prefaced: Tulsa.

41 **Years later she retained a distinct memory:** SS, *I&S*, pp. 68–69; and Snow.

she became devoted: see her remark, "I love the fairy stories so much I still read them" (Williams, p. 109).

"When I think of Stevie's": Woodcocks, int A.

"She transmogrified": Marghanita Laski, int A.

in the books she acquired: see inscriptions in her library, Tulsa.

"Peggy" she remained: DD, int A.

Mrs Hoster's College was: AB, int A.

"the famous Mrs Hoster's": Horder.

two fingers: EL, int A. See also *NYP*, p. 27.

consulting engineer: Snow.

"the soul of probity", "We must get rid": and

"a kind man, a man of kind": "Too Tired". SS gave the quotation about the horrible child to the character Rackstraw, *H*, p. 154. remained in the employ . . . for a year: Snow. But, according to MS, SS spent only *six* months "with some nameless tycoon" before becoming Sir Neville Pearson's secretary (MS to HF, HFP).

42 whose father had risen: see Sheridan Morley, *Gladys Cooper: A Biography* (London: Heinemann, 1979), pp. 122–23.

"like something out of Dickens": Margaret Ralph, undated l to A, por.

died in November of 1924: notation in diary of MAS, Tulsa. Confirmed by death certificate of M. H. Clode, GRO.

salary of £100: MS to HF, HFP.

"this darling Lion": *NYP*, p. 77.

described by Stevie's friends: AB, Racy Buxton, Dr Curley, HF, Marghanita Laski, Sir John Lawrence, OP, Woodcocks—int A.

when watching Aunt enjoy: *NYP*, pp. 88–89. Molly Everett, int A, notes that MS told her it was "all true" about the game pie and Aunt.

did not have a literary aunt: *I&S*, p. 66–67.

borrowing from Santayana: *MA*, pp. 296–97.

"two English animals": *H*, p. 176.

"But most women, especially": *H*, p. 28.

43 In 1926 she and Molly: MS, pc to Miss Orange, 3 May 1926, Tulsa.

another Cornish holiday: Joan Prideaux, l to A, 2 Aug. 1981, por.

Molly became, in 1928, a convert: MS to HF, HFP.

grieved the Lion Aunt: AB and Woodcocks, int A.

Molly and Stevie were never: AB, DD, AO, Woodcocks—int A.

"Pearl came to breakfast": *H*, pp. 60–61.

44 "When I left school, at seventeen": SS, l to HH, 17 Oct. 1957, HHP.

read too much: *NYP*, p. 12.

Journals Stevie began: now in Tulsa.

"what Shelley read": SS, "Out of the Uproar", *O*, 22 May 1960, p. 19.

"suddenly became a real swot": Horder.

45 never intellectually snobbish: AB, int A.

"Read what you like, capture": script of BBC radio broadcast, "World of Books", 18 Mar. 1961, BBCa.

"A poet reading books is hungry for food": ibid.

a quick reader, one who: Marghanita Laski and AO, int A.

46 **"lost to us, ahem"**: *NYP*, p. 180.

at St John's Church in 1910: see *The Sentinel*, 15 July 1910.

exchanged them, for instance, with John Hayward: *MA*, p. 286.

appears in the typescript: p. 52, Hull.

"if Miss Gertrude Stein": Douglas West, "Thoughts at Random", *Daily Mail*, 17 Sept. 1936, p. 16.

47 **He is the "P."**: MS key to *NYP*, HFP.

"Well and who is this 'P.'": *NYP*, p. 174.

a form she later tried: e.g., "The Parklands", "Eulenspiegelei", "The Boat", "Lady 'Rogue' Singleton", "The Warden", "Saint Anthony and the Rose of Life", and "The Singing Cat".

"about 1924": SS, l to HH, 17 Oct. 1957, HHP. See also Williams, p. 107.

exercises from the 1920s exist in typescript: at Tulsa. Dates of these unpublished poems, and SS's comments about them, are from her notations on the typescripts.

48 **His function in the firm**: AB, int A.

49 **Hester Raven-Hart**: Hester Raven-Hart presented SS with a copy of Conrad's *Victory* inscribed "Hester Raven-Hart to Margaret Smith—22 Feb 1927," and SS also had among her books a copy of Blake's poems given to her in the 1920s and inscribed, "Bouquet from Hester". These books are now in Tulsa. During SS's childhood she sometimes visited the Raven-Harts, who lived at Fressingfield Vicarage in Harleston, Norfolk, where Hester's father was rector. According to AB, int A, Hester's father, Canon Raven-Hart, provided the inspiration for the kindly rector, Uncle Heber, in *H*. Canon Raven-Hart died nine months after the death of SS's mother: see *The Suffolk Chronicle and Mercury*, 28 Nov. 1919, p. 6.

met Joan Prideaux and: Joan Prideaux, l to A, 2 Aug. 1981, por.

six pages in her diary: Tulsa.

50 **"in the rich full way"** and **"cultured gentlewomen"**: *NYP*, pp. 116–17.

Sir Neville first met and **"That they were both divorcees"**: Sheridan Morley, *Gladys Cooper* (cited above), p. 122–4. The interpolation in the second quotation is by Morley.

51 **"She knew all about that"**: SJ, int A.

"Peggy from Palmers Green": phrase used by EL: "Don't you think we could do a musical on Novel on Yellow Paper—or Peggy of Palmers Green? incorporating lots of the poems??" (l to SS, 3 Dec. 1956, PGP).

Chapter Five

52 **On 13 September, 1929:** SS, pc to MS, Tulsa.
in Germany for two weeks: for SS's visits to Germany see *MA*, pp. 296 and 301; *I&S*, p. 68; and SS, l to HH, 17 Oct. 1957, HHP.
especially for *Novel on Yellow Paper*: *NYP*, pp. 83–113 and elsewhere.
did once send a copy of Isherwood's: now in Tulsa.
the first "last train": *I&S*, p. 68.
"everything in the novel is true": SS, l to DJ, 4 Sept. 1936, por.
"I was grateful and relieved": Terrence Pim, l to SS, 6 March 1938, PGP.
In the typescript: Hull.

53 **sees Pompey off, *Schokolade*, *Spieglein*, and last time she was in Berlin:** *NYP*, pp. 85, 94, 101, and 97.
"staying with a dotty family" and "I travelled about with Frau. S.": SS, l to HH, 17 Oct. 1957, HHP. Tolstoy's play is discussed in *NYP*, pp. 85–86 and 241–50.
"*männlicher Protest*", meets Karl, and Swiss-German: *NYP*, pp. 99–100, 32.
"drove me deutsch-wards": SS, l to HH, 17 Oct. 1957, HHP.
"*I met Karl* (!)": SS, pc to MS, 12 July 1931, Tulsa.

54 **"In 1931 I went again to":** SS, l to HH, 17 Oct. 1957, HHP.
"*ganz platonisch*": *NYP*, p. 100.
owned many of these: see her library, Tulsa.
Stevie reports she has had: SS, pc to MAS, 16 July 1931, Tulsa.
Stevie introduced her childhood friend . . . is lifelike: DD, int A.
Stevie has Pompey recall and Six feet two: *NYP*, pp. 30–32, 36.
his calling card: we found it inscribed "To Margaret with my best wishes" in SS's copy of *Faust*, Tulsa. MS's key to *NYP* identifies Karl as Karl Eckinger. See *NYP*, p. 44, where Pompey says Karl gave her a translation of *Faust* as a gift.
Eckinger studied history: These details are from the *curriculum vitae* which Karl Eckinger placed in his published dissertation, approved by the faculty of Friedrich-Wilhelms-Universität, *Lord Palmerston und der Schweizer Sonderbundskrieg* (Berlin: Ebering, 1938). See the parallel details in *NYP*, p. 44. Eckinger's *vita* notes that he was born 9 August 1904, which made him two years younger than SS.
discussing the struggles of Luther . . . the nightmare it became by 1936?: *NYP*, pp. 31–49.

55 **Sally Chilver, remembers:** SC, int A.
"pedantic self-righteous": SS, l to HH, 17 Dec. 1957, HHP.

56 **"feelings were all nicely":** SS, l to HH, 17 Oct. 1957, HHP.
Suzannah Jacobson: SJ, int A.
The only "goy" and "Jew friends": *NYP*, pp. 10, 103.
Cruelty was always terrifying: Lady Lawrence, int A.

Pompey's loathing: *NYP*, pp. 103–4.

begins to weep for them: *NYP*, pp. 108–9.

unflattering conclusions: *NYP*, pp. 102 and 104.

only when she boarded the Channel ferry: unedited recording of SS, int KD, 7 Nov. 1970, in the possession of KD.

to discourage the playwright: *MA*, p. 269. In this letter SS refers to Vienna as part of the "larger German Reich".

Helen Fowler remembers: HF, int A.

[Footnote] See O. F. Fritschi, "Zum Gedenken an Karl Eckinger", *Neue Zürcher Zeitung*, 14 Apr. 1983, p. 53; and "Chefredaktor Dr Karl Eckinger tritt zurück", *Der Zürcher Oberländer*, 30 June 1972, p. 1. Also Gertrud Häusermann-Häusermann, l to A, 28 Nov. 1983, for the name "Elfrieda".

57 **"I never want to go" and made synonyms of:** *MA*, pp. 296 and 345 ("how cruel, how *German*").

she criticized Dean Inge: SS, "A Poem and a Review", *Frontier*, 4 (Spring 1961), 41–42.

"Have Done, Gudrun": *The Times*, 9 Jan. 1970, p. 9.

In 1931, probably before her second: this paragraph is based on a l of Gertrude Wirth to A, 2 Feb. 1980, por, and her int A.

The local paper reported: see *PGSG*, 20 Nov. 1931, p. 8, and 18 Dec. 1931, p. 8.

58 **"This vicar—the Rev. Humphrey":** *H*, p. 76. See also *MA*, p. 76.

"At that period" she said: *I&S*, p. 72.

"quite catch successfully": *MA*, p. 302.

"love is everything, it is the only thing": *H*, p. 50.

"Stevie Smith shows that she loved him": David Garnett, "Current Literature", *NS* (5 Sept. 1936), p. 321.

according to Anna Browne: AB, int A.

Eric Armitage: MS's key to *NYP*, HFP.

"I've been curious as to how": June Nethercut, l to A, 18 Nov. 1980, por.

[Footnote] Woodcocks, l to A, 26 March 1984, por.

59 **Rosemary Cooper . . . remembers:** Rosemary Cooper, int A.

Doreen Diamant remembers: DD, int A. Also see SS's response to the question, "Did you nearly marry . . . ?": "Oh no—well yes, I suppose I did really" (*I&S*, p. 72).

"Now we are engaged to be married": *NYP*, p. 216.

In these deleted sections: see typescript of *NYP*, Hull. Several passages not deleted and not referring to Freddy do have SS's query "libel?" in the margin.

"maybe never had a meaning": *NYP*, p. 240.

60 [Footnote]: AB, l to A, 15 July 1982, por.

61 **gather silver and gold:** See Iona and Peter Opie's *Children's Games in Street and Playground* (Oxford Univ. Press, 1969), pp. 84–86.

tour d'ivoire and "a hearth-rug's ivory tower": *NYP*, pp. 37 and 206.

In appearance Eric was: This paragraph is based on June Nethercut, l to A, 18 Nov. 1980, por.

"foot-on-the-ground" and **in his extreme moods:** *NYP*, pp. 220–21, 232.

Anna Browne says: AB, int A.

62   **"Freddy's heart":** *NYP*, p. 234.

"I can see why": June Nethercut, l to A, 18 Nov. 1980, por.

"He wasn't anything like": June Nethercut, l to A, 7 Oct. 1980, por. In this l Mrs Nethercut verifies the authenticity of the line in the film, taken from *NYP*, about the commonness of people in Bottle Green, and the other about a home where one was allowed to do anything except smack the dog—both favourite sayings of her grandmother (who was Eric's mother).

Talking with young friends years later: Piers Russell-Cobb, int A.

she told Anna Browne and **Helen Fowler recalls:** AB and HF, int A.

flirted "with my escort": Jeni Couzyn, " 'But at Heart You Are Frightened' ", *Poetry Dimension* (London: Abacus, 1973), p. 83; and George MacBeth, int A.

"I don't feel happy in love. I think I'm": *I&S*, p. 77.

"probably not": MS to HF, HFP.

Mrs Fowler said . . . impunity: HF, int A.

"When my generation was young" and **"took the last fence":** EL, int A.

63   **as an experiment:** Lady Lawrence, int A.

said to Sally Chilver: SC, int A.

"a bit of a sentimentalist in his idea": *MA*, p. 139.

"falling in love with themselves when young": SS, "New Novels", *WR*, NS 28, June 1951, p. 80.

"It is a matter of painful experience": "New Novels", *WR*, NS 32, (Oct. 1951), p. 76.

good at discussing the tangles: SC, int A.

like a schoolgirl about love: SS, "Books", *Queen*, 16 Jan. 1963, p. 9.

Stevie admitted she wasn't much good: *MA*, p. 294.

64   **Love mapped a country . . . in their novels:** HF, int A.

"quelque chose de Lesbos": *OTF*, p. 131.

according to Elisabeth Lutyens: EL, int A.

a day when Elizabeth Sprigge: Ronald Orr-Ewing, int A.

Stevie noted a woman: HF, int A.

"But there was this love between them": *MA*, p. 192.

"the repulsive current euphemism": SS, "New Novels", *S*, 16 Oct. 1953, p. 438.

"How I have enjoyed sex": *NYP*, p. 121. See also SS's poem, "Conviction (iv)".

65   **"for many, marriage is a chance clutch":** introduction to SS's poem, "The Jungle Husband", typed on a sheet of introductions to poems, Tulsa. See also the variant, "love is a desperate chance clutch", *H*, p. 49.

"I love life," she said . . . as a spouse: *I&S*, pp. 70, 72–73.

"The Herriots": see *MA*, pp. 74–79.

exactly what a man said: *MA*, p. 282.

66 "how awful marriage can get": ibid.

told friends how at a "stuffy black-tie dinner": Racy Buxton, int A.

"warm her hands at other": AB, int A.

her review of a book whose heroine: SS, "Read and Relax", *MW*, Sept. 1941, p. 18. For the adage in French see *NYP*, p. 43, and *MA*, p. 197.

"married to her Aunt": Gertrude Wirth, int A.

"You want to get right away" and "The unmarried girls have an idea": *NYP*, pp. 65 and 149.

"the witch-craft spider-web": SS, "New Novels", *WR*, NS 28, June 1951, p. 78.

"In some ways I'm romantic": *I&S*, p. 75.

"People think because I never married": Neville Braybrooke, "Poet Unafraid", *Sunday Telegraph*, 14 March 1971, p. 11.

67 "But when it is over": *NYP*, p. 236.

"no man ever again shall scour": *OTF*, p. 19.

"with the rights and wrongs": *OTF*, p. 137.

"Nothing that produces suffering": *H*, p. 65.

characteristically loving letter: MAS, undated l to SS and MS, on a view letter-card of Eastbourne (Tulsa). We date the holiday by a dated pc SS sent to MAS (Tulsa).

68 In 1933 . . . commuter belt: David Pam, *The New Enfield* (London Borough of Enfield, 1977), p. 17.

not "excessive": Snow.

comfort themselves with tea and Pompey has speeches: *NYP*, pp. 16, 203–04.

Once Sir Neville's wife: OP, int A.

"the same Beaver who": Mrs Robert Hardy, l to A, 18 Feb. 1981, por. Mrs Hardy wonders if she is not the "Baby N." (although her name is Sally) mentioned in the subtitle of "To the Dog Belvoir" whose life, according to the poem, was saved by a canine cognate of Chaucer's Knight—a "perfect gentle quadruped". ("Belvoir" is also the name SS gave to the Labrador retriever in her poem, "The Hound of Ulster".)

based on Sir Neville's mother: MS's key to *NYP*, HFP.

Pompey describes her visits: *NYP*, pp. 26–27.

"It didn't prove easy": Sheridan Morley, *Gladys Cooper: A Biography* (London: Heinemann, 1979), p. 163.

69 "by no means intended for your eye": Curtis Brown, l to SS, 28 June 1934, PGP. E.B. is almost certainly the late Eleanor Brockett according to Juliet O'Hea, a former director of Curtis Brown, Ltd. E.B.'s "Reader's Report," which is quoted in the next few pages of the text, is also part of the PGP. SS's coments are written on the report.

70 **later said she too disliked:** SS, l to Mr Baynes (sic) [Jocelyn Baines], 28 Apr. 1961, PGP.

71 **"Ignes Fatui":** We quote the title and the lines of this poem from the Tulsa version.
when she wrote in 1951: Boulton.
"fear and pain and": Williams, p. 107.

72 **writing writing writing:** *NYP*, pp. 203–4.
answered other letters sent to the agony columns: HF, int A.
"lovely mixture": *MA*, pp. 276–77.

Chapter Six

73 **"A pity you were not saying":** *NYP*, p. 72.
into a "go-ey girl": SJ, int A.
"in this world of catch-as-catch-can": *NYP*, p. 183.
filled with single ladies: Colin Amery, int A.
isn't at all chic: *NYP*, p. 65.

74 **prefers tiring travel:** *NYP*, p. 209.
company more than the fame: *MA*, p. 286.
Bottle Green and the pub-lub life: *NYP*, p. 240.
the most compelling thing about her: Jonathan Mayne, int A.

75 **a young editor and junior partner:** Ian Parsons, int A; also Williams, p. 107.
based a character (Topaz): MS's key to *NYP*, HFP; also Ian Parsons, int A.
rhythm which got into her head and **"pseudo-American":** *I&S*, p. 73.
"go away and write a novel" and **"He also said":** Williams, p. 107. See also Horwell.
"in a dream state": Audax, "What I Hear", *JO'L*, 38, 11 Feb. 1938, 782.
completed in six weeks: MS to HF, HFP. It was about ten weeks according to Andax, ibid.
"the depressing, and to me humiliating": Ian Parsons, l to SS, 12 Feb. 1936, PGP.

76 **she complained in 1955:** *MA*, p. 299.
regale him with imitations: Ian Parsons, int A.
"As Lady Godiva said": *MA*, p. 281.
"very grateful to the old N.S.": SS, l to HH, 15 Apr. 1958, HHP.
that of Hamish Miles, went to Stevie's office, and **"amusingly described":** RHD, l to A, 13 Sept. 1980, por.
"You must imagine a very": The proof-correcting sequence is found in *OTF*, pp. 108–13.

78 **inscribed in Miles' copy:** now in the possession of his son, Professor Hamish Miles.
"Miss Smith was . . . diseased": Professor Hamish Miles, l to A, 2 Oct. 1979, por.

"I think it is much the best": Raymond Mortimer, l to SS, 17 Sept. 1936, PGP.

invited Stevie to lunch: Diana Oakeley, l to A, 19 Sept. 1980, por. As JH's sister, she is in possession of his engagement calendars for 1926–50. The lunch party is JH's calendar entry for 29 Oct. 1936.

"Since most of my time": JH, l to SS, 14 Sept. 1936, PGP.

she had already written half of a new novel: SS, l to JH, 16 Sept. 1936, King's.

79 **"the talk of the town"** and **"Reminiscent of *Tristram Shandy*":** JH, "London Letter", *The New York Sun*, 26 Sept. 1936, p. 17.

Stevie was grateful and mentioned: SS, l to JH, 16 Sept. 1936, King's.

"rarely endorses a book": Frances Phillips, l to SS, 12 Jan. 1937, PGP.

"I think it is a brilliant": Noel Coward, copy of l to Mr Hunt, 7 Jan. 1937, PGP.

New Yorker **advertisement:** Frances Phillips, l to SS, 12 Jan. 1937, PGP.

"got a bit tipsy": SJ, int A.

"publication nerves" and **"very sad-dog":** SS, l to DJ, 4 Sept. 1936, por.

"far little behind": *NYP*, p. 56.

"had a warped way", "you can't get drunk", "she didn't think",
80 **took Stevie to parties,** and **ask Suzannah to lunch:** SJ, int A.

"ladylike boy": *NYP*, p. 54. "Larry" is identified in MS's key to *NYP*, HFP.

"fabulous parties where one met": SJ, l to A, Jan. 1981, por.

Stevie talked all through: SJ, int A.

"we've got to mind our P's and Q's": SC, int A.

"my vents have been blocked": SC, int A.

"Good-bye to all my friends": *NYP*, p. 9.

"none of her beautiful friends": Marie Scott-James, "A Sentimental Journey", *London Mercury*, 34, Oct. 1936, 560.

Jacobsons didn't invite: SJ, int A.

Naomi Replansky ... twice visited Palmers Green: Naomi Replansky, int A.

81 **"The little bits":** Naomi Replansky, undated l to SS, PGP.

"only a consciousness": AO, int A.

inclusion among books she owned: SS's copy of Belloc's *The Jews*, Tulsa.

a charge he denied: see Belloc's "Preface To The Second Edition", *The Jews* (London: Constable, 1928), p. xvi.

"an alien body" and **"Jewish problem":** ibid., pp. xix and xiv.

"renowned among Jews": Noel Annan, "The joys of bigotry", *TLS*, 27 April 1984, p. 467.

"although I was never anti-semitic": Stephen Spender, *World Within World* (New York: Harcourt, Brace and Company), 1951, p. 12.

82 **neighbours were "hurt":** Woodcocks, int A.

"Charles is grieved": Barbara Flower, l to SS, 29 Oct. 1940, PGP.

Aunt was furious . . . that too helped: MS to HF, HFP.

"Darling Auntie Lion": *NYP*, p. 225.

"most utter rubbish": *MA*, p. 254. Uncle James is based on SS's elderly cousin, Alfred Hook, according to MS's key to *NYP*, HFP.

"beastly" letter and "These are all": *MA*, p. 256.

Pompey says that she will fight: *OTF*, pp. 157–59.

83 **"Very essentially civilized" and "*Barbare* is always":** *OTF*, p. 197.

"what he and his race": *OTF*, p. 198.

"I didn't want to write novels": Frank Entwisle, "'Such A Nervy Business . . . '", *ES*, 24 Nov. 1960, p. 5.

"the poems with the beastlies": *MA*, p. 255.

It was Hart-Davis who had suggested: SS, l to HH, 17 Oct. 1957, HHP.

"We shall obviously have to cut": RHD, l to SS, 27 Oct. 1936, PGP.

"this time next year" and "A lot of the poems": *MA*, p. 256.

84 **"two rather grand poems by":** *MA*, p. 256.

"We still have something over 200 copies": RHD, l to SS, 28 Nov. 1936, PGP.

"The sales of N.O.Y.P. have been": RHD, l to SS, 1 Jan. 1937, PGP.

"What a splendid letter": RHD, l to SS, 3 Dec. 1936, PGP.

"Stevie my dear, You write like": RHD, l to SS, 7 Apr. 1937, PGP.

"What a little tiger": RHD, l to SS, 30 Apr. 1937, PGP.

85 **"This week a book has made me":** David Garnett, "Current Literature", *NS*, 5 Sept. 1936, p. 321.

Paul Bailey: "Stevie Smith: Sometimes Awful, Mostly Marvellous", *Nova*, Nov. 1969, pp. 25 and 28.

[Footnote] **"That awful Bloomsbury" and "I have only one fault":** *MA*, p. 316; and Clive Bell, l to SS, 11 Sept. 1936, PGP.

86 **"One summer's day":** David Garnett, l to A, 27 Oct. 1980, por.

"I stayed once": SS, l to Dr Polly Hill, 13 Nov. 1969, por.

"During the last eighteen months": Arnold Palmer, "Winter-Flowering Novelists", *Yorkshire Post*, 19 Jan. 1938, p. 6.

artist wants attention, assessment, and employment: SS, "Do Academies Foster Art?", *JO'L*, 43, 21 June 1940, 340.

Chapter Seven

87 **"impossible to classify":** "Without Title", *TLS*, 12 Sept. 1936, p. 727.

acquired her copy of Sterne's novel: now in Tulsa.

88 **"One begins reading, a bit annoyed":** Patrick Monkhouse, "Good and Bad People", *ES*, 10 Sept. 1936, p. 10.

"compellingly exasperating": Edwin Muir, "New Novels", *L*, 16, 16 Sept. 1936, 547.

"young Joyce out of Anita Loos": Peter Burra, "Fiction", *S*, 157 (18 Sept. 1936), 474.

complained to a friend: *MA*, p. 254.

"For it is not subject to the laws": Ian Parsons, "Novel on Yellow Paper", *Now and Then*, Summer 1936, pp. 39–40.

89 **"In these days of Fascism and Communism"**: Marie Scott-James, "A Sentimental Journey", *London Mercury*, 34 (Oct. 1936), 560.

"the sort of book I must have by me": Leonard Isaacs, l to SS, 22 Feb. 1941, PGP.

she included her press cuttings: *MA*, p. 254.

"(if you will pardon the liberty, but": Kenneth Clark, undated l to SS, PGP.

90 **"Recently I re-read [*Novel on Yellow Paper*]"**: Mary Renault, l to SS, 7 Feb. 1958, PGP.

wrote contemptuously of the magazine: *MA*, p. 281.

"I have discovered something original": Richard Church, "New Books", *JO'L*, 36 (3 Oct. 1936), 23.

"After watching your things": NM, l to SS, "Christmas Day" [1936?], PGP.

91 **met Gertrude Hermes**: Gertrude Hermes, telephone int A.

"Les Sitwells", along with: *NYP*, p. 185.

"I am so much enjoying": Osbert Sitwell, l to SS, 3 Feb. 1938, PGP.

"fan letter": Osbert Sitwell, l to SS, 17 Feb. 1938, PGP.

interest in Edith Sitwell was slight: *MA*, p. 269.

"sinister": SS, "Sitwell Panorama", *JO'L*, 44 (6 Dec. 1940), 243.

"I do not know who Miss Stevie Smith is": Robert Nichols, "Stevie Smith", *Now and Then*, Winter 1936, p. 36.

had been baptized "Patience": *NYP*, p. 20.

92 **rather like a crumbling statue**: *NYP*, p. 20.

in the acerbic view of her friend: EL, int A.

"lovely to stay with": *NYP*, p. 20.

"It is based on the same": SS, l to RM, 11 Jan. 1960, RMP.

93 **"a cool thou"**: Ian Parsons, int A.

Sir Phoebus asks Pompey: *OTF*, p. 118.

Inez Holden . . . once described: Celia Goodman, l to A, 22 Oct. 1980, por.

who suggested to Stevie the title: AB, int A.

"I don't mind saying I am a lucky girl": *NYP*, p. 20.

"I am *toute entière* visitor": *NYP*, p. 212.

94 **a quatrain John Hayward called "Peevy"**: PGP.

"basically English": C. V. Wedgwood, int A.

"She couldn't talk about anything": HF, int A.

one tooth left in his head: SS, "Be Merry," *NS*, 12 Apr. 1963, pp. 525–26.

"That's a difficult child": Lady Lawrence, int A.

"a sort of gale" and the "Fairy Queen" anecdote: AB, int A.

95 **"affable" circumstances**: *MA*, p. 111. See also *NYP*, p. 144. In her essay, "Too Tired for Words" (not to be confused with her BBC script

of the same title), SS further illustrates her faculty for invention released by fatigue: *MA*, pp. 111–18.

"she invented witticisms so complete": SC, int A.

Norah Smallwood remembered: Norah Smallwood, int A.

"and gossip is what novels": SS, "New Novels", *O*, 19 Sept. 1954 p. 13.

"high-powered conversation" and **"unnecessary":** Horwell."

"nothing thicker than a knife's blade": Virginia Woolf, *Orlando* (New York: New American Library, 1960), p. 28.

"a girl for the glooms": *OTF*, p. 265.

"he would say 'No' as the chance": SS, "New Novels", *O*, 27 June 1954, p. 9.

96 richly compostly loamishly sad": *NYP*, p. 13.

"sickly green gaslight": *OTF*, p. 37.

"Gothic castles, preferably": SS, "Woman in Man's Dress", *JO'L*, 56 (13 Dec. 1946), 146.

"a new way of looking at the world": Irving Howe, "Messages from a Divided Man", *New York Times Book Review*, 29 March 1981, p. 7.

she can't seem to acquire: *NYP*, p.231.

"when we are writing the life of a woman": Virginia Woolf, *Orlando* (New York: New American Library, 1960), p. 175. Of course Woolf may have intended irony.

admired . . . Maugham: *MA*, p. 285.

97 **"seems to have been genuinely typed, direct":** Douglas West, "Thoughts at Random", *Daily Mail*, 17 Sept. 1936, p. 16.

"the Old Interior Monologue, come home": Williams, p. 114.

"the natural run of the voice": SS's copy of *Orlando* is in Tulsa. We are quoting from the New York, New American Library edition, 1960, p. 138.

[Footnote] **"I will tell you who he is":** "Too Tired". SS's copy of the 1832 Lempriere's *Classical Dictionary* is in Tulsa.

98 **"You are Stevie Smith":** See *Leave The Letters Till We're Dead*, eds. Nigel Nicolson and Joanne Trautmann, Vol VI (London: The Hogarth Press, 1980), p. 75.

had read Dorthy Richardson: SS's reading journals, Tulsa.

Fielding to whom she refers: *NYP*, p. 77.

"Reader, do you" and **"But first, Reader":** *NYP*, pp. 28 and 38.

About Sinclair Lewis's: SS's reading journals, Tulsa.

once praised Edmund Wilson's: SS, "Puzzled Young Man", *JO'L*, 61 (28 March 1952), 331.

"reminiscent of Stevie Smith": Muriel Spark, "Amphibian Man", *O*, 20 Oct. 1957, p. 19.

"apologized for having more or less fallen": Gertrude Wirth, l to A, 2 Feb. 1980, por.

"I'd been hoping . . . that all this enforced": pc to SS, 1 Jan. 1956, PGP—signed "Montague".

people unfairly referred to Stevie: Francis Wyndham, int A.

99 **"has hung like a millstone"**: Thomas.
   Marghanita Laski found: Marghanita Laski, int A.
   it was a far better work: SS, l to DJ, 22 Aug. 1969, por.
   "What are you doing with *that*": Molly Everett, int A.
   "I am already sick of my first book": *OTF*, p. 107.
   "long comic-tragic poem": Inez Holden, "Stevie Smith" [obituary notice], *Royal Society of Literature Report*, 1971–72, p. 52.
   "the beauty and subtlety of unuttered thought": SS, review of *Love and Death* by Llewelyn Powys, *Life and Letters To-day*, 22 (July 1939), 138.

Chapter Eight

100 **larger publishing house**: Mrs Ralph, undated l to A, por.
   in 1936 or early in 1937: date inferred from addresses on SS's letters.
   concealed from the baronets: AB, int A.
   Stevie had a bell: KF, int A.
   first floor: *I&S*, p. 82.
   Tower House: address on l of SS to RHD, 24 Aug. 1937, por.
   A small, dark . . . her drawings: *I&S*, p. 82. That it was dark and had volumes of *Country Life* we know from our int Lady Georgina Coleridge.
   wrote a rum account: SS, l to JH, 12 Nov. 1942, King's.
   friendly to everyone and neat but not fashionable: Mrs Ralph, undated l to A, por.
   "fidgets": SS, l to JH, 10 Sept. 1943, King's.
   twice what she was worth and **inquired after her health**: *MA*, p. 262.
   Sir Neville had been . . . amused by it: Ronald Orr-Ewing, int A.
101 **"I'm sorry you work in a Gothic ruin"** and **"such who entering"**: V. S. Pritchett, l to SS, 21 Jan. 1941, PGP.
   some sympathized with the baronets: Rosemary Cooper and EL, int A.
   "faithful dog-o": *MA*, p. 293.
   "reading bloody manuscripts": *MA*, pp. 260 and 264.
   "the scourings of about ten years": NM, *You May Well Ask* (London: Gollancz, 1979), p. 153.
   "like a bird": ibid.
   "I have written a quite different": NM, l to SS, "Tues" [1937], PGP.
   "Stevie Smith . . . stands outside any": Mark Storey, "Why Stevie Smith Matters", *Critical Quarterly*, Summer 1979, p. 42.
   There are always certain people who are aware": NM, "Bouncing With Blake", *Now and Then*, Winter 1937, p. 27.
   Other reviewers pointed: see *The Granta*, 46 (5 May 1937), 390; and "Humour", *TLS*, 8 May 1937, p. 365.
102 **"absolute confidence"**: *MA*, pp. 266 and 285.

"brilliantly funny and intimate": G. W. Stonier, "The Music Goes Round and Round", *NS*, 13 (17 Apr. 1937), 640 and 642.

"lyrical-sardonic" and **"shocks of pain"**: Anon, *London Mercury*, 36 (May 1937), 107.

"when you sign your contract": RHD, l to SS, 8 Dec. 1936, PGP.

"We will find out about libel laws here": Noel Wiren, l to SS, dated "Nov 23rd," PGP. For still more evidence of SS's lifelong concern about libel consider her question to an editor at Longmans before they published her *Selected Poems*: "I wonder if 'Lord Say and Seal' had not better come out on account of possible libel???" (SS's copy of her l to Mr Baynes [sic], 28 April 1961, PGP). The editor agreed that the poem "certainly ought to come out" (Jocelyn Baines, l to SS, 8 June 1961, PGP).

"Major Raven-Hart has chosen": SS, l to NM, "Monday" [1937], Edinburgh. Published in *You May Well Ask* with the surname deleted.

"I am pretty sure that Raven-Hart was": NM, l to A, 13 March 1981, por.

103 **"I wonder if Major R. H. has recognized"**: NM, l to SS, dated "17th," PGP.

"Major R. H. has pottered": SS, l to NM, 20 May 1937, Edinburgh. Published in *You May Well Ask* with the Major's surname deleted.

"Stevie had a rather peculiar flirtation": NM, l to A, 3 Jan. 1981, por.

"You mightn't be so popular with Jews": NM, l to SS, dated "17th," PGP.

"No no,—no Yiddisher" and **"Rupert, Hamish and even"** (p. 104: SS, l to NM, 20th May 1937, Edinburgh.

[Footnote] **"everlastingly writing"**—*MA*, p. 300; **The Major's military experience**—see editor's note to Raven-Hart's translation of *Germans in Dutch Ceylon* (Colombo, Ceylon: Colombo National Museum, 1953), p. iii; the three books whose titles are given in the footnote were all published in London by F. Muller. That "L. A. C. Errant" was Raven-Hart's pseudonym is clear from a page, listing Raven-Hart's other writings, in *Ceylon: History in Stone* (Colombo, Ceylon: Daily News Press, 1964).

104 **"I am not at all sure"**: RHD, l to SS, 31 March 1937, PGP.

"I am sorry to have to tell": *MA*, p. 262.

"As for *Over the Frontier*, really": DJ, l to SS, 19 Nov. 1937, PGP.

"about General Ironside. I": RHD, l to SS, 13 Oct. 1937, PGP.

"I have forgotten to tell": RHD, l to SS, 31 March 1937, PGP.

105 ***Married To Death***: *MA*, p. 263.

"I am not a very good girl": *MA*, p. 264.

very anaemic: SS, l to NM, 20 May 1937, Edinburgh.

"two friends of mine" and **"vicariously the"**: SS, l to DJ, 28 June 1937, por.

"I never rode again": SS, "Mosaic", *Eve's Journal*, May 1939, p. 107.

"**There are parts round the Norfolk**": SS, l to DJ, 28 June 1937, por.

"**She has been reading this grand old play**": *MA*, p. 265.

on whom she had based "Harriet": MS's key to *NYP*, HFP; and Jane Stockwood, int A.

London Gate Theatre: the typescript of *NYP* (Hull) has "Gate" deleted and the query "libel?" written in the margin, p. 142.

106 "**just inside the door by the fireplace**": SS, one of two letters to DJ dated "Sunday", por.

"**5 ft. 2 1/2 inches**": SS's actual height, according to her passport, PGP.

wanted to take Johnston's play: SS to DJ, 28 June 1937, por.

"**to be in the way and be in the way**" and "**I do not make**": *MA*, p. 260.

"**You will have to come again and meet more Reds!**": NM, l to SS, dated "Tuesday," PGP.

"**I like Naomi but of course**": *MA*, p. 259.

[Footnote] "**to have a pro-Russian**"—carbon copy of SS, l to Pearl Binder, 14 Feb. 1951, PGP; "**relying on the pledge**"—SS, to Hermon Ould, 29 Jan. 1951, PEN.

107 **safer, much safer for mankind**": SS, "Mosaic", *Eve's Journal*, Apr. 1939, p. 106.

"**Well read in the political game**": *MA*, p. 105.

"**more talkie from Naomi**": *MA*, p. 259.

"**I can so much more be interested in Helen**": SS, "Mosaic", *Eve's Journal*, April 1939, p. 107.

"'**No, I am not interested to concentrate upon politics**'": *OTF*, p. 256.

"**all my rich friends are communists**": SS, "Mosaic", *Eve's Journal*, April 1939, p. 107.

Margaret Gardiner . . . had a son: Margaret Gardiner, int A.

the character "Tengal": "Not Believing", p. 26. See also *H*, pp. 85–86.

108 "**Stevie transformed herself**": Margaret Gardiner, int A.

"**She had asked whether she could stay**": Margaret Gardiner, "Louis MacNeice Remembered", *Quarto*, No. 6 (May 1980), p. 13. See also *I&S*, p. 84: "could she be sure of . . . a glass of hot milk at bedtime?"

"**very highly trained intelligences**": SS, "Mosaic", *Eve's Journal*, April 1939, p. 107.

"**mauled**" **Shakespeare** and "**speaks English like a parakeet**": see Paul Levy, "A Genius For Comedy", *O*, 14 June 1981.

"**Darling darling I love you so much**": SS's two monologues in typescript, and her l to Lydia Keynes, 20 May 1937, are at King's.

109 "**possibilities in Charles, but**": Lydia Keynes, l to SS, 2[?] June 1937, PGP. Gordon Square refers to the house Lydia and Maynard Keynes shared with Virginia Woolf's sister Vanessa and her husband, Clive Bell.

"**I was in Cambridge last weekend**": SS, fragment of l to Lydia Keynes, King's.

"**a family . . . at a garden party**" and "**The David Garnetts**": RM, 1 to SS, dated "June 27", PGP.

"**This is a painful letter to write**": David Garnett, 1 to SS, 21 June 1937, PGP.

110 "**I rather think she may have to drop Pompey**": Desmond Shawe-Taylor, "How Now, Pompey?", *NS*, 15 (22 Jan. 1938), 136 and 138.

and sent to Hamish Miles: the book is now in posssession of his son, Professor Hamish Miles. Hamish Miles left Cape for *The Times* late in 1937 and soon thereafter suffered a nervous breakdown: Malcolm Muggeridge, 1 to SS, 11 Nov. 1937, PGP.

[Footnote] "**you were so very kind to me**": SS, 1 to David Garnett, 1 June 1949, in possession of Richard Garnett.

111 "**Look out for choppy weather in January**": *MA*, p. 262.

"**Phoebus is getting hot-cattish**": SS, 1 to NM, 1937, Edinburgh.

"**I think it is an amusing situation**": *MA*, p. 261.

"**I am no martyr**": *MA*, p. 262.

"**Don't go running down the safe and ordinary**": *NYP*, p. 119.

112 "**Out of desperation over her lonely 'life in Death',**": HH, "Stevie Smith", *Neue Zürcher Zeitung*, 15 Dec. 1957, p. 4.

"**I may say I was shanghaied**": *OTF*, p. 267.

"**wars, oppressions and cruelties are**": SS, "New Novels", *WR*, NS 39 (May 1952), 66–69.

"**There is nothing worse**": *OTF*, p. 25.

"**It is horrible, I am so ashamed**" and "**was nothing to be proud of**": *MA*, pp. 287 and 285. In a copy of *OTF* SS presented to Roger Senhouse in 1970 (now in Tulsa) she wrote: "I do not *very* much like this one."

"**not always rationally comprehensible**": Tangye Lean, "H. G. Wells As Good As Ever", *News Chronicle*, 24 Jan. 1938, p. 4.

"**'Over the Frontier' is not . . . a novel**": "Miss Pompey Casmilus", *TLS*, 15 Jan. 1938, p. 43.

"**Every variety of groupismus**": Malcolm Muggeridge, "New Fiction", *DT*, 18 Jan. 1938, p. 6.

"**deliciously inventive**": Frank Swinnerton, "New Novels", *O*, 23 Jan. 1938, p. 6.

"**strong sensitive intelligence**" and "**fancy surface**": RL, copy of 1 to Ruth Atkinson, c/o Jonathan Cape Ltd., 4 Jan. 1938, PGP.

113 "**took criticism very badly**": RL, 1 to A, 11 Aug. 1981, por.

presumably at the lunch party: this paragraph is based on an entry of 29 Oct. 1936 in JH's engagement calendar, in possession of Diana Oakeley; RL, 1 to SS, dated "Saturday", PGP; and RL, 1 to A, 11 Aug. 1981, por.

"**I revelled in the first part**" and "**I felt it has worried *you***": (p. 114) RL, 1 to SS, dated "Jan 19th", PGP.

114 "**perhaps gone to her head**": RL, 1 to A, 11 Aug. 1981, por.

"**cross and snappish**" and "**About the baby talk**": RL, 1 to SS, dated "Thursday" [1938], PGP. Probably not long after she received this 1, SS mocked the phrase used by Miss Lehmann, "lets the side down", by having the speaker in her poem "Girls" declare: "I will let down the side if I get a chance."

fuelled the controversy by telling Stevie: RL, fragment of l to SS, PGP.

"When I got your subsequent letters": *MA*, pp. 265–66.

115 **fragile truce:** over two decades later SS said she was "not very fond" of RL: SS, l to HF, dated "Sunday" [1961?], por.

"brilliantly sustained" and **"a foolish mask":** Edwin Muir, "New Novels", *L*, 19 (19 Jan. 1938), 159.

"brilliant clowning": "New Novels: The Eternal Bourgeois", *Glasgow Herald*, 27 Jan. 1938, p. 4.

had read Kafka's *The Castle*: SS's reading diaries, Tulsa.

"curiously of Kafka": Edwin Muir, "New Novels", *L*, 19 (19 Jan. 1938), 159.

"shadowy world, part Kafka": Marie Scott-James, "News from Nowhere", *London Mercury*, 37 (Feb. 1938), 456.

"most comfortable as a guest" and **"the danger of compulsion":** HH, "Stevie Smith", *Neue Zürcher Zeitung*, 15 Dec. 1957, p. 4.

"understand the last third": Storm Jameson, copy of l to Miss Atkinson (of Jonathan Cape, Ltd), 6 Jan. 1938, PGP.

"the author tries to land too big a fish": Arnold Palmer, "Winter-Flowering Novelists", *Yorkshire Post*, 19 Jan. 1938, p. 6.

"There is another frontier faced": Victoria Glendinning, "Sturm in a teacup", *TLS*, 18 Jan. 1980, p. 54.

116 **"power and cruelty are":** *OTF*, p. 272.

"a mixed bag of brick-bats": *MA*, pp. 266–67.

"Alfred Lord Tennyson (Alfie to me)": H. Mayor, *NS*, 15 (28 May 1938), pp. 930 and 932.

Chapter Nine

117 **"To keep or not to keep out of war":** *MA*, p. 105.

"a favourite character throughout": Adele C. Friedman, "The Broadside Ballad Virago: Emancipated Women In British Working Class Literature", *Journal of Popular Culture*, 13 (Spring 1980), 469. Patriotism and love, or some blend of the two, motivated the military women in these ballads, according to Friedman, who attributes to Hyder Rollins the conclusion that the ballads represent a historical and psychological truth that departs from the conventional one of woman's role in society.

"now and then things get so frightful": *MA*, p. 266.

"Friendship . . . is the stuff of life" and **"I am so grateful to my darling friends":** *MA*, p. 105.

"I am glad . . . that my friends don't": SS, "In My Dreams", *CP*, p. 129.

"rather off the handle quite nice": *MA*, p. 267.

"so amusing, and not groupy at all": *MA*, p. 271.

118 **"academic gairls":** KF, l to A, 15 Dec. 1982, por.

based on Inez: Francis Wyndham, int A.

"a forehead fringe" and **"It was as if I had":** Inez Holden, *It Was Different at the Time* (London: John Lane, The Bodley Head, 1943), pp. 9 and 10.

"a fringe awry that is": *NYP*, p. 74.

Inez was a friend: our account in this and the following paragraph is based on our int Barbara Clutton-Brock, except for SS's wry attitude toward the cats, for which see *MA*, p. 313.

119 **"Private Views":** *MA*, pp. 130–33.

"Making her own" and **"My friend is *still*":** AO, int A.

"jerky jabs at the attention": Ivor Brown, "Proper Words and Proper Places", *Manchester Guardian*, 14 May 1938, p. 9.

"idiot" and **"if he is out to find":** Osbert Sitwell, l to SS, 22 May 1938, PGP.

"The pressure of money is": SS, "Mosaic", *Eve's Journal*, Jan. 1939, p. 112.

admired her writings and published: see SS's 1937 comment about R. A. Scott-James, "he really is very nice indeed" (*MA*, p. 262).

Armide Oppé, who became: AO, int A.

120 **"The problem now is how can I say":** SS, l to RHD, 27 Aug. 1937, por.

"Reviewing is a difficult ring": SS, copy of undated l to Reginald Clarke, PGP.

"may prove too much a shadow": SS, "These Are These", *London Mercury*, 37 (April 1938), 654.

pleased Osbert Sitwell: Osbert Sitwell, l to SS, 22 April 1938, PGP.

Less responsive in 1938: this paragraph is based on *MA*, pp. 270–71.

121 **"Why put cups on the table":** *MA*, pp. 267–68.

"the wraith of Edith [Sitwell]": SS, l to DJ, 4 July 1938, por.

efforts to discourage: *MA*, p. 269.

Stevie wrote in September . . . "Horrible times": *MA*, pp. 271–72.

"my voice is awful": *MA*, p. 272.

"My poems will be out": SS, l to DJ, 4 July 1938, por.

122 **"a perpetual hollow . . . the misunderstandings":** RL, l to SS, 6 Nov. 1938, PGP.

asking that he delete: SS, l to RHD, 4 Aug. 1938, por. The other poems were "Salon d' Automne" and "Henry Wilberforce", which are now published in *MA*, pp. 240 and 218; and "Bed" and "Father Damien Doshing", which are now published in Tracy Warr's "Unpublished Stevie Smith", *Poetry Review*, 74 (Sept. 1984), 23–24.

"I think the selection is the best": RL, l to SS, 24 Nov. 1938, PGP.

123 **"all the romantic, self-deluding":** V.H.F., *Country Life*, 85 (1 April 1939), 334.

"when it comes to poetry": RL, l to SS, 24 Nov. 1938, PGP. See also *MA*, p. 266.

"exploits a distraught" and **"She writes alternately"**: George Stonier, "Five Poets, Five Worlds", *NS*, 16 (3 Dec. 1938), 930.

"How I let it get in" and **"blue for bad"**: *MA*, p. 266.

Brown's piece: Ivor Brown, "Proper Words and Proper Places", *Manchester Guardian*, 14 May 1938, p. 9.

"political" poem: see the script for SS's BBC Radio Third Programme broadcast of 21 July 1952, "Poems and Drawing III", BBCa; and SS's introduction to "O Happy Dogs of England" on a sheet of poem introductions, Tulsa.

"well up in Church history" and **"could correct"**: SC, int A.

124 **"Why doesn't the Church"**: SS, "Mosaic", *Eve's Journal*, April 1939, p. 105.

"Mr.—.": SS, "Mosaic", *Eve's Journal*, May 1939, p. 106.

"and I don't know that I": SS, l to DJ, 26 Oct. 1938, por.

125 **"We are not innocent"** and **"I wish for innocence"**: *H*, pp. 103 and 143.

126 **"There was pity"**: *NYP*, p. 251.

"What of the heart of Pompey": *OTF*, p. 18.

127 **"a desperate character"**: *NYP*, p. 193.

"very depressed and deathly": *MA*, p. 272.

she joined PEN: SS, l to Storm Jameson, 5 Jan. 1939, Austin.

German measles: Hermon Ould, l to SS, 9 Feb. 1939, Austin.

disliking the "noise" and an article she published: *MA*, pp. 105–7.

128 **became a fire-watcher and . . . met Norah Smallwood:** Norah Smallwood, int A.

renting a bedsitter, "an atmosphere of security" and **she took her first novel:** Isobel English [June Braybrooke], Introduction to Olivia Manning's *The Doves of Venus*, (London: Virago, 1984), p. xii.

"my dearest Stevie": Olivia Manning, l to SS, 8 Sept. 1939, PGP.

for whom she grieved: Francis King, "Olivia Manning: 1915–1980", *S*, 245 (2 Aug. 1980), 21.

"the people I like best": Olivia Manning, l to SS, 11 Oct. 1939, PGP.

129 **a miscarriage:** Francis King, "Olivia Manning" (cited above).

sent some short stories . . . but he rejected: SS, l to John Lehmann, 12 July 1939; and John Lehmann, l to SS, 22 Aug. 1939, Austin.

"Surrounded by Children": *MA*, pp. 26–27.

"Storm back through the gates": SS, "A Dream of Comparison," *CP*, p. 314.

"I don't want to see your back": Racy Buxton, int A.

"the impishness which at one party": Jacynth Ellerton [Lady Lawrence], "Stevie Smith", *The Tablet*, 10 April 1971, p. 360.

130 **"She would *demand* company"**: EL, *A Goldfish Bowl* (London: Cassell, 1972), p. 313.

"man in the poem is meant": SS, l to TK, 12 May 1966, files of O.

"The Herriots": *MA*, pp. 74–79.

review of **Llewelyn Powys's** *Love and Death*: SS, "Reviews of Books", *Life & Letters Today*, 22 (July 1939), 138–39.

"a horrible new world": *MA*, p. 273.

131 **a visit to Cambridge where:** entry of 17 Aug. 1940, in engagement calendar of JH, in possession of Diana Oakeley.

"Stevie began to have a look round": Norah Smallwood, int A.

typescript of her third novel: in possession of Janet Watts when we examined it, now at Tulsa.

"He had never *been in a raid"* **and "Yesterday morning":***MA*, p. 278–79.

132 **"I often have tea with Inez":** *MA*, p. 279.

Chapter Ten

133 **"How long will the war last":** typescript of *H*, Tulsa.

"stunning": *MA*, p. 283.

"If you'll excuse me": Osbert Sitwell, l to SS, 17 Feb. 1938, PGP.

"I've known and loved your poems": John Gabriel, 1 to SS, 31 Dec. 1940, PGP.

"very much" and **"grave-yard note of mine":** MA, p. 277.

134 **"missing believed drowned.":** Clothilde Gabriel, l to SS, 8 April [1941], PGP.

"not a very popular author" and **"the way that war":** SS, l to Clothilde Gabriel, 18 April 1941, George Orwell Archive, University College, London.

reviewed for *Aeronautics*: SS, "Finer Fiction", *Aeronautics*, April 1941, p. 74.

"penurious author" and **"I wouldn't really be out of London":** *MA*, pp. 275–6.

PEN World Congress lunch: Bernard Crick, *George Orwell: A Life* (London: Secker & Warburg, 1980), p. 395.

"cities and races" and **"a stranger reading Miss Brittain's":** *MA*, pp. 176–77.

"that is not steeped in blood": SS, "An Amusing Thriller", *JO'L*, 43 (14 June 1940), 319.

"British lion getting really off the ground": *MA*, p. 177.

"Our times have been upon the rack of war": *OTF*, p. 94.

though the Nazis were hell: *I&S*, p. 74.

135 **"keep your anger to yourself":** SS, "For and on Behalf of Miss Jones", *T*, 23 April 1943, p. 19.

"those in close immediate touch": *OTF*, pp. 94–95.

"family Christmas", "if I had it always to do", and **"never very certain":** *MA*, p. 280.

"the *New Statesman* **has been wonderful":** *MA*, p. 276.

"there isn't an earthly chance": RHD, l to SS, 19 March 1940, PGP.

"We should be sorry to see": JC, l to SS, 14 Feb. 1941, PGP.

136 **"Will you get"** and **sales figures:** ibid

paper covers and **"a larger public"**: JC, l to SS, 29 April 1942, PGP.
"so many drawings" and **art department**: *MA*, pp. 280–81.
choose 80 out of 200, **"I wish I wasn't cursed"**, **C. V. Wedgwood . . . compounding her problem** and **"The office looks like a paper chase"**: SS, l to JH, 2 Jan. 1942, King's.
"obscene poem", **"I take your opinion"**, **"London is looking perfect"** and **"I can't stand 'em"**: *MA*, p. 283.

137 **Hayward dissuaded her** and **Rothschild thanked her:** JH, l to SS, 26 March 1942, PGP; Lord Rothschild, l to SS, 18 May 1942, PGP.
"was very impressed by the sherry": SS, l to JH, 19 Aug. 1942, King's.
wrote to Aunt from the Lizard: SS, pc to MAS, 2 July 1942, Tulsa.
Holden and . . . Gollancz and **"it pours out":** SS, l to JH, 19 Aug. 1942, King's.
Stevie's story "In the Beginning of the War": *MA*, pp. 28–30.
"both Got Out in Time" and **"Olivia doesn't mean to be patronising":** *MA*, pp. 282–83.
"Stevie survived on her friends": Mrs Cleverdon, int A.

138 **"the most persistent liar":** *MA*, p. 284.
other links: Rosemary Cooper and Barbara Clutton-Brock, int A.
used to fire-watch with Orwell: Wallace Finkel, int A.
Orwell claimed that his secretary: Eric Blair [George Orwell], copy of l to SS, 17 Oct. 1942, BBCa.
"you never gave me a date": *MA*, p. 284.
"sick-man fancy of a pool": SS, "Books", *Other Voices*, 28 Jan. 1955, n.p. See also SS's passing remark, "There is no Orwellian hankering in this book for extreme future cruelty" ("New Novels", *O*, 28 Nov. 1954, p. 11). And see her even earlier remark, "does not the author [of *Nineteen Eighty-Four*] seem to believe that Evil is stronger than Good? And almost to enjoy that belief?" ("Stevie Smith's Guide to Good Reading", *MW*, Sept. 1949, p. 127.)

139 **her attitude as romantic:** KF and Lettice Cooper, int A.
Norah Smallwood insisted: Norah Smallwood, int A.
repeats an anecdote: Bernard Crick, *George Orwell: A Life* (London: Secker & Warburg, 1980), p. 423.
Kay Dick . . . derides such speculation: KD, int A.
"Orwell's leg-pulls": SC, l to A, May 1984.
"knew George and his first wife" and **"all this comes into *The Holiday*"** (p. 140): *MA*, p. 315.

140 **"is rather a fool . . . girls can't play":** *H*, p. 145.
Tom Fox, and these are the two: C. V. Wedgwood, int A.
"Basil knelt down": *H*, pp. 67–68.
"that rich nursery of English talent": SS, "New Novels", *WR*, NS 45 (Nov. 1952), 70.
"Basil and [his] chums are": *H*, pp. 104–5.

141 **praised exactly the middle-class virtue:** SS, "Plomer's Own Good", *T*, 17 Dec. 1943, p. 18.

an actress friend: Sheila Rayner, int A.

Tom becomes mute: *H*, p. 35.

"went very deep": C. V. Wedgwood, int A.

"a friend of mine": *MA*, p. 285.

"I have read the poems": C. V. Wedgwood, l to SS, 28 Sept. 1942, PGP.

whose title comes: SS, l to HH, 17 Oct. 1957, HHP.

142 **"a feat of modern letters":** E. C. Bentley, "Books of the Day", *DT*, 1 Jan. 1943, p. 3.

"the jungle theme": H. E. Degras, *International Women's News*, 37 (Jan. 1943), 62.

143 **George Stonier said:** George Stonier, "Fun and Poetry", *NS*, 24 (12 Dec. 1942), 393.

"they are an integral part of the verse": Mark Benney, "The Mocking Muse". We found this cutting among the PGP but have been unable to identify where it was published.

144 **when she met Sally Chilver:** this paragraph is based on SC, int A.

Helen Fowler thinks and **Elisabeth Lutyens recalled:** HF and EL, int A.

met . . . Kay Dick: this paragraph is based on *I&S*, pp. 81 and 83.

Stevie often was lonely: this paragraph is based on *MA*, pp. 285–86.

145 **"disturbed by the dismay":** JH, l to SS, 3 Sept. 1943, PGP.

"Maugham's fault of pride": "A Choice of Kipling's Prose", *WR*, *NS* 47 (Jan. 1953), 66–68.

"I feel an absolute 'case',": SS, l to JH, 10 Sept. 1943, King's.

146 **"a perfectly awful operation":** SS, l to JH, 10 Sept. 1943, King's.

account of the horrors: *H*, pp. 58–59.

the staff of Westonbirt: K. M. Warburton, obituary of "Miss Ward-Smith", p. 5 from an old Westonbirt School Magazine in the files of Margaret Evett, the School Secretary.

"so happy at her school now": *H*, pp. 75–76.

a bay wreath and **"ineffectual people"** (p. 147): SS, l to RM, 28 Feb. 1944, RMP.

147 **received a novel by Friedl Benedict:** C. V. Wedgwood, l to SS, 10 Feb. 1944, PGP.

"like two slaphappy clowns": *I&S*, p. 84.

"look well, a bit tired maybe" and **"They talked more easily":** SS, "Helen Comes to Town", *MW*, July 1944, pp. 46–47.

"the war within": a phrase from the poetry of Gerard Manley Hopkins.

"condemned" her feelings of loneliness: *MA*, p. 115.

"disused railway track" and **"a she-ass licks":** script for BBC radio Third Programme broadcast of 18 Oct. 1951, "Poems and Drawings II", BBCa.

148 **"Plum self-confident":** SS, "Helen Comes to Town", *MW*, July 1944, p. 47.

Chapter Eleven

149 **"won't visit for breakfast":** KF, l to SS, 6 April 1945, PGP.
suffering from cancer: Gertrude Wirth, int A.
"two pints of milk a day": Daniel T. Da[nly?], l to SS, 12 June 1945,
PGP.
organic lesions: Dr L. Leventon, l to SS, 18 June 1945, PGP.
off to Scotland: DD, int A.
"This ideal of happiness in work": SS, "Women at Work", *T*, 23
Sept. 1949, pp. 20–21.

150 **met Stevie for lunch:** this paragraph is based on T. R. Fyvel, l to A, 15
Aug. 1981.
an undated letter to Kay Dick: *MA*, pp. 287–88.

151 **"Blake was full":** SS, "Blake's Vision", *T*, 26 Dec. 1947, p. 14.
"an unbeliever with a religious temperament": "Not Believing",
p. 23.
"Life and Letters": this paragraph is based on five letters of Reginald
Moore to SS which are part of the PGP. They are dated 8 May 1945, 15
June 1945, 1 July 1945, 26 July 1945 and 24 Feb. 1947.
"A Very Pleasant Evening": *MA*, pp. 31–34.

152 **"a child's ruthless absence":** Desmond MacCarthy, "Mirrors-I", *The
Sunday Times*, 27 April 1947, p. 3.
remembers her at a wartime party: Jane Stockwood, int A.
"The Story of a Story": *MA*, pp. 50–59.
wrote a key: Margery Hemming's copy is now in the possession of
Judith Hemming.
"Sunday at Home": *MA*, pp. 44–49.
originally titled "Enemy Action": see letters to and from SS, Dec.
1948–April 1949, BBCa, especially SS, l to H. N. Bentinck, 16 Feb.
1949.
accepted in early 1946: John Singer, l to SS, 12 April 1946, PGP.
he returned it: John Singer, l to SS, 15 May 1946.
kept trying: Mary Louise Aswell (Fiction Editor of *Harper's Bazaar*), l
to SS, 3 June 1946, PGP; Cyril Connolly (Editor of *Horizon*), l to SS, 1
April 1947, PGP; and Denys Kilham Roberts (Editor of *Orion*), l to SS,
31 March 1948, PGP.
describing the changes she had made: *MA*, pp. 288–89.
he did sensibly repair to cupboards: Judith Hemming, int A.
"a plain simple symbol of betrayal": *MA*, p. 290.

153 **"storm in a tea-cup":** *MA*, p. 288.
"personal war" and "You want it both ways": *MA*, pp. 51 and 53.
"a strong communication": *MA*, p. 126.
Other literary projects: H. van Thal, letters to SS, 13 May 1946 and 16
May 1949, PGP; and SS, copy of l to H. van Thal, 17 May 1949, PGP.
met Antoinette Watney and "Taking Stevie Home": Antoinette
Watney, int A.
complained Elisabeth Lutyens: EL, int A.

Helen Fowler sometimes resented: HF, int A.

back in touch: SS, l to HF, 7 Aug. 1947.

"I understood only too well": Kenneth Barrow, *Flora* (London: Heinemann, 1981), p. 164.

154 "which the BBC has had": SS, l to HF, 7 Aug. 1947.

"Is There a Life Beyond the Gravy?": *MA*, pp. 60–73.

"high and successful note": "Realism and Fantasy", *TLS*, 15 Nov. 1947, p. 589.

Stevie told Kay Dick that "I should never": SS, l to KD, 2 Oct. 1945, St L.

according to Sally Chilver: SC, l to A, 25 Aug. 1980, por.

"Had he not written" and "We're all dead": *MA*, pp. 62 and 73.

"and those of riper years": SS, "Short Stories of All Kinds", *JO'L*, 56 (28 Nov. 1947), 714.

155 "This was a fascinating exercise": *MA*, pp. 289–90.

reminiscent of Yeats: Ronald Orr-Ewing, int A.

"other people will get the right rhythm": Orr, pp. 226–27.

neglected books in favour of music: DD, int A.

music doesn't interest intellectuals: Rosemary Cooper, int A.

Neither Helen Fowler nor Suzannah Jacobson: HF and SJ, int A.

"Stevie wasn't interested in the other arts": Sir John Lawrence, int A.

professional actresses did: Rosalinde Fuller, telephone int A. See also SS's comments in interviews: "I don't really want other people to read them" (Orr, p. 226); and, in response to a question about how she feels when other people read her poems, "I don't like it at all especially when they're rather good actresses" (script of BBC Woman's Hour radio broadcast of 16 Feb. 1970, "Poet Talking"—Tulsa and BBCa).

"Last night some of [my poems]" and "Does this suggest" (p. 156): *MA*, p. 291.

156 Elisabeth Lutyens set ten and bickering with Sally: EL, int A.

Jonathan Miller learned in his twenties: this paragraph is based on Jonathan Miller, telephone int A.

"Beside the Seaside": *MA*, pp. 13–25. As the story is short and we quote it often, we do not cite page numbers for the quotations.

157 Sally Chilver remembers a conversation: SC, l to A, 27 Dec. 1983, por.

"the people in these stories": SS, "In Roman Tenements", *DT*, 16 Nov. 1956, p. 11.

158 "When I am with people": *H*, p. 7.

Betty Miller who wrote to say: Betty Miller, undated l to SS, PGP.

159 Olivia Manning rang Francis Wyndham: All of the memories of Francis Wyndham in this chapter are from his int A.

a "spiky" one: HF, KF, and Neville and June Braybrooke, int A.

160 "I am uncertain of his surname": KF, l to A, 13 Dec. 1982, por.

"I have an absolute loathing of death": Francis King, "Olivia Manning: 1915–1980", *S*, 245 (2 Aug. 1980), 21.

"I have none of Stevie's sentimental": KD, *Friends & Friendship* (London: Sidgwick & Jackson, 1974), p. 44.

In 1947 she sent: SS, l to KD, 31 July 1947, St L.

"Dear Olivia says": *MA*, p. 295.

"I am sure they loved each other": KF, l to A, 15 Dec. 1982, por.

a review of *The Doves of Venus*: SS, "New Novels: Youth and Age", O, 23 Oct. 1955, p. 12.

deeply offended: Neville and June Braybrooke, int A.

"much of Stevie in her" and **"obvious attempts at disguise"**: Isobel English [June Braybrooke], Introduction to Olivia Manning's *The Doves of Venus*, (London: Virago, 1984), pp. xi–xii.

"the world's most excruciating bore": ibid, p. 158.

161 **"an odd love/hate relationship"** and **"Once at a Hampstead party"**: KF, l to A, 13 Dec. 1982, por.

Francis King did not know: this paragraph is based on Francis King, l to A, 11 March 1985, por.

"A funny thing happened": Olivia Manning, l to KF, 16 Nov. 1971, por.

"No, I have not seen 'Stevie'": Olivia Manning, l to KF, 27 March 1977, por.

162 **Oliva's animosities were legendary**: KD, int A.

"feared below": not an exact quote but an echo of SS, "Dirge", *CP*, p. 186.

Chapter Twelve

163 **"a great woodland"** and **"The most beautiful"**: *MA*, pp. 101–2.

164 **20 February, 1949**: death certificate of Charles Ward Smith, GRO.

pre-deceased him and **turned up in Avondale Road**: MS to HF, HFP.

administrators of his will: Manager of Midland Bank Ltd, l to MS, 13 Feb. 1950, PGP.

£470 1s: will of Charles Ward Smith, records at Somerset House, London.

did not attend her father's funeral: MS to HF, HFP.

more suited to the eye: Ronald Lewin, copy of l to SS, 15 April 1947, BBCa.

rejected that story for the identical reason: Ronald Lewin, copy of l to SS, 8 April 1948, BBCa.

"a bit above the heads" and **might care to enter**: Rayner Heppenstall, l to SS, 9 Dec. 1948, PGP.

"to encourage writers": SS Author's Note with description of Third Programme Short Story Competition, BBCa.

"came to the conclusion that it was hopeless": D. F. Boyd, memo to Mr Bentinck, 3 March 1949, BBCa.

165 **"allow me to say how *enormously*"** and **"improved 100%"**: H. N. Bentinck, copy of l to SS, 14 March 1949, PGP.

"even more nervous and shy": AK, memo to Chief Producer, 29 March 1949, BBCa.

paid twenty guineas: E. M. Layton, l to SS, 18 March 1949, PGP; and BBC Contracts Department confirmation sheet, 18 March 1949, PGP.

"I think radio may, in the end, intrigue": AK, l to SS, 3 May 1949, PGP.

Glenda Jackson reported: Glenda Jackson, int A.

"One can argue interminably": Douglas Cleverdon, "Stevie Smith Recollected", undated script of a BBC radio tribute [May 1971], BBCa.

Anna and Stevie met: this paragraph is based on AB, int A, and AB, l to A, 23 May [1980], por.

"talking so hard": AB, l to A, 23 May [1980], por.

166 **the Lawrences estimate:** George and Olga Lawrence, int A.

"I have never myself bought a volume": SS, copy of l to JC, 17 Feb. 1948, PGP.

had been negotiating with Peter Nevill: Neville Armstrong, l to SS, 7 Oct. 1948, PGP.

"I wrote it during the war": *I&S*, pp. 73–74.

"I have not been happy about": SS, copy of l to Leo Kahn, 13 Oct. 1948, PGP.

"cannot be said that it is war", "I do not know that we can bear" and **"crisis and exile":** *H*, pp. 13, 8 and 9.

"drawn toward the Grecian" and **"Persephone sub-theme":** Boulton.

"knew how brief that beauty was": Edith Hamilton, *Mythology* (New York: Mentor, 1942), p. 54.

167 **"She was stolen away by Pluto":** SS, script titled, "Books, Plays, Poems: Poems by Living Poets", broadcast on the BBC radio Home Services (Schools), 15 June 1966, Tulsa.

Armide Oppé remains surprised: AO, int A.

168 **"Stevie Smith's Victorianism":** Hermione Lee, "Fits & Splinters", *NS*, 4 May 1979, p. 652.

"There's something of the Mrs Humphry Ward in me": *NYP*, p. 30.

"the conflicts commonly endured by Victorian": Hilary Spurling, *Ivy When Young* (London: Hodder and Stoughton, 1981), p. 78.

"Oh what a pity it is": SS, "Mr and Mrs Browning", *T*, 28 Nov. 1947, p. 21.

169 **we have come further than the apes:** *H*, pp. 64–65.

"I was oppressed by such a sense of melancholy": *H*, p. 102.

"gives a clear, bitter picture": Hermione Lee, "Fits & Splinters", *NS*, 4 May 1979, pp. 652–53; and SC, undated letter to SS, PGP.

"I am a middle-class girl": *H*, p. 92.

"I least know how to deal with [it]": P. H. Newby, "New Novels", *L*, 42 (11 Aug. 1949), 244.

170 **Joan Robinson once sent:** Joan Robinson, l to A, 8 Nov. 1980, por.

"pick-rag mind", "long summer days" and "that brings with it
so much glee": *H*, pp. 117, 122 and 124.
a long composite review: Inez Holden, "Some Women Writers",
Nineteenth Century and After, 146 (August 1949), 130–36.

171 "God knows why I should be as the old thing": SS, l to HH, 6 April
1959, HHP.
"nervy, bold and grim" and "all his Elks": *H*, pp. 191–92.
"this humane person": *H*, pp. 191–2.
"It's wickedly clever of you": SS, l to HH, 17 Dec. 1957, HHP.
"And how persistent are your sallies": SS reading notebooks, Tulsa:
H, pp. 192–93; F. Dostoevsky, *Notes from Underground*, trans. Constance
Garnett (New York: Heritage Press, 1967), p. 56.

172 "a piece of astringency", "I detest your Christianity" and "says
that human beings are our concern": *H*, pp. 193, 199 and 202.
"Some of us had begun to think": R. A. Scott-James, "The Return of
Stevie Smith", *Britain Today*, Oct. 1949, p. 46.
"most mature work": Richard Church, "Probes of the Spirit", *JO'L*,
58 (22 July 1949), 457.
"extraordinarily of our time": Howard Spring, "New Books",
Country Life, 106 (8 July 1949), 136–37.
"We are vacillating": *H*, p. 90.

173 "no longer quite current": Frances Phillips, l to SS, 4 Aug. 1949,
PGP.
"parts of the book immensely": Blanche Knopf, l to SS, 10 Nov.
1949, PGP.
"it is sold out": SS, l to HF, 24 Oct. 1949, por.
"I like *The Holiday* very much better": *I&S*, pp. 73–74.
"We shall most certainly respect": Alexander Greu[nts?], l to SS, 27
June 1949, PGP.
"The train of death": *H*, p. 155.
"three heavenly weeks" and "all windmills": SS, l to HF, 24 Oct.
1949, por.

174 her reading notes: Tulsa.
"When I am happy I live and despise writing": SS, "My Muse",
CP, p. 405.

Chapter Thirteen

175 sometimes visit . . . Mary Lee Settle: quotations and memories of
Mary Lee Settle are from her int A.
had appeared in a contemporary revue: HF, int A.

176 come to the Lansdowne Club: Rose Macaulay, l to SS, 26 Oct. 1951,
PGP.
one held in January 1950: SC, int A. Also see SS, l to HF, 20 July 1951,
por, for references to HF and James MacGibbon being at the party. See
entry of 6 Jan. 1950 in JH's engagement calendar (in possession of his
sister, Diana Oakeley), for his presence at the party.

"So enormous was this party": quotations and memories of Dr Hill are from her letters to A, 18 Feb. 1984 and 14 Feb. 1986, por.
the firm was "coming up": SS, l to JH, 6 Feb. 1950, King's.

177 **that autumn:** Chapman & Hall press release of Oct. 1950, PGP.
"probably already ceased": SS, l to Hermon Ould, 13 Dec. 1950, PEN.
"kindness", "the rain falls" and **"fallen prey"**: SS, l to Daniel George, 9 Feb. 1951, Hull.
"*perhaps* simplified": Boulton.
"to shift [my] feelings outwards: "Too Tired."
Elsewhere she amplified: undated script at Tulsa which seems to be for a school poetry reading.
Stevie explained, she was criticizing: text of BBC broadcast for schools, 8 June 1966, PGP.
or tiredness: "Too Tired".

178 **According to Stevie:** "Too Tired".
"a wicked poet who": SS, "The Living Poet", Third Programme BBC radio broadcast of 4 Oct. 1963, script at BBCa.
a friend's copy: Dr Polly Hill's copy, in her possession.
In a similar vein she once commented: SS, "Poems and Drawings III", Third Programme BBC radio broadcast of 21 July 1952. Script at BBCa. Also see *MA*, pp. 113–14.
The poem's last line: The words "in death", which appear in the last line of "Do Not!" in *our* text are not present in the version of the poem in *CP*. SS added the clarifying words "in death" to the last line of the poem in KF's copy of *Harold's Leap*, in the possession of KF.

179 **a piece she wrote in 1949:** SS, "Women at Work", *T*, 23 Sept. 1949, p. 20.
She once prefaced: "Too Tired".
"The feeling of wanting to get away": SS, "Poems and Drawings II", Third Programme BBC radio broadcast of 18 Oct. 1951. Script at BBCa.

180 **"the pages continued"** and **"Stevie liked to travel out"**: Nigel Dennis, l to A, 20 Dec. 1980, por.
"marriage is not always a solution": SS, "Poems and Drawings III", cited above.
"There are a great many hats": "Too Tired".

181 **"toys with the thought"**: SS, "Poems and Drawings I", Third Programme BBC radio broadcast of 28 March 1951. Script at BBCa.

182 **"ticklish comic element"**: Boulton.
Stevie cited the commentary: Boulton.
"I am sure something must have": SS, copy of l to Minina Mesquita (of Jonathan Cape Ltd), 29 Jan. 1948, PGP.
The firm tried and **Stevie wrote:** SS, copy of l to Messrs Jonathan Cape Ltd, 10 Feb. 1948, and copy of l to Minina Mesquita, 11 Feb. 1948, PGP.
"so long as the books are with us": JC, l to SS, 16 Feb. 1948, PGP.

thanked Cape: SS, copy of l to JC, 17 Feb. 1948, PGP.

[Footnote] **ten guineas:** SS, l to Niouta [AK], 30 Jan. 1952, BBCa; **over £60:** E. M. Layton, memo to AK, 24 Oct. 1951, BBCa; **"Financially this is charming":** SS, l to RM, 29 Oct. 1951, RMP.

183 **"I would like to call it a happy letter":** JC, l to SS, 25 July 1950, PGP.

"invited Stevie to contact her": AK, l to SS, 3 May 1949, PGP.

By November of 1950: SS, l to Niouta [AK], 20 Nov. 1950, BBCa.

"changed the order round": SS, l to Niouta [AK], 2 Jan. 1951, BBCa.

a lot of thought and time: SS, l to AK, 30 Jan. 1952, BBCa.

"it is not easy": SS, l to AK, 3 April 1952, BBCa.

"readings à thèse,": SS, l to AK, 3 Dec. 1952, BBCa.

"Hazards and temptations": SS, "Poems and Drawings III", cited above.

"it's like mixing a cake": SS, l to AK, 10 Sept. 1951, BBCa.

184 **"batted away":** SS, l to RM, 29 Oct. 1951, RMP.

"an angel though I think": SS, l to AK, 19 Oct. 1951, BBCa.

"nerve-racking": SS, pc to KD, 23 July 1952, St L.

In the margin of the letter: SS, l to AK, 16 Sept. 1952, BBCa.

In the past, the composers: SS, l to AK, 29 Sept. 1952, BBCa.

"You say I have had": SS, l to AK, 3 Dec. 1952, BBCa.

185 **"You really are a comfort", "I did ring"** and **"they always say":** SS, l to KD, 21 May 1953, St L.

"horrifying figures": Jack McDougall, l to SS, 2 Feb. 1953, PGP.

"languid & triste" and **"There's no two ways":** SS, l to KD, 12 Nov. 1953, St L.

"I now address": SS, undated l to NM, Edinburgh. Our dating is by internal evidence. Rpr. in *You May Well Ask*, p. 157, but our version is according to the original.

186 **"It is shockingly sad":** SS, l to D.J. Enright, 27 May 1953, por.

only listened to her Muse when: SS, "My Muse", *CP*, p. 405.

in the *Observer*: "Not Waving but Drowning" was never published in *Punch*. It appeared in *O*, 29 Aug. 1954, p. 8.

"too low for words," and ***Punch* like it":** *MA*, p. 294.

"I thought that in a way": "Schools Broadcast" notes, PGP. This is probably a draft of her BBC broadcast of 8 June 1966.

[Footnote] **among her papers:** PGP.

187 **a painful knee condition:** SS, letters to KD, 22 Jan., 22 and 26 Feb., and 23 Sept. 1953, St L.

just won unaided a battle: SS, l to David Carver, 22 May 1953, PEN.

"if my dear baronets can loose me": SS, l to HF, 30 May 1952, por.

Stevie had to cancel: *MA*, p. 294.

188 **reading notebooks:** Tulsa. See also *NYP*, p. 42.

"I always said": *I&S*, pp. 63–64.

slashing her wrists: SC and AB, int A.

wrote to Anna Browne and **"all those women had to go to work":** AB, int A.

"It's that old long long office job": SS, l to HH, 15 Aug. 1958, HHP.

In another letter: SS, l to HH, 17 Oct. 1957, HHP.

189 **"I am a Nervous Wreck":** *MA*, pp. 295–96.

"general debility and anaemia,": L. M. Leventon, LRCP, LRCS, l to Sir Neville Pearson, Bart., 8 Aug. 1953, PGP.

"I . . . am so glad": SS, undated pc to RM, RMP.

went on holiday: hotel bill for 12–26 Aug. 1953, with jottings on back about MS's share, PGP.

190 **"reasonably comfortable,":** Sir Neville Pearson, l to SS, 26 Aug. 1953, PGP.

"If I could afford" and **the difficulty in singling out:** SS, l to KD, c. Sept. 1953, St L.

repetitive thoughts: SS, l to KD, 26 Oct. 1953, St L.

"oh if only that was all": *MA*, p. 295.

"It is what had to be written": SS, l to HF, 30 May 1952, por.

share this enthusiasm: SS, "New Novels", *WR*, NS 45 (Nov. 1952), 66.

a good neglected poet: Andrew Feiling, "A Second Course of Modern Verse", *Time and Tide*, 34 (24 Oct. 1953), 1398.

slip lines from her poems: See, for example, her reviews in *WR* of Oct. 1952 and Jan. 1953, for quotations from "Happiness" and "The Lads of the Village". See her reviews in *O* of 24 April 1955, 19 Feb. 1956, and 7 Oct. 1956, for quotations from "Le Revenant", "A Dream of Comparison", and "Will Man Ever Face Fact and not Feel Flat". These quotations appear without author or title citation.

many poems came to her: *MA*, p. 129.

"Perhaps England": SS, "New Novels", *O*, 5 Sept. 1954, p. 11.

"I would not call": SS, "Family Affair", *Books of the Month*, 72 (March 1957), 9.

191 **"Miss De Beauvoir":** "The Devil's Doorway", *S*, 191 (20 Nov. 1953), 602.

"Those who eschew": SS, "A Novel of Feeling", *O*, 2 Feb. 1958, p. 15.

"Satirists must never": SS, "New Novels", *WR*, NS 36 (Feb. 1952), p. 79.

"gives a perfect impression": SS, "New Novels", *O*, 7 March 1954, p. 9.

staying with the Brownes: our account of this day and of "The Old Sweet Dove . . ." is based on AB, int A.

Chapter Fourteen

192 **Chapman & Hall decided:** Jack [McDougall], l to SS, 2 Feb. 1953, PGP.

She first approached: SS, l to KF, 4 March 1953, in possession of the London booksellers, Bertram Rota.

"But then so often": *MA*, p. 298.

"**your presence and voice to recite**": Mervyn Horder, l to SS, 4 Jan. 1955, PGP.

"**the keenest pleasure**": L. P. Hartley, l to SS, 3 Jan. 1951, PGP.

"**I would love you to see**": *MA*, pp. 299–300.

193 "**writing today**", "**brittle laughter**" and "*à la*": *MA*, p. 300.

193 **Dylan Thomas once complained**: EL, int A.

"**actually finding it impossible**": Jonathan Barker, "David Wright In Conversation", *PN Review*, 6 (1980), 51.

his request for drawings: David Wright, undated l to SS (accepting 13 poems for *Nimbus*), PGP.

"**They all move me**": L. P. Hartley, l to SS, 11 Feb. 1955, PGP.

not shared by the firm of André Deutsch: SS, l to DA, 27 April 1955, Deutsch.

"**brought about by having**": SS, l to DA, 16 May 1955, Deutsch.

"**They have a Lear-like**": L. P. Hartley, l to SS, 28 Aug. 1955, PGP.

194 **sent Stevie pulls** and "**oh dear oh dear**": SS, l to DA, 22 Nov. 1955, Deutsch.

"**I absolutely am riveted to them,**": *MA*, p. 301.

"**brought into VG**" and "**in working out**": Hilary Rubinstein, l to SS, 21 Dec. 1955, PGP.

Diana Athill returned Stevie's advance: SS, l to DA, 9 Jan. 1956, Deutsch.

"**I am not a trained drawer**": Orr, p. 229.

To a friend she wrote of her weakness: SS, l to KF, 4 March 1953, in possession of the London booksellers, Bertram Rota.

although on occasion: When *Punch* asked for a drawing to go with "The Celts" SS did fifty (she said) to give them a choice (SS, l to KF, 4 March 1953, in possession of Bertram Rota); and she once remarked that looking through her drawings inspired her to write more poems (Williams, p. 113).

She doodled: Williams, p. 113; Orr, p. 229; *I&S*, p. 70.

195 "**I take a drawing**": Williams, p. 113.

very much "part of": SS, l to DA, 22 Nov. 1955, Deutsch; and *MA*, p. 298.

196 "**a reminder of the banality**": Mark Storey, "Why Stevie Smith Matters", *Critical Quarterly*, 21 (Summer 1979), 44.

"**A Saturday afternoon**" and "**a butcher**" (page 197): *MA*, p. 341.

197 "**This does not break down**": SS, *Some Are More Human Than Others* (London: Gaberbocchus, 1958), n. pag.

In one of her novels: *H*, pp. 65 and 68.

198 "**depends upon its being always**": Christopher Ricks, "Stevie Smith", *Grand Street*, 1 (Autumn 1981), 148.

199 **The features editor**: W. S. Aitken, l to SS, 16 March 1955, PGP.

"**To School in Germany**": *MA*, pp. 35–38.

Stevie wrote another fictional story . . . in August: On a l of Harold M. Harris to SS, 10 Aug. 1955, inviting her to submit another story, SS jotted " 'Getting Rid of Sadie' sent off 15/8/55", PGP.

"**Getting Rid of Sadie**": *MA*, pp. 39–43.

200 "**Even modern writers**": "Angels and Horrors", *Time and Tide*, 39 (11 Jan. 1958), 49.

"**the poem about the lady**": L. P. Hartley, l to SS, 28 Aug. 1955, PGP.

"**she has a criminal abortion**": SS, "New Novels", O, 6 March 1955, p. 8.

"**bitchy**", "**highbrow**" and "**caviar**": TK, int A.

"**utterly devoted**": TK, l to SS, 4 May 1955, PGP.

"**completely flummoxed**": TK, l to SS, 5 Nov. 1955, PGP.

Kilmartin remembers: TK, int A.

"*Faites attention,*": SS, "New Novels", *WR*, NS 29 (July 1951), 78. See also SS, "Stevie Smith's Guide to Good Reading", *MW*, Sept. 1949, p. 127.

201 "**a kind of shabby chic**," **remembered**: Hester Marsden-Smedley, int A.

"**another genius**": KD, int A.

Anna Browne once, According to Mrs Browne, "**middle-class aristocracy**", **and Ivy . . . felt too**: AB, int A.

Orr-Ewing said: this paragraph is based on Ronald Orr-Ewing, int A.

she told Naomi Replansky: Naomi Replansky, int A.

"**frosty**": Rosemary Cooper, int A.

"**desperate . . . at being told**": SS, l to TK, 4 Jan. 1956, files of O.

202 "**for all the sensibility**": John Rosenberg, l to SS, 9 Jan. 1956, PGP.

"**I see with much regret**": Arthur Hansen, l to SS, 2 Feb. 1956, PGP.

prefer to speak about . . . her poetry: see Arthur Hansen, l to SS, 12 Feb. 1956, PGP.

"**Time, mercifully**": script of SS's talk on Modern English Literature, Tulsa. SS also said this in "New Novels", O, 26 Dec. 1954, p. 7.

" '**lecturing' on Modern**": SS's report on *The Compassionate Lady*, files of The Bodley Head, Archives, The Library, University of Reading.

"**I do so hate lecturing**": SS, l to RM, 8 May 1956, RMP.

203 "**if you wanted to bring a libel action**": "Question Time", BBC radio Light Programme, broadcast 18 March 1956. Script at BBCa.

did not recommend: Kay Fuller, memo to A.H.O.T.F., Langham, 13 April 1956, BBCa.

" '**S' is difficult**": SS, l to RM, 8 May 1956, RMP.

bad news from Diana Athill: DA, l to SS, 31 May 1966, PGP.

did not want the advertising . . . to stir up Palmers Green: *MA*, p. 301.

The Towers of Trebizond . . . **and wrote to tell her so**: SS's l is mentioned in Rose Macaulay, l to SS, 1 Sept., 1956, PGP.

204 "**noveletish**": *MA*, p. 307.

"**You are very sagacious**": Rose Macaulay, l to SS, 1 Sept. 1956, PGP.

enthusiastic reviews: For SS's reviews of Llewelyn Powys's books see: *Life & Letters To-day*, 22 (July 1939), 138–39; and "Farewell to Dorset", *JO'L*, 45 (29 Aug. 1941), 342. For her reviews of John Cowper Powys's books see: "Stevie Smith writes about Reading", *MW*, Dec. 1946, p. 6; "New Novels", *WR*, NS 42 (Aug. 1952), 70–72; "New Novels", *O*, 31 Oct. 1954, p. 7; "Books of the Year", *O*, 23 Dec. 1956, p. 6; and "Powys and Homer", *O*, 8 March 1959, p. 22.

John Cowper Powys . . . wrote to tell her so: John Cowper Powys, letters to SS, 1 Nov. 1951 and 31 Oct. 1954, PGP; and his l to "Mr White", 27 Sept. 1952, PGP. See also *H*, p. 124.

"lonely widower" and "with acclamation": Littleton Powys, *Still the Joy of It* (London: Macdonald, 1956), pp. 33–34.

"bravely unusual": SS, "Heard of a Good Book", *MW*, Nov. 1943, p. 81.

writing to John Hayward: *MA*, pp. 285–86.

205 **"Our wedding night":** *Still the Joy of It*, (cited above).

"a happy love poem": "Too Tired"; and SS, "The Living Poet", script at BBCa of BBC radio Third Programme, broadcast 4 Oct. 1963.

"have a boss shot": *MA*, p. 303.

"gleeful macabre": Janice Thaddeus, "Stevie Smith and the Gleeful Macabre", *Contemporary Poetry: A Journal of Criticism*, 3 (1978), 36–49. SS told John Horder that "I Remember" is "terribly sad really although it's also very funny . . . One feels sort of sorry for them . . . though one can't help roaring with laughter of course, but still . . . I mean, it's an expression of the loving kindness of mankind, you know. Two lots of bombers are setting out to bomb each other. There's this poor old man in bed with a girl who's got T.B. and obviously won't last very long and neither will he. I don't know, it's terribly sad and it's terribly ridiculous too, I think. (Unedited script at Tulsa of BBC radio "World of Books", broadcast 2 Feb. 1963.)

"I altered the setting": SS, l to Derek Parker as related by his l to A, 19 Sept. 1979, por.

206 **"I must look west":** AO, int A.

"Wonderful to see a sunset": SS, pc to MS, 24 Sept. 1956, PGP.

proposed to the BBC in June: Dorothy Baker, l to SS, 11 June 1956, PGP.

"cri de coeur": *MA*, p. 116.

suggested that some . . . be sung and **changed her script:** SS, l to Douglas Cleverdon, 9 Dec. 1956, BBCa.

"was difficult to do": SS, l to RM, 13 March 1957, RMP.

"angel of a producer" and "gloomy drawl": SS, l to HF, 28 March 1957, por.

"I do not myself think that this broadcast": SS, l to Miss Wakeham, 27 Feb. 1957, BBCa.

207 **"For three long weeks, and long weeks":** *I&S*, p. 68. For our account of this trip to Milan we also consulted the unedited recording of KD's int SS, in possession of KD. Our other sources are AB and Barbara Clutton-Brock, int A; SS, l to HH, 19 Aug. 1959, HHP; and a note attached to a l to A from David Wright (concerning the British Council's £10 payment), 27 Nov. 1980, por. The name of the Brazilian Consul in

Milan was supplied by the Brazilian Consulate-General in New York.
208 **Stevie referred to it:** SS, l to HH, 19 Aug. 1959.
   "soft at the centre": Roy Fuller, *The London Magazine*, 5 (Jan. 1958), 64.
   "It sounds awful": SS, l to DA, 12 Dec. 1957, Deutsch.
   "curious chit-chat rhythms": Muriel Spark, "Melancholy Humour", *O*, 3 Nov. 1957, p. 16.
   "a poet capable of writing": "Light Fantastic", *TLS*, 4 Oct. 1957, p. 588.
   "*Punch*'s reviewer": P.D. [Peter Dickinson], *Punch*, 233 (13 Nov. 1957), 582.
209 **"to de-whimsy the funny-peculiar":** Anon., *The Listener*, 58 (17 Oct. 1957), 623.
   drawings with no poems: *MA*, p. 280.
   "dashing publisher": SS, l to HH, 15 April 1958, HHP.
   "owes much to Thurber": John Betjeman, "Something Funny", *DT*, 12 Dec. 1958, p. 7.
   "not random jottings": Oswell Blakeston, "Universal Predicament", *Art News and Review*, 3 Jan. 1959, p. 8.
210 **"artistic analyst":** "Picture Books", *S* (21 Nov. 1958), p. 730.
   "so I asked him what": [Clive Cullerne-Brown], "Letter From Anywhere", *Hampstead and Highgate Express*, 14 Nov. 1958, p. 6.
   Natal Witness: F.M., "Among the New Books", *Natal Witness Saturday Magazine*, 25 April 1959, p. 2.

Chapter Fifteen

211 **Hans Walter Häusermann:** this paragraph is based on HH's handwritten notes of his first meeting with SS, item 35 of the HHP.
212 **highly favourable review:** HH, "Zeichnungen von Stevie Smith", *Neue Zürcher Zeitung*, 16 Aug. 1959, p. 4.
   finishing a lecture: SS, l to HH, 17 Oct. 1957, HHP.
   vivid but far too long: Charles Davy, l to SS, 20 Nov. 1957, PGP.
   "hawking it round": SS, l to HH, 30 Dec. 1957, HHP.
   "the kiddy-press": SS, l to HH, 23 Jan. 1958, HHP.
   "rationalists" and **"Goya etchings":** SS, l to David Wright, 10 May 1959, Manuscripts Department, Lilly Library, Indiana University.
   "I do not think there is any harm" and **"the nature of God and Goodness":** *MA*, p. 153.
   "the nerve of cruelty" and **"I could not forget Hell":** *MA*, pp. 154–5.
213 **"Some people may not find the tug":** "Not Believing", pp. 20–21.
   she balanced Lecky: "Not Believing", p. 25.
   "The Gospels, from which as a pacifist,": *MA*, p. 193.
   pamphlets of The Catholic Truth Society: "Not Believing", p. 25.
   "My feelings fly up": *MA*, p. 153.

"a determination to make words mean", In "Why d'You Believe?" a poem and "is fraught with the perils": "Not Believing", p. 21–23.

cheered up by the degree: *MA*, p. 160.

214 **"One may like the Catholics":** SS, "The Personal Devil", *O*, 19 March 1961, p. 31.

paradoxes of Christ: *MA*, p. 161.

215 **"icy indifferent wind"** and **"hurrying back to religion":** "Not Believing", p. 24.

crack about the Assumption: SC, 1 to A, 3 March 1985, por.

Gemini **published lengthy comments:** the responses of Stockwood and Barrington Ward appeared under the heading, "The Necessity of Belief" in a subsequent issue of *Gemini* (pp. 41–43), and the 1 of John Pinnington appeared in the correspondence section (p. 63) of the same issue. These pages were sent to us by Dr Elisabeth Schnack, to whom SS sent them along with the issue of *Gemini* containing "Not Believing". We have been unable to locate the complete issue of *Gemini* with the responses to SS's essay, and so are unable to cite its date and volume number.

subsequently engaged Stevie: see John Pinnington's letters to SS, 10, 17 and 27 May 1960, and 21 and 24 Oct. 1962, PGP.

"It is with great pleasure and curiosity": SS, "Unbelieving", *S*, 199 (6 Dec. 1957), 800.

216 **"All that the Christians write":** ibid.

"was a home in life": SS, "Happy and Holy", *O*, 23 Sept. 1956, p. 12.

In 1957 she reviewed: for SS's reviews mentioned in this paragraph see "Spokesman for God", *O*, 3 Feb. 1957, p. 13; "Caritas", *O*, 8 Dec. 1957, p. 19; "Educated Silliness", *S*, 200 (3 Jan. 1958), 23; "All Saints", *S*, 201 (5 Dec. 1958), 829.

217 **"it's not atheism either":** Mary Renault, 1 to SS, 7 Feb. 1958, PGP.

Stevie told Helen Fowler: SS, 1 to HF, 7 March 1958, HFP.

letters . . . passed between Diana Athill and Stevie: DA, 1 to SS, 17 March 1958, PGP; SS, 1 to DA, 22 March 1958, Deutsch.

218 **"Tengal [her name for Bernal] was a very pious boy":** *H*, pp. 85–86.

"these things . . . are being taught today": "Not Believing", pp. 26–27.

"anti-religious poem," she called it: scrap of paper, Tulsa.

"has as loud a sneer": SS, 1 to HH, 1 July 1959, HHP.

"Our temperament rules us" and **"I daresay my own":** "Not Believing", p. 31.

219 **"who could not let God go":** Father Gerard Irvine, "Stevie Smith", unpublished address given at Saint Matthew's Church memorial service for SS, London, 31 March 1971, copy in possession of A.

Helen Fowler felt: HF, int A.

Teresa of Avila's "reasonable fear": SS, "St Teresa of Avila", *O*, 12 June 1960, p. 27.

told **Helen Fowler, "No,":** *MA*, p. 305. See also SS's reviews, "Double, Double", *O*, 17 Feb. 1957, p. 14; "In Defence of Witches", *O*, 15 Feb. 1959, p. 21; and "Miching Mallecho", *O*, 8 July 1962, p. 25.

"I am not a really proper": *MA*, p. 292.

"how closely connected belief": HH, "Zeichnungen von Stevie Smith", *Neue Zürcher Zeitung*, 16 Aug. 1959, p. 4.

"seen so much of what I was thinking": SS, l to HH, 17 Dec. 1957, HHP.

220 **"caked in ashes & cross":** this paragraph is based on SS, letters to HH of 30 Dec. 1957, 23 Feb. 1958, 7 March 1958, and 15 April 1958, HHP.

lasted through December: SS, l to HH, 10 Dec. 1958, HHP.

"They say it only means going": SS, l to HH, 15 April 1958, HHP.

"a little extra money": SS, l to HF, 11 May 1958, por.

"the stuff one has to read": SS, l to RM, 11 May 1958, RMP.

Helen Fowler recalls: HF, int A.

221 **"awfully racketty":** this paragraph is based on SS, l to HH, 28 May 1958, HHP.

"but on the whole" and **"I am typing this":** ibid.

"pretty dull": SS, l to HH, 24 July 1958, HHP.

"I adore these trips": SS, l to HH, 28 May 1958, HHP.

"I have a passion": SS, l to HF, 20 May 1958, por.

222 **"Papa in the Norfolk":** SS, l to HH, 28 May 1958, HHP.

"The dear boys do an *awful*": SS, l to RM, June 1958, RMP.

took Stevie to see an historic house: HF, int A.

"odd Scottish sunburn": SS, l to HF, 22 June 1958, por.

"Don't really think", "silly programme" and **"the old hands":** SS, l to HH, 15 Aug. 1958, HHP.

"more social life", "coming back from the Point" and **new young author:** SS, l to HF, 15 Sept. 1958, por.

223 **guests came in togas:** Maria Browne, int A.

"Books of the Year": SS, *O*, 27 Dec. 1959, p. 8.

a book Laurence Fowler: SS, letters to HF, 22 June and 24 July 1958.

"Bliss, isnt it?": SS, l to HH, 22 Oct. 1958, HHP.

"I do not think the batch": J. T. C. Hall, l to SS, 17 March 1959, PGP.

German publisher and **"turned from Lit":** SS, l to HH, 22 Oct. 1958, HHP.

wrote detailed letters: SS's letters to Ladislav Horvat are por, and his letters to her are among the PGP. The sums mentioned in this paragraph are from Mr Horvat's l to SS, 12 Jan. 1959.

"a beautiful red petticoat": SS, l to HH, 14 Nov. 1958, HHP.

an Anglican canon: Canon Scrutton, l to SS, 22 July 1957, PGP.

"little cracks": J. Mangold, l to SS, 7 Aug. 1957, PGP.

former Catholic priest: John V. Simcox, l to SS, 15 Aug. 1957, PGP.

224 **Mary Tudor:** SS, "Elizabeth . . .", *O*, 12 Oct. 1958, p. 21.

also elicited protests: SS, l to HH, 14 Nov. 1958, HHP.

"boyish tastes": SS, "Strange & Sickly", *O*, 9 Nov. 1958, p. 22.

"the sickness of states": *MA*, p. 148.

"I gather that you want to say": Neville Braybrooke, l to SS, 11 March 1957, PGP.

"a very un-birthday present": SS, l to HH, 10 Dec. 1958, HHP.

225 **"a remarkable evocation"**: this short essay appears in *MA*, pp. 148–52, and page citations for specific quotations will not be given.

"the sharp edge of our own opinions": *MA*, p. 173.

226 **"My reading of the priest's"**: Flannery O'Connor, *The Habit of Being*, ed. Sally Fitzgerald (New York: Farrar Straus Giroux, 1979), p. 354.

advanced once by Dorothy Sayers: In a review, SS spoke of Sayers' "good essay" on Dante's *Divine Comedy*, but wondered "how she can bubble over quite in that way about the sufferings of the damned". Sayers replied to what she took to be SS's "childish notions about the doctrine of Hell", and SS responded in part by accusing Sayers of "fascinating heresy". See *T*, 30 April (p. 19), 7 May (pp. 14–15) and 14 May (p. 12), 1948.

"You may think it is curious": *MA*, p. 165.

"went up the steps": SS, pc to HF, 3 Feb. 1961, in possession of A.

the *Guardian* commissioned: the poem was published in the *Guardian* on 16 May 1964.

227 **who asked (and got)**: Francis Stevenson, O.S.B., l to SS, 4 June 1964, PGP.

Stevie wrote warmly: see Francis Stevenson, O.S.B., l to SS, 11 June 1964, PGP.

several times visited: HF, int A.

told her of the letter: JL, l to SS, 18 April 1964, PGP.

Chapter Sixteen

228 **felt lonely and** and **"getting rather sick"**: SS, l to HH, 10 Dec. 1958, HHP.

"hardly Christmas fare", **"misery alone"** and lift **"the glumness"**: SS, l to HH, 17 Dec. 1958, HHP.

Who Is The Devil?: SS, "Saints and Suffering", *DT*, 2 Jan. 1959, p. 12.

"dodgy . . . politically": SS, l to HH, 17 Dec. 1958, HHP.

"forgetting it was Gregory": SS, l to HH, 6 April 1959, HHP.

229 **first printed in November**: in *X: A Quarterly Review*, 1 (Nov. 1959), 55–57.

"a new idea": this and the following two paragraphs are based on SS's poem and her l to HH, 6 April 1959, HHP.

"very nice & a wonderful financier": SS, l to HH, 6 April 1959, HHP.

230 **"neatly ruled pages"**: Ladislav Horvat, l to SS, 12 April 1959, PGP.

"cat card" and **"Swiss sporting cats"**: SS, l to HH, 6 April 1959, HHP.

"a book for Stevie Smith fans": Muriel Spark, "Top Cats on Show", *O*, 29 Nov. 1959, p. 4.

well in England and sold out in America: Sam Carr (Editor,

Batsford Ltd), l to SS, 25 Sept. 1959, PGP; Mary Velthoven (Editor, Viking Penguin Inc), l to A, 11 May 1979, por.

"even the most illustrious": Philip Larkin, "Frivolous and Vulnerable", *NS*, 64 (26 Sept. 1962), 416.

"sweet little catsy-watsies", and **"what mind have animals"**: *MA*, pp. 134–35. The entire essay is on pp. 134–147. We do not cite page numbers for subsequent quotations from it in this chapter.

231 **"Help! But it *was*"**: SS, l to HF, 19 Nov. 1959, por.
wrote saying so to Stevie's publisher: Sam Carr, l to SS, 26 Nov. 1959, PGP; and Sam Carr, copy of l to G. D. Fisher, 26 Nov. 1959, PGP.
" 'a good armful of flowers' ": *MA*, p. 180.
"Stevie was shocked": AB, int A.

232 **Sally Chilver remembers**: SC, int A.
The title of a review: SS, "Keeping cool about cats", *O*, 19 March 1967, p. 26.
"I am not a Cat Lover": SS, "Cats as gods & devils", *O*, 2 June 1963, p. 19.
irritated with her sister's cat: Donald and Molly Everett, int A; *MA*, p. 319.
"three top cats": *MA*, p. 313.
A Turn Outside: MA, pp. 335–58.
since the late 1940s: the Cleverdons, int A.
"romantic about old Death": *MA*, p. 305.
hitherto unpublished poem: St. L.

233 **went to Oxford**: EL, int A. All of EL's memories in this paragraph are based on this interview.
"sailing as near the wind": SS, "Who's who on Mount Olympus", *O*, 23 Dec. 1962, p. 20.

234 **"He brought me back"** and **"just right"**: SS, l to David Wright, 10 May 1959, MSS Department, Lilly Library, Indiana University.
"possessive and capricious furies": SS, "Odd Man Out", *O*, 2 Aug. 1959, p. 8.
"the old Turk" and **"dragooned escort"**: SS, l to Roland Watson, 25 May 1959, Birmingham.
"Pompey Casmilus in her teens": HH, l to SS, 10 July 1959, PGP (a copy is among the HHP).
with the George Lawrences: signature of SS in their guestbook, 26 July 1959, in their possession; and SS, l to HH, 19 Aug. 1959, HHP.
"Have just had a heavenly day" and **"feeble little ass"**: this paragraph, except for the last sentence, is based on SS, l to HH, 19 Aug. 1959, HHP.
commit suicide: John Gardner, l to A, 19 Oct. 1980, por.
As the 1950s . . . and Birmingham: SS, l to HH, 12 Dec. 1959, HHP.
"in a sort of *palace*": SS, l to HF, 26 Nov. 1959, por.

235 **rattled and whistled**: SS, l to RM, 11 Jan. 1960, RMP.
"there doesn't seem much to say": SS, l to HH, 12 Dec. 1959, HHP.
It was around . . . *Selected Poems* came about: John Guest, int A.

"feel too keen on any": SS, l to HH, 12 Dec. 1959, HHP.

"I know they're *something*": SS, l to Roland Watson, 8 Nov. 1960, Birmingham.

"a lot of new poems", **"write a new one"** and **"far too much"**: SS, l to RM, 11 Jan. 1960, RMP.

"a Hell File": SS, letters to DA, 2 May and 4 July, 1960, Deutsch.

"this neurotic attitude to work": SS, l to DA, 2 May 1960, Deutsch.

"fascinating yet a painful figure": SS, "Facades of Rome", O, 8 May 1960, p. 19.

236 **the Oxford student:** John Pinnington, letters to SS, PGP.

in school hockey games: Dr James Curley, int A.

a surgeon removed: SS, l to Ladislav Horvat, 7 Jan. 1961, por.

"lazy comfort": SS, l to Ladislav Horvat, "Sun" [Jan. 1961], por.

"spending a fortune on stamps": MAS, l to SS, 29 Jan. 1961, PGP.

"wonderful" mid-Victorian: reading list from SS, l to HF, 16 Jan. 1961, por.

the BBC invited her and **recuperating:** SS, l to Miss Pearce, 2 Feb. 1961; and Miss Pearce, copy of l to SS, 28 Feb. 1961, BBCa.

Nancy Hodgkin: SS, l to Miss Pearce, 24 Feb. 1961, BBCa. See also MA, p. 313.

In March . . . poet's reading and **"code iddy head!"**: SS, letters to Miss Pearce, 9 and 19 March 1961, BBCa.

"awful nasalness" and **"lame dogs"**: SS, l to RM, 20 March 1961, RMP.

"horrid little knee", **"a deserter"** and **"*They* at least"**: SS, l to RM, 14 April 1961, RMP.

a brief contretemps: this paragraph is based on Victor Gollancz, letters to SS, 16 and 20 June 1961, PGP; and SS, "Laws of the wild", O, 2 July 1961, p. 26.

237 **But one friend she saw often:** this paragraph is based on James Cubitt, int A.

"rather downcast" and **"to be easy & happy"**: SS, letters to Ladislav Horvat, 26 April and 5 July 1961, por.

a guest of the Watneys: the remainder of this paragraph and the following paragraph, are based on Antoinette Watney, int A; SS, l to Jocelyn Baines, 10 Aug. 1961, Archives, The Library, University of Reading; and SC, l to A, 30 Mar. 1985, por.

238 **"wonderful, but fierce"**: SS, l to TK, 10 Aug. 1961, files of O.

"why cant people be honest?": MAS, l to SS, 11 Aug. 1961, PGP.

"There is also a Voice: SS, l to Jocelyn Baines, 10 Aug. 1961, Archives, The Library, University of Reading.

far more cordial: MAS, l to SS, 19 Aug. 1961, PGP.

Anna Browne recalls: AB, l to A, 30 March 1985, por.

239 **"groundwork for a novel"**: MAS, l to SS, 19 Aug. 1961, PGP.

on 10 August: SS, l to TK, 10 Aug. 1961, files of O.

"Don't be silly, it's about": Antoinette Watney, int A.

the removal of a tumor and "BENIGN LUMP": SS, letters to

Jocelyn Baines, 11 and 19 Oct. 1961, Archives, The Library, University of Reading.

turning a cartwheel: Dr James Curley, int A.

In December Stevie broadcast: this paragraph is based on Peter Orr, int A; and Peter Orr, letters to SS, 27 Nov., 11 and 14 Dec. 1961, PGP.

Wright organized a poetry reading: this paragraph is based on David Wright, l to A, 1 Nov. 1980, por.

240 **"*I* (rather late in life)":** SS, l to HH, 15 Dec. 1962, HHP. Also SS, l to Roland Watson, 5 July 1962, Birmingham.

Dr Curley speaks: Dr James Curley, int A.

she had struck Aunt: Rosemary Cooper, int A.

"I fancy things are not so easy": MS, l to SS, 7 Aug. 1962, PGP.

a fortnight in Venice: the memories of the Lawrences in this paragraph are based on their int A, and on Jacynth Ellerton [Lady Lawrence], "Stevie Smith", *The Tablet*, 10 April 1971, p. 360.

241 **"a heat wave and all the fixed":** *NYP*, p. 29.

"streaming out from the hand of God": Thomas.

"That's me": Lady Lawrence; int A.

"the feeling one has sometimes in Italy": SS, "Private Lives", *Time and Tide*, 38 (4 May 1957), 552.

her friend declined: Lady Lawrence, int A.

They speak of terrible things": J. E. [Lady Lawrence], *Frontier*, (Winter 1962), 614.

"My poor publishers were trying": Williams, pp. 112–13.

"simply nowhere to put her": Kathleen Nott, "Tickling The Muse", *O*, 23 Sept. 1962, p. 28.

Anthony Thwaite: Anthony Thwaite, "Peculiar Talent", unpublished [?] review in typescript, PGP. According to Thwaite, l to A, 6 July 1985, this review may have been broadcast on the BBC World Service.

"The animus . . . a witch": Hugh Gordon Porteus, "Sibyl", *S*, 209 (19 Oct. 1962), 610.

242 **"enormously under-valued":** John Coleman, "Books", *The Queen*, 221 (30 Oct. 1962), 19 and 20.

"who attract addicts": "Light But Not Slight", *TLS*, 28 Dec. 1962, p. 1006.

"I am an addict": *MA*, p. 6.

"The poets I delight in": Sylvia Plath, "Context", *The London Magazine*, 1(Feb. 1962), 46.

"a deliciously Smithish letter": Plath's phrase in a l to Peter Orr, late Nov. 1962—according to Peter Orr, int A.

"Dear Sylvia Plath": SS, l to Sylvia Plath, 22 Nov. 1962, Rare Book Room, Smith College Library, Northampton, Massachusetts.

243 **"I am not aware":** Philip Larkin, "Frivolous and Vulnerable", *NS*, 64 (26 Sept. 1962), 416.

"'break through' with a new name": JL, l to SS, 28 Aug. 1963, PGP.

Among those who sent Laughlin: JL, letters to SS, 4 Dec. 1963 and 13 May 1964, PGP.

left Stevie grateful but puzzled: SS, l to JL, 12 Dec. 1963, por.

Ogden Nash's praise: Ogden Nash, l to JL, 7 Feb. 1964, por.

Robert Lowell also sent . . . a puff: Robert Lowell, l to JL, 31 Oct. 1963, por.

a "sad fate": SS, "Boccaccio heroines", *O*, 5 July 1964, p. 26.

244 **fearful of being described** and **The English Poet Miss Stevie Smith:** SS, l to JL, 25 Oct. 1963, por.

Stevie also quoted to Laughlin: SS, letters to JL, 23 and 25 Oct. 1963, por.

"I unreservedly say: Admire Stevie Smith": Joseph Bennett, "Each in His Own Fashion", *New York Times Book Review*, 5 July 1964, p. 4.

Marianne Moore, who wrote: Marianne Moore, l to JL, 9 April 1964, por.

"Isn't it super": SS, undated l to JL, postmarked 19 April 1964, por.

stayed with the Fowlers: this paragraph is based on HF, int A; and HF, l to A, 5 April 1985, por.

245 **Robert Lowell met Stevie:** see SS, l to JL, 2 Sept. 1963, por.

The Poetry Center: Galen Eberl (Secretary of The Poetry Center), l to SS, 17 April 1964, PGP.

Academy of American Poets: JL, l to SS, 10 April 1964, PGP.

"as long as I don't come to America": SS, undated l to JL, postmarked 19 April 1964, por.

246 **"The young man turned up":** *MA*, pp. 309–10.

"mixed up with Jazz": SS, l to JL, 9 Nov. 1964, por. Michael Horowitz, l to A, 16 Aug. 1985, identifies the "actress", who read Sylvia Plath's poems as Glenda Jackson.

"She came up to me": Glenda Jackson, int A.

Chapter Seventeen

247 **a long letter to Rachel Marshall:** this and the subsequent paragraph are based on *MA*, pp. 311–13.

248 **a letter she wrote to an editor:** this paragraph is based on SS, l to Brocard Sewell, 8 Jan. 1965, Manuscripts Department, Lilly Library, Indiana University.

wrote to thank Ladislav Horvat: this paragraph is based on SS, l to Ladislav Horvat, 31 Jan. 1965, por.

"I don't want to write": this paragraph is based on an unedited script at Tulsa of SS's int John Horder on the BBC radio "World of Books", broadcast 2 Feb. 1963.

249 **"I have been writing quite a lot":** this paragraph is based on SS, l to HH, 9 March 1965, HHP; and *MA*, p. 175.

several interviews: engagement diaries of SS, PGP.

a 'Monitor' film: the rest of this paragraph is based on *I&S*, p. 67.

250 **On the telecast:** this paragraph is based on our viewing of a tape of the telecast.

Giles Gordon: this paragraph is based on Giles Gordon, int A; *I&S*,

p. 62; John Horder, "Diary", *L*, 12 April 1979, p. 512; and SS, letters to Giles Gordon, 11 and 13 Jan. and 1 and 5 Feb. 1965, por.

"I twisted": this paragraph is based on SS, l to Barbara Clutton-Brock, 18 June 1965, por.

251 **Clifford Doyle . . . remembered:** Clifford Doyle, int A.

Among other readings: SS's engagement diaries, PGP.

less nervous before a reading: AB, int A.

Helen Fowler noticed: HF, int A.

"don't think she liked": this paragraph is based on Colin Amery, int A.

"You sounded marvellous": Jonathan Williams, pc to SS, 28 Aug. 1965, PGP.

"I have been doing a lot": SS, pc to Jonathan Williams, 29 Dec. 1965, the Poetry Collection of the University Libraries, State University of New York at Buffalo.

a film crew: see Jonathan Vickers' description of the film in his "Stevie Smith—Her Broadcasts and Recordings", *Recorded Sound* (July 1982), p. 18.

252 **"I don't think Auden liked":** unedited recording of SS, int KD, 7 Nov. 1970.

"novels for girls": HH, l to SS, 10 July 1959, PGP (copy, HHP).

tell Stevie of their marriage: see SS, l to HH, 12 Jan. 1967, HHP.

Mrs Häusermann-Häusermann recalls: the description of her visit to SS in 1965, in this and the next four paragraphs, is based very closely on Mrs Häusermann-Häusermann's account (written in German) included in her l to A, 28 Nov. 1983, por. SS was unprepared for lunch guests because she had forgotten it was a Bank Holiday weekend and they could not dine out.

253 **"Sorry you are feeling":** TK, copy of l to SS, 27 March 1965, files of O.

an evaluation of a symposium: *MA*, pp. 199–201.

"Faith's best friend": SS, *L*, 74 (30 Dec. 1965), 1083.

"it is because man is so lonely": unedited script at Tulsa of SS's int John Horder on the BBC radio "World of Books", broadcast 2 Feb. 1963.

Simone Weil: *MA*, pp. 196–97.

Calvin: SS, *L*, 74 (9 Dec. 1965), 968–69.

Isherwood's book: SS, *L*, 73 (22 April 1965), 607–08.

254 **"remarkably feeble":** SS, l to HH, 7 Feb. 1966, HHP.

"I wish your mother was alive": *I&S*, p. 67.

"too marvellous": SS TK, 24 May 1966, files of O.

Seventy to eighty people: this paragraph is based on AB, int A.

in the Albert Hall: This paragraph is based on SS, l to TK, 15 June 1966, files of O; Alasdair Clayre, int A; Christopher Logue, l to A, 27 Oct. 1980; and Michael Horovitz, l to A, 1 July 1985, por.

Stevie wrote to Professor: this paragraph is based on SS, l to HH, 14 Aug. 1966, HHP.

255 **"the most original poet"** and **"occurs throughout"**: John Horder, "Poet Exposed", *Catholic Herald*, 6 Jan. 1967, p. 6.

The *TLS* reviewer: "Waving and Drowning", *TLS*, 19 Jan. 1967, p. 48.

256 **one galley of the poem:** Tulsa.

"a religious poem": SS's notes, Tulsa.

257 **In her poem "How do you see?"**: A poem sent to SS by someone who must have been responding to "How do you see?" perhaps fed the inspiration which led to "The Frog Prince". The correspondent's poem, "To Stevie Smith—A Prayer", asks in its first line why SS wants people to be good without enchantment, and it concludes that without enchantment there is no hope for people to be good. The closing lines of "The Frog Prince" seem to be SS's direct reply. See A. N. Pospieszalski, l to SS, 16 May 1964, PGP.

"The nerviness" and **"Sweet Death"**: see SS's poem "Why do I . . . ", *CP*, p. 508.

Anna Browne remembers: AB, int A.

"cheerful as it is a word-play": SS's notes, Tulsa.

258 **"Stevie just made it up"**: Lady Lawrence, int A.

Sally Chilver, Elisabeth Lutyens: SC and EL, int A.

she felt as well a link: SS was among the signers of a l, printed in *The Times* on 13 Jan. 1967, on the subject, "Moves to Ban Export of Animals". This protest against cruelty to animals was organized in the main by Olivia Manning. See *The Letters of J. R. Ackerley*, ed. Neville Braybrooke (London: Duckworth, 1975), note 1, p. 313.

once told Anna Browne's daughter: Alice Browne, int A. In a BBCa script for a BBC broadcast of "The Lively Arts", dated 5 Nov. 1965, SS said of "The Best Beast": "I wrote this poem which, as a matter of fact, is all in monosyllables. Though I didn't realise that until I got to the last verse but one. And so I thought I'd better keep it up for the last verse. It was quite a feat I think really."

259 **used this introduction** and **"I am haunted"**: SS's notes, Tulsa.

260 **a holograph of her poem, "Longing for Death"**: Tulsa.

"a literary criticism": SS's notes, Tulsa.

he does not do so in her poem: SS's notes, Tulsa.

"The poem deals with the price": the rest of this paragraph, and the following paragraph, are based on SS, l to HH, 28 July 1958, HHP.

261 **"the melancholy note is struck"**: SS's notes, Tulsa.

262 **"a poem that is really a novel"**: SS's notes, Tulsa.

262 **"Nobody writes or wishes to"**: from SS's poem, "Mrs Arbuthnot", *CP*, p. 492.

263 **"a little child who has"**: SS's notes, Tulsa.

in a public way: BBCa script of BBC "World of Books", dated 6 Dec. 1966.

Chapter Eighteen

264 **Stevie wrote to Rachel:** this paragraph is based on SS, undated fragment of l to RM, RMP; and Bernard Bergonzi, "Tones of Voice", *Guardian*, 16 Dec. 1966, p. 7.

"So often when evening comes": ibid.

writing to another friend: the remainder of this paragraph is based on SS, l to Polly Hill, 25 March 1967, por.

265 **Stevie confessed . . . in English papers:** SS, l to JL, 27 Feb. 1966, por.

telling David Wright: the remainder of this paragraph is based on *MA*, pp. 313–14; and SS, "Deafness Without Disguise", *L*, 83 (19 Feb. 1970), 255.

"the poem is a parable": SS, undated script at Tulsa which seems to be for a school poetry reading.

published that summer: in *TLS*, 17 Aug. 1967, p. 742.

266 **"my favourite smell":** SC, int A.

In June 1967: this paragraph is based on Geoffrey Handley-Taylor, l to A, 1 Jan. 1981, por.

attended the Royal Garden Party: this paragraph is based on SS's engagement diary, PGP; and Norah Smallwood, int A.

last time Elisabeth Lutyens met: this paragraph is based on EL, int A.

267 **The peculiar flavour:** this paragraph is based on Anthea Hall, "As Mrs Lutyens said to Miss Smith . . . ", *The Journal* [Newcastle upon Tyne], 23 Sept. 1967, p. 8.

to read . . . in Brussels: the account of the reading in this and the following paragraph is based on Alasdair Clayre, int A; Michael Horovitz, memoir of the 1967 Brussels poetry reading in a lengthy footnote to his editorial "of absent Friends", *New Departures*, No. 7/8 and 10/11, p. 15, and letters to A, 1 July and 16 Aug. 1985, por; and the following cuttings from the PGP: J.D., "Show in PvSK to Brussel lokt massa "Psychedelic Feast", *Het Nieuwsblad* [Brussels], 9 Oct. 1967; K.V.A., "Poetic Show: Britse dicthers met eigen werk", in both *Het Volk* [Ghent] and *De Nieuwe Gros* [Brussels], 9 Oct. 1967; and D.C., "Veel Geblatt, Weinig Wol Poetic Show in Paleis voor Schone Kunsten" [paper and date not given]. In his l to A, Horovitz notes that the poets were "more or less forced to enjoy ENORMOUS liquid hospitalities" and states: "It wasn't, you know, my or our idea to bill it a 'Psychedelic Feast' & I don't think WE mentioned 'flowers, love and peace, that much—the organisers in Bruxelles Palais voor Schoone Kunsten kept on & on abt. this, & were v. hooked on the whole 'swinging London'/hippy image as selling line for the show—I did try to disabuse them, somewhat."

268 **When Stevie returned:** this paragraph is based on AB, int A; and SS's engagement diary, PGP.

offered her £250: Julia Wootten [?] (Society of Authors), l to SS, 14 Dec. 1967, PGP.

Miss Spear made a will: will of MAS, 2 Nov. 1967, records at Somerset House, London.

"**The heart-block**": SS, l to JL, 30 Dec. 1967, por.

"**just getting up**": SS, l to Barbara Clutton-Brock, 4 Jan. 1968, por.

269 "**contradictions of the New Testament**": SS, "Can't Have It", *L*, 77 (4 May 1967), 596.

Ste. Thérèse: SS, "The Little Way", *L*, 77 (16 March 1967), 369.

Thurber and Vicky: *MA*, pp. 184–85.

"**Miss Manning is apt to be**": SS, "Our Love Provoked", *The Sunday Times*, 26 Nov. 1967, p. 55.

"**I don't know . . . about cats**": SS, l to Ann Thwaite, 22 May 1968, por.

270 "**our poor dumb-wits**": SS, "Keeping cool about cats", *O*, 19 March 1967, p. 26.

"**to the point**": SS, l to TK, 15 Jan. 1968, files of *O*.

Worrying over the illness: this paragraph is based on SS, "Consolations", *L*, 79 (4 Jan. 1968), 23–24.

271 "**prim orthodox—churchy**": RL, l to A, 11 Aug. 1981, por.

"**with its raw note**" and "**I do not see the dead**": RL, Epilogue, *The Swan in the Evening* (London: Virago, 1982), pp. 162–63.

272 "**'Michael Field', that odd amalgam**": *MA*, p. 181.

religion for children: SS, "Old story, revised versions", *O*, 1 Dec. 1968, p. 30.

ninety-sixth birthday: SS's engagement diary, PGP.

"**going downhill**": Woodcocks, int A.

"**shattered**": AB, int A.

in her diary: PGP.

"**We saw Aunt again**": SS, l to HH, 11 Feb. 1968, HHP.

"**before the poor darling**": SS, l to Marghanita Laski, 24 Feb. 1968, por.

weekend in Kent with Norah: this paragraph is based on SS's engagement diary, PGP; and Norah Smallwood, int A.

"**Aunt is slipping away**": SS, l to Marghanita Laski, 5 March 1958, por.

273 **dined** and **Granada TV**: SS's engagement diary, PGP.

"**Darling Aunt died**": ibid.

"**mixture of dialects**": SS, l to TK, 11 March 1968, files of *O*.

The Vicar was ill: Rev. Clifford Doyle, int A.

her diary: PGP.

"**to lie on the bosom**": Norah Smallwood, int A.

"**don't be silly, it's Aunt**": Antoinette Watney, int A.

Stevie and Molly to lunch: Woodcocks, int A.

Cleverdons gave a dinner party: the account of this party is based on the Cleverdons, int A.

thank Doreeen Diamant: SS, l to DD, 9 April 1968, por.

one neighbour: Wallace Finkel, int A.

274 **fresh flowers**: SS's engagement diary, PGP; and Woodcocks, int A.

inspired the poem, "O Pug": AB, int A. SS mentions the ride in the

Brownes' car with Pug in her 1968 engagement diary, by her 21 April entry.

"I wish Aunt were here": Woodcocks, int A.

Helen Fowler remembers and **Anna Browne believes:** HF and AB, int A.

On the eve of her departure: this paragraph is based on *MA*, p. 316.

275 **Stevie told Hans Häusermann:** the remainder of this paragraph is based on SS, l to HH, 12 Aug. 1968, HHP; and HH's notes of his telephone conversation with SS, item 38 of the HHP. SS introduced "So to fatness come" by noting "This is what may be said when human beings lose people they love. I like it because, though it is stern, it is quite truthful and so it is comforting. Also the metric is very lithe and subtle. This is something I give great care to and find pleasure in." (SS, l to James Gibson, 9 June 1970, Austin.)

a brief and moving essay: this and the next paragraph are based on Manelle Bernstein [?] (O), l to SS, 18 Sept. 1968, Tulsa; and SS's unpublished and untitled 4 page typescript (including a last page with the poem "So to fatness come") which begins: "It must be obvious, I think, that those who have a firm belief" (Tulsa).

276 **Sally Chilver thought:** SC, int A.

One of Stevie's neighbours: the remainder of this paragraph is based on Mrs Bradford (int A) and the poem, "There is an Old Man", (*MA*, p. 220).

277 **In 1967 Ladislav Horvat:** this paragraph is based on Ladislav Horvat, int A.

"I have decided": SS, l to HH, 12 Aug. 1968, HHP.

"heavenly": this paragraph is based on *MA*, p. 317.

poetry readings: SS's engagement diary, PGP.

Eddie Linden remembers: Eddie Linden, int A.

278 *Python:* Sebastian Barker, *Who Is Eddie Linden* (London: Jay Landesman Ltd, 1979), p. 215.

Eddie Linden went: Eddie Linden, int A.

"impassioned Apologia": RL, Epilogue, *The Swan in the Evening* (London: Virago, 1982), p. 164.

Stevie's Christmas letter: this paragraph is based on *MA*, p. 318.

Chapter Nineteen

279 **negotiations were proceeding:** Robert Gottlieb, copy of l to John Guest, 19 Feb. 1969, PGP.

Laughlin wrote . . . and Stevie said: SS, l to JL, 7 April 1969, por.

280 **"a most profound impression":** this paragraph is based on JL, l to SS, 18 June 1969, PGP.

Rosemary Cooper remembers: Rosemary Cooper, int A.

"I wasn't able somehow": the rest of this paragraph is based on SS, l to JL, 7 April 1969, por.

In **"What Poems Are Made Of"**: this paragraph is based on *MA*, pp. 127–29.

281 **some comic relief**: this paragraph is based on SS's review, *MA*, pp. 187–91; SS, l to TK, 6 Feb. [1969], files of O; as well as SS's poem "Seymour and Chantelle".

A review of *Against All Reason*: SS, "God's Will", *NS*, 75 (25 April 1969), 590–91.

In the late winter and spring: this paragraph is based on SS's engagement diary, PGP; and *MA*, pp. 318–19.

282 **"detestable cat" and "two 'Busy Lizzies'"**: *MA*, pp. 318–19.

wrote . . . to Lucinda Fowler: this paragraph is based on SS, l to Lucinda Fowler, 29 May 1969, por; HF, int A; and SS's engagement diary, PGP.

283 **a reading with Phoebe Hesketh**: this paragraph is based on SS to RHD, 15 June 1969, PGP; and RHD, l to SS, 30 June 1969.

delivered the manuscript . . . to Sam Carr: this paragraph is based on SS's engagement diary, PGP; SS, l to Jean MacGibbon, 1 Oct. 1970, por, and printed on the endpapers of the hardback edition of *MA*; and SS, Preface to *The Batsford Book of Children's Verse* (London: Batsford, 1970), p. 3.

In June of 1969: this paragraph is based on SS's 1969 diary, PGP, and her l to Anthony Thwaite, 27 June 1969, por.

284 **"weeping with boredom"**: SS, l to AB, 29 June 1969, por.

In July Anthony Powell: this paragraph is based on SS, l to Anthony Powell, 10 July 1969, por; *MA*, pp. 203–5; and SS, l to Terence de Vere White, 15 July 1969, Special Collections, Mugar Memorial Library, Boston University.

Most of the summer: SS's diary, PGP.

"well known now for *not* being able": SS, l to Dr Audrey Insch, 15 July 1969, por.

Writing a few days later: this paragraph is based on SS, l to Dr Audrey Insch, 19 July 1969, por.

285 **In August Stevie heard**: this paragraph is based on DJ, l to SS, dated "Sunday, 17th", PGP, and SS, l to DJ, 22 Aug. 1969, por.

been to see her own physician: Dr James Curley, unsigned copy of l to SS, 26 Aug. 1969, in possession of Dr Curley.

"You must not dream": Aylet Hyams, l to SS, 28 Aug. 1969, PGP.

a Mrs McCleod: Aylet Hyams, l to SS, 16 Sept. 1969, PGP.

stayed in Norfolk: SS's diary, PGP.

That autumn she did: this paragraph is based on SS's diary, PGP, and Professor Bruce Berlind, l to A, 21 July 1980, por.

286 **Stevie told Suzannah**: SS, l to SJ, 8 Nov. 1969, por.

"Poor old Pug": SS, l to Dr Audrey Insch, 12 Nov. 1969, por.

Evening Standard: Thomas.

Christmas came: this paragraph is based on SS, undated l to AB, por.

for a reading: SS's diary, PGP. It was at the London Graduate School of Business Studies.

"a photographic portrait": Roy Strong (Director, National Portrait Gallery), l to SS, January 1970, PGP.

one from William Plomer: William Plomer.

delighted to have a portrait: SS, undated l to William Plomer, Special Collections, University Library, Durham.

"I do not think it is like me.": *MA*, p. 323.

287 **In March the *New Statesman*:** this paragraph is based on *MA*, pp. 206–9; and Sir John Lawrence, int A.

In April Stevie "skidded": this paragraph is based on *MA*, pp. 321–22.

288 **the Arts Council, with whom:** this paragraph is based on Eric W. White, l to SS, 17 April 1970, PGP; and Hull.

It was probably in these weeks: this and the following paragraph are based on Bernard Stone and John Morley, int A.

"I do claim all the same": SS, undated l to Anthony Thwaite, por.

"Did you see Scorpion": SS, l to KD, 3 Oct. 1970, St L. In this l SS complains about stomach trouble.

290 **On 21 June:** this paragraph is based on SS, l to Jean MacGibbon, 1 Oct. 1970, por, and printed on the endpapers of the hardback editions of *MA*; and HF, int A.

In October Stevie read: this paragraph is based on SS's diary, PGP; SS, l to Dr Audrey Insch, 22 Oct. 1970, por; and SS, l to Piers Russell-Cobb, 1 Nov. 1970, por.

291 **On 6 November Stevie recorded:** this paragraph is based on SS, letters to Lord Moyne, 10 and 18 Nov. 1970, por; *I&S*, pp. 61–63; and SS, l to KD, 10 Nov. 1970, St L.

Stevie went to stay: this paragraph is based on SS, l to Lord Moyne, 22 Nov. 1970, por; SS, l to John Guest, 24 Nov. 1970, por; and *MA*, pp. 323–24.

292 **she was worse off:** SS, l to Dr James Curley, 26 Dec. 1970, por.

"Doctors in command": SS, l to Robert Nye, 27 Dec. 1970, in possession of the bookshop, About Books, Toronto, Ontario.

between peeling potatoes and passing out: SS, l to Lord Moyne, 27 Dec. 1970, por.

"how am I going to stop Molly": this paragraph is based on the Everetts, int A.

Stevie was still at her sister's: this paragraph is based on SS, letters to J. W. Arnold, 1 Dec. 1970 and 5 Jan. 1971, por; and the Everetts, int A.

293 **"I'm afraid I've been snatched":** SS, undated l to Robert Nye, in possession of the bookshop, About Books, Toronto, Ontario.

"I'm so glad Laurence": HF, int A.

"Well, if she's really dying": James MacGibbon, int A. But see our letter, "Old Friends," *London Review of Books*, 20 Mar. 1986, p. 4.

Guest kept a log: John Guest, in a l to A, 25 Nov. 1980, explains that after the merger of Longmans and Penguins in 1972, the SS files could not be found. However, as the result of an enquiry made by Derek Parker of the BBC, he had already copied meticulously parts of her last letters from the *Scorpion* file. The log is part of his transcription, from a document titled "Stevie Smith's Illness", and material in brackets is Guest's interpolations.

295 **death mask of John Donne:** James MacGibbon, int A.

"shocking pink" and "The gamine child": extract from a memoir by HF, sent to A, 11 March 1985, HFP.

"Olga, this is": Olga Lawrence, int A.

Lady Lawrence . . . have tea: Lady Lawrence, int A.

had been moved: addresses on SS's letters to friends.

a brain tumour: Dr James Curley, int A.

the word "Death" encircled: James MacGibbon in a BBC 4 radio broadcast of 23 June 1974. Also see *CP*, p. 9.

Helen Fowler remembers: HF's memoir, HFP.

296 **"Dear Anthony":** *MA*, pp. 325–26.

During one of their visits: this paragraph is based on HF's memoir, HFP.

297 **"it is only on his deathbed":** SS, "The Converted Poet", *T*, 23 Jan. 1948, p. 19.

Chapter Twenty

298 **Five days after Stevie's death:** this paragraph is based on HF's memoir, HFP; and a cutting from a Buckfast [?] paper, "Miss Florence M. Smith, Buckfast", supplied to us by the Everetts.

Molly Ward-Smith was not happy: this paragraph is based on HF and the Everetts, int A.

Evening Standard: "Tea Party for Stevie," *ES*, 26 March 1971.

Helen Fowler has suggested: HF, int A.

299 **The Venerable Edward Carpenter:** this paragraph is based on cuttings about the memorial service from *The Times* and *DT* of 1 April 1971, and *PGSG* of 9 April 1971; Father Clifford Doyle, int A; and Father Gerard Irvine's eulogy, "Stevie Smith", a copy of which he kindly gave to us.

Some of the people . . . also attended the small tea-party: this paragraph is based on EL, *A Goldfish Bowl* (London: Cassell, 1972), pp. 312–13; MS to HF, HFP; and the Everetts, Brownes and Woodcocks, int A.

300 **In her will:** this paragraph is based on SS's will, 26 April 1966, a copy of which was supplied to us by her solicitor, A. J. Davey; "Stevie Smith Will", *DT*, 18 Aug. 1971, p. 7; and, for the detail about the house, HH's notes of his telephone conversation with SS, HHP.

Molly lived on: this paragraph is based on the Everetts, HF, Father Jerome Gladman, and AO, int A; and MS's will, Somerset House, London.

newspaper reported the "row": "'Plaque for Stevie' row", *PGSG*, 12 Jan. 1978, p. 1; "Plaque for Stevie Smith", *Weekly Herald*, 13 Jan. 1978.

301 **an account of their visit:** "Stevie as USA sees her", *PGSG*, 17 Aug. 1978, p. 1.

Index